God and Evolution

God and Evolution

Protestants, Catholics, and Jews Explore Darwin's Challenge to Faith

Jay W. Richards, Editor

Discovery Institute Press Seattle 2010

Description

This book seeks to analyze "theistic evolution," to critique various attempts to reconcile Darwinism with traditional theistic religion, and to respond to criticisms of intelligent design. Contributors include William Dembski, Logan Gage, David Klinghoffer, Stephen Meyer, Denyse O'Leary, Jay Richards, Jonathan Wells, John West, and Jonathan Witt. Edited by Jay W. Richards.

Copyright Notice

Publisher's Note

This book is part of a series published by the Center for Science & Culture at Discovery Institute in Seattle. Previous books include *Signature of Controversy*, edited by David Klinghoffer; *The Deniable Darwin & Other Essays*, by David Berlinski; and *Darwin's Conservatives: The Misguided Quest*, by John West.

Library Cataloging Data

God and Evolution: Protestants, Catholics, and Jews Explore Darwin's Challenge to Faith Edited by Jay W. Richards.

388 pages, 6 x 9 x 0.86 inches & 1.25 lb, 229 x 152 x 22 mm. & 0.57 kg

BISAC Subject: REL106000 RELIGION / Religion & Science
BISAC Subject: SCI027000 SCIENCE / Life Sciences / Evolution
BISAC Subject: REL113000 RELIGION / Essays

ISBN-13: 978-0-9790141-6-1 ISBN-10: 0-9790141-6-6 (paperback)

Publisher Information

Discovery Institute Press, 208 Columbia Street, Seattle, WA 98101
Internet: http://www.discoveryinstitutepress.com/
Published in the United States of America on acid-free paper.
First Edition, Second Printing. October 2010.

Contents

Introduction

Squaring the Circle

Jay W. Richards

When someone asks me: "Can you believe in God and evolution?," I always respond: "That depends. What do you mean by 'God' and what do you mean by 'evolution'?" No one seems to be very satisfied with this retort, which seems evasive; but it's the honest answer, since the initial question, as it stands, is hopelessly ambiguous. Without more detail, it's susceptible to almost any answer.

Asking whether one supports so-called "theistic evolution" has the same problem. Unless you define "theistic" and "evolution" very carefully, it might refer to positions that, on closer inspection, are more different than they are alike. One version might be an oxymoron, one a triviality, one an interesting proposition, and another, a complete muddle.

Besides being vague, these questions, and practically every answer to them, are controversial. Perhaps no subject now inspires more heated arguments at family reunions and cocktail parties. Whether in religious or secular, scientific or literary circles, giving the "wrong" answer can put you on the fast track to being labeled a heretic. A scientist in an academic setting who expresses any doubt about Darwinism, for instance, may find himself in a battle for tenure and funding. In his church, the same scientist may be suspected of creeping liberalism because he doesn't think the word "evolution" means atheism. Or he may be thought a "fundamentalist" because he thinks his faith has something to do with his science, and vice versa.

Such countervailing social pressures don't encourage clear thinking or clear speaking. So when they encounter the question, many people, especially academics, choose obfuscation over clarification. If pressed, they may attempt to stake out a moderate both-and position: "I think evolution is God's way of creating."[1] For the conflict-averse, this may be a reassuring response, but what does it mean?

In the century and a half since Charles Darwin first proposed his theory of evolution, Christians, Jews, and other religious believers have not only pondered its truth—or lack thereof—they have grappled with how to make sense of it theologically. So far, they haven't reached a consensus and tend, instead, to argue among themselves. It can be quite confusing. In fact, the whole subject of God and evolution, and especially what is called "theistic evolution," is an enigma wrapped in a shroud of fuzz and surrounded by a blanket of fog.

The purpose of this book is to clear away the fog, the fuzz, and the enigma.

Getting Our History Right

ONE OF THE FIRST patches of fog that needs to be cleared away is pop culture's caricature of the historical relationship between evolution and religion. In America, that caricature is epitomized by the old film *Inherit the Wind,* which reduces the debate over Darwin to a battle royal between intolerant Bible-thumpers and enlightened champions of free speech. In England, the caricature is epitomized by an exchange between a scientist and a clergyman. On June 30, 1860, a mere seven months after Charles Darwin released his *Origin of Species,* Oxford's Museum of Natural History hosted a famous debate on Darwin's theory of descent with modification, or what would later be called his theory of evolution. Among its distinguished participants were "Darwin's Bull Dog," Thomas Henry Huxley, and the Anglican Bishop Samuel Wilberforce. During the course of the debate, it is said that Wilberforce asked Huxley if it was through his grandmother or grandfather that he supposed that he was descended from apes. Huxley purportedly said that he "was not ashamed to have a monkey for his ancestor; but he would be ashamed to be connected with a man who used great gifts to obscure the truth."

The Huxley-Wilberforce repartee is often portrayed as a decisive victory for good science over bad religion. J. R. Lucas cheekily summarizes the received account:

> Huxley's simple scientific sincerity humbled the prelatical insolence and clerical obscurantism of Soapy Sam; the pretension of the

> Church to dictate to scientists the conclusions they were allowed to reach were, for good and all, decisively defeated; the autonomy of science was established in Britain and the Western world; the claim of plain unvarnished truth on men's allegiance was vindicated, however unwelcome its implications for human vanity might be; and the flood tide of Victorian faith in all its fulsomeness was turned to an ebb, which has continued to our present day and will only end when religion and superstition have been finally eliminated from the minds of all enlightened men.[2]

It's a memorable story. But as Lucas's arch tone suggests, this trope, like so many stories drawn from the hallowed pages of Darwiniana, is mostly mythology.

First of all, careful historians suspect that the grandparent-ape exchange probably never happened. Even though it is the most widely recounted detail from the famous debate, there was no contemporaneous report of it.

Second, it misrepresents how Darwin's theory was initially received. To judge from the story of the debate, you would think that objections to Darwin's theory came mainly from clerics and religious believers for religious reasons, while being quickly embraced as good science by scientists.[3] In truth, Darwin had a number of scientific critics, and even Bishop Wilberforce focused on scientific rather than religious objections to Darwin's theory.

Third, the Christian response to his theory was diverse from the very beginning. For instance, Charles Hodge and B. B. Warfield, two nineteenth century Presbyterian theologians at Princeton Theological Seminary, initially disagreed with each other on the merits, and theological implications, of Darwin's theory. Both were pillars of conservatism. But while Hodge saw the theory as atheism masquerading as science, Warfield thought it could be reconciled with Christian orthodoxy (though he later came to agree more with Hodge[4]). There was similar ambivalence among Catholics. Although most traditional Catholics opposed the implications of Darwin's theory for human beings, the *Origin of Species* was never placed on the Church's Index of Prohibited Books.

Fourth, the Huxley-Wilberforce debate is often used to illustrate a larger myth about an innate war between science and religion. Tales of that warfare usually include Copernicus, Galileo, Giordano Bruno, and William Jennings Bryan. Though many science textbooks still spread this "warfare myth," most historians of science recognize that this simplistic trope distorts a much more complicated and interesting history of interaction between science and religion in the West. Many have argued, in fact, that the Judeo-Christian tradition actually helped give rise to natural science.[5]

There is no innate war between natural science and theistic religion; nevertheless, the question of God-and-evolution remains complicated, controversial—and confused.

A Range of Views

There are practically as many views of how God relates to evolution as there are people who have pondered the subject. Still, most views fall into one of several categories. Unfortunately, before defining the categories, you have to overcome a terminological hurdle. What do you do with the troublesome word "creationist"? The word is usually used pejoratively, to bring to mind "**young earth creationists**" who believe that God created the universe in six, twenty-four hour days sometime in the last ten thousand years. Critics assume that the young earth view is so disreputable that anyone associated with it will likewise be tarnished. However you judge that uncharitable assumption, you can't use the word "creationist" these days without carrying some of this baggage along for the ride.

This is an accident of history. In a less complicated world, the word "creationist" would not be a put-down but simply a way to refer to people—Christians, Jews, Muslims, Sikhs, and other theists—who believe in a doctrine of creation. Regrettably, that's not how the world or the word works. Like it or not, the entire discussion about God and evolution takes place in a rhetorical context designed to misdirect and misrepresent certain views, especially those views that take God seriously.

Since we're stuck with the word creationist, though, we'll just have to slog ahead.

Besides "young earth creationists," there are folks who refer to themselves as "**old earth creationists**" and others who call themselves "**progressive**

creationists." Old earth creationists generally hold to mainstream scientific views of the age of the Earth and the universe, but believe that God worked directly in nature (as a "primary" or "efficient cause"[6]) to create some things. These might include heavenly bodies like galaxies and the solar system, the first reproducing cell, various forms of life, human beings, human souls, and so forth. Old earth creationists disagree among themselves on the *loci*—the places—where they think that God acts directly, but all agree that, sometimes at least, God acts directly in natural history to bring about things that nature would not produce if left to its own devices.

Progressive creationists also believe that God acts directly at various points in cosmic history, but they tend to see more evolutionary development between the seams of God's specific acts.

Then there are those who don't fit simply on the "creationist" spectrum, but do challenge materialistic theories of evolution. For example, "**intelligent design**" or "ID" theorists argue that nature, or certain aspects of nature, are best explained by intelligent design. On this view, repetitive, law-like or mechanistic explanations that invoke, say, the gravitational force and natural selection, explain some aspects of nature, but a full explanation of the natural world will include intelligent agency as well.

Moreover, ID theorists have argued that physical laws are themselves the result of intelligent design, even if they are not, arguably, adequate to explain everything in nature.

At the same time, ID theorists focus on the detectable effects of intelligence, rather than on the specific locations or modes of design within nature. As William Dembski, one prominent ID proponent, puts it: "Intelligent design (ID) is the study of patterns in nature that are best explained as the result of intelligence."[7] Since ID is minimal, it is logically compatible with almost any creationist or evolutionist view that allows for intelligent agency as an explanation within nature. (The contributors to this volume fall into the ID camp.)

Finally, there are **theistic evolutionists**, who would appear to subscribe to a hybrid position that combines both "theism" and "evolution." Most theistic evolutionists contrast their view with "special creationism," which would include any view that suggests that God has acted directly in natural history. However, logically speaking, a theistic evolutionist could also be an ID proponent (in fact, there are many such people). Nevertheless, most self-described theistic evolutionists distinguish themselves from intelligent design proponents, and are, in some cases, harsh critics of ID. So, like the word creationist, "theistic evolution" tends to have a meaning different from what its etymology alone would suggest.

So what exactly is theistic evolution? It would be nice to open Webster's, find the definitions of "theism" and "evolution," stick the definitions together, and be done with it. Alas, it's not that simple. Behind the phrase "theistic evolution" lurks a lot of mischief and confusion.

A Dilemma

When dealing with God and evolution, most people have an intuitive feeling that there's some contradiction lurking in the neighborhood, some dilemma that has to be resolved. Even children, at some point, begin to sense this. Most probably ask their parents what my eleven year old daughter asked me recently: "So why did God make dinosaurs that all died out millions of years before Adam and Eve?" Several years earlier, she had asked, obviously garbling the kindergarten evolution lesson: "Did we used to have tails?" Perhaps you'd have ready answers to these questions. But if you're like millions of other parents, you might try to punt.

For punters, theistic evolution (or "evolutionary creationism" as it's sometimes called) might seem to promise some relief. But eventually, if you tell an attentive child that evolution is just God's way of creating, she's going to ask you what you mean. It would be nice to have something more than a pat answer accompanied by some hand waving.

The difficulty begins when we start to dig into the common textbook definitions of the term "evolution." Here, evolution is often *defined* by its opposition to creation. Consider just two academic sources among legion: "That organisms have evolved rather than having been created is the single most important and unifying principle of modern biology."[8] And here's the Harvard paleontologist George Gaylord Simpson: "Man is the result of a purposeless and natural process that did not have him in mind."[9] Darwin himself understood his theory this way. As he said, "There seems to be no more design in the variability of organic beings, and in the action of natural selection, than in the course which the winds blow."[10]

These descriptions of (Darwinian) evolution don't leave a lot of wiggle room. And notice that the idea of organisms evolving *rather than* being created is not presented as a side-light, as the private opinion of a few scientists. In the first quote, as in many others, evolution is described as the "single most important and unifying principle of modern biology." It would be hard to put the point any more strongly.

Surely, for the sake of truth and sanity, it's better to ask and answer the follow up questions directly than to avoid them indefinitely.

THE DILEMMA WITH DILEMMAS

BEFORE DIVING IN, HOWEVER, let's step back and think about dilemmas in general. Whenever you're trying to hold together ideas that seem to contradict each other, you have a *dilemma*—or trilemma if there are three ideas involved. (We don't have a word for apparent contradictions that involve more than three ideas, perhaps because most of us just give up or think about something else when things get that complicated.)

Anyone who has studied theology or philosophy will be familiar with one famous trilemma, called the problem of evil. The problem of evil involves three ideas believed by most theists (that is, people who believe in a personal God who transcends the universe). The problem is that it seems at first blush like these three beliefs can't all be true:

1. God is all powerful.
2. God is perfectly good.
3. There is evil in the world.

There's no obvious contradiction here, as there is if you claim that 2 plus 2 equals 4 and 2 plus 2 does not equal four, or you say that your best friend is a married bachelor. In fact, the three claims above aren't even about the same things. The first two are about God. The third is, strictly speaking, about the world. So we're not dealing with what is called a formal contradiction. Still, most college freshmen sense trouble when they first see the problem of evil presented in Philosophy 101.

And the freshmen are surely right. The basic intuition motivating this trilemma is that if God is really all good and all powerful, it seems that he would not create a world with evil in it. Whatever world he created, he would know how things are going to turn out and would prevent evil from popping up anywhere in his creation. Since there obviously is evil in the world (premise three), something must be wrong with at least one or both of the other premises. Either God is not all powerful or he's not perfectly good. Or maybe he's neither.

Now there are a few basic ways to resolve any real dilemma (or trilemma in this case). If you want to be consistent, you need to drop one of the premises. Since contradictions describe impossible situations, at least one of them has to be false.

To solve the problem of evil, for example, you might decide that God is good but not all powerful, so he just can't keep all the evil out. Or maybe he's all powerful but not perfectly good, so he's not really that concerned if evil turns up. Or maybe we're mistaken and there is no evil in the world, meaning there's no reason to doubt that God is all-powerful and perfectly good. If you go in any of these three directions—problem solved.

Regrettably, each of these solutions requires that you abandon one of the beliefs that, if you're a theist, you'd prefer to retain. If you believe that God is all powerful, perfectly good, *and* that, nevertheless, there is evil in the world, then you want to find a way to reconcile the three beliefs, not sacrifice one for the others. So you may hold out hope that what looks like a contradiction is not really one at all, but just an unfortunate misunderstanding.

The traditional response to the problem of evil, for instance, is called the free will defense. According to this rejoinder, in choosing to create the world, God chose to create free beings such as humans and angels, beings so free that they could choose against him, could do evil. Even though God is all powerful, he couldn't create a world with free beings and no evil, any more than he could create a square circle. No matter how powerful God is, there is no possible world with free beings and no evil. So God had to accept a trade-off.

Usually the free will defense is paired with the greater good defense, which says that a world with free beings and some evil is better, all things being equal, than a world that lacks both free beings and evil. This assumes that the existence of free beings has intrinsic value. And since God knows how things are going to turn out, he can bring about a greater good by allowing free beings to exercise evil than by creating a world of robots that always do what they're told. As Joseph (in the book of Genesis) explained to his brothers, who had sold him into slavery, "You meant it for evil, but God meant it for good."

There is, of course, much more that could be said about this thorny issue (for example, the problem of evil in the natural world), but this short summary should give you some sense of the basic moves for solving a dilemma. When it comes to the question of God and evolution, we're dealing with a different subject but the same basic options.

Presumably, a theistic evolutionist is someone who claims that both theism *in some sense* and evolution *in some sense* are true, that both God and evolution somehow work together in explaining the world. But of course, all the real interest is hidden behind the phrase "in some sense." So let's lay out the main senses of the two words in question.

"Theism"

Although different people understand God differently, the word "God" has a pretty stable meaning in ordinary conversation. If I tell Christopher Hitchens, an atheist, that I believe in God, he has some sense of what I mean. In ordinary English and other Western languages, "God" usually refers to a Creator, a personal being who has chosen to create the world, who is powerful and perfect in whatever ways such a being could be powerful and perfect, and who transcends the universe. That is, God would exist, would be, whether or not he had chosen to create the world. The world, in contrast, exists as the result of his free choice, for his purposes and at his discretion.

Of course, "God" doesn't refer to just any old being like a bunny resting on a down or the guy in Mumbai who answers your questions when you call Dell tech support. God, though a "being" in the sense that he "exists" (or, more precisely, is), is himself the source of other beings, and in that way, he is qualitatively different from all other beings. Classical theists often say that God is "Being itself." That way of speaking is a bit obscure to the uninitiated. At the very least, however, what this means is that God doesn't participate in some more fundamental reality called "being" along with everything else. He is the Source of all being. Moreover, unlike you, me, and the burrito I had for lunch, God necessarily exists. He exists in every possible world.

Technically speaking, you could believe that such a God exists, and be either a theist or a deist. A **theist** believes that God both created the world and continues to conserve and interact in and with it.[11] In fact, God is so intimately related to the world that, while being separate from the world, he still wholly pervades it. So theists speak of God as both transcendent and immanent.[12] What the theist will never do is identify God with the world.

A **deist** holds a more minimal view, believing that God created the world but doesn't really keep up with the day-to-day activities on the ground. Or even if he keeps up, he doesn't get directly involved. He maintains a strictly hands-off policy.

Besides theism and deism, and leaving aside polytheism, the other main options are **pantheism** and **panentheism**. Pantheism identifies God and nature. For the pantheist, God doesn't transcend the world nor is he independent of it. He's not really even immanent in the world. Rather, God is the world and the world is God. For most pantheists, moreover, God is not really personal either. After all, the universe just doesn't look much like an agent with purposes and a will. So for the pantheist, "God" might be thought of as a rational principle or a life force that somehow pervades the universe; but God, for the pantheist, most certainly is not a transcendent Creator.

A hybrid position is called panentheism, which holds that God has some transcendent qualities but is nevertheless in the world, or, to put it differently, the world is in him. The world, we might say, is part of God. God and nature may be distinct but they're inseparable. A panentheist might think of God as a Creator, but not in an absolute sense. God might push or pull or persuade or cajole things to go in a certain direction. He might have purposes. But he won't call everything into existence from nothing simply by his free choice. God will evolve along with the world.

Though there are a few Christian academics who identify with panentheism,[13] the vast majority of Christians, Jews and Muslims, and the historic thinkers in these traditions, are theists. That's because the basic tenets of their religions hold that God is a transcendent Creator who at least occasionally acts directly in the world. All three of these Abrahamic faiths believe that God specially communicated with Abraham and Moses, for instance.

In addition, Christians believe that God became a man, Jesus, at a particular time and place; that Jesus was conceived by the Holy Spirit in the Virgin Mary rather than by ordinary means; and that after Jesus died, he was raised from the dead and ascended into heaven. All this implies the Christian belief that God is triune. He exists eternally as three "persons" while still being one God. Though less central to Christian doctrine, most Christians also believe that Jesus worked certain extraordinary miracles, such as calming a storm and raising a girl from the dead.

Take away all beliefs about God acting in history, and you have at best only a shadow of theistic belief.

Of course, theists don't believe that God is aloof from the world except when he acts directly in nature. For theists, God transcends the world, is free to act directly in it, and always remains intimately involved with it.

At the same time, the theist need not believe that God always acts directly in the world. Traditionally, Christian theologians have argued that God can act in the world in two different ways. He can act directly or "primarily," such as when he creates the whole universe or raises Jesus from the dead. It's God's world, so that's his prerogative. He's not violating the universe or its laws when he does this, or invading alien territory, since he's the source of both the universe and whatever "laws" it might have.

He also can act through so-called "secondary causes." These include the choices or tendencies of the creatures he has made. For instance, he can work through the evil choices of Joseph's brothers to achieve a greater good of getting the descendants of Abraham to Egypt so that they don't die from famine.

God can also bring about his purposes through natural processes and laws that he has established, such as the electromagnetic force. An event might be both an expression of a physical law and the purposes of God. It's not as if atheists appeal to gravity while theists appeal to miracles. Gravity is as consistent with theism as are miracles. But for the theist, gravity is a creature, or rather, it describes creatures. It's like a mathematical description of how God has ordained physical objects to act in ordinary circumstances; it's not an eternal law governing God's behavior.

Christians, Jews, and other theists recognize that God can act through secondary causes when they thank God for their food, even though they know that God normally provides our food, not as manna from heaven, but through natural causes like rain, spring, and soil, and through human actions like sowing and reaping. God is so free and so powerful that he can act either directly or through secondary causes. He's like a doting gardener who creates his own sun, seeds, water, nutrients, and dirt. And he's perfectly happy to have "flowers" who can make their own decisions.

Therefore, for theists, God, while acting either directly or through secondary causes, continually upholds, oversees and superintends his entire creation in "providence," even as he allows his creatures the freedom appropriate to their station.[14]

We've just scratched the surface, but we've probably said enough about theism for our purposes.

"Evolution"

THOUGH GOD IS THE grandest and most difficult of all subjects, the meaning of the word "evolution" is actually a lot harder to nail down.

In an illuminating article called "The Meanings of Evolution," Stephen Meyer and Michael Keas distinguished six different ways in which "evolution" is commonly used:

1. Change over time; history of nature; any sequence of events in nature.

2. Changes in the frequencies of alleles in the gene pool of a population.
3. Limited common descent: the idea that particular groups of organisms have descended from a common ancestor.
4. The mechanisms responsible for the change required to produce limited descent with modification, chiefly natural selection acting on random variations or mutations.
5. Universal common descent: the idea that all organisms have descended from a single common ancestor.
6. "Blind watchmaker" thesis: the idea that all organisms have descended from common ancestors solely through unguided, unintelligent, purposeless, material processes such as natural selection acting on random variations or mutations; that the mechanisms of natural selection, random variation and mutation, and perhaps other similarly naturalistic mechanisms, are completely sufficient to account for the appearance of design in living organisms.[15]

Meyer and Keas provide many valuable insights in their article, but here we're only concerned with "evolution" insofar as it's relevant to theology.

The first meaning is uncontroversial—even trivial. The most convinced young earth creationist agrees that things change over time—that the universe has a history.[16] Populations of animals wax and wane depending on changes in climate and the environment. At one time, certain flora and fauna prosper on the earth, but they later disappear, leaving mere impressions in the rocks to mark their existence for future generations.

Of course, "change over time" isn't limited to biology. There's also cosmic "evolution," the idea that the early universe started in a hot, dense state, and over billions of years, cooled off and spread out, formed stars, galaxies, planets, and so forth. This includes the idea of cosmic nucleosynthesis, which seeks to explain the production of heavy elements (everything heavier than helium) in the universe through a process of star birth, growth, and death. These events involve change over time, but they have to do with the history of the inanimate physical universe rather than with the history of life. While this picture of cosmic evolution may contradict young earth creationism, it does not otherwise pose a theological problem. The generic idea that one form of matter gives rise, under the influence of various natural laws and processes, to other forms of matter, does not contradict theism. Surely God could directly guide such a process in innumerable ways, could set up a series of secondary natural processes that could do the job, or could do some combination of both.

In fact, virtually no one denies the truth of "evolution" in senses 1, 2, or 3. And, pretty much everyone agrees that natural selection and random mutations explain some things in biology (number 4).

What about the fifth sense of evolution, universal common ancestry? This is the claim that all organisms on earth are descended from a single common ancestor that lived sometime in the distant past. Universal common ancestry is distinct from the mechanism of change. In fact, it's compatible with all sorts of different mechanisms or sources for change, though the most popular mechanism is the broadly Darwinian one. It's hard to square universal common descent with some interpretations of biblical texts of course; nevertheless, it's *logically* compatible with theism. If God could turn dirt into a man, or a man's rib into a woman, then presumably he could, if he so chose, turn a bacterium into a jellyfish, or a dinosaur into a bird. Whatever its exegetical problems, an unbroken evolutionary tree of life *guided and intended by God,* in which every organism descends from some original organism, sounds like a logical possibility. (So there's logical space where both intelligent design and theistic evolution overlap—even if ID and theistic evolution often describe people with different positions.[17])

Besides the six senses mentioned by Meyer and Keas, there is also the metaphorical sense of evolution, in which Darwinian Theory is used as a template to explain things other than nature, like the rise and fall of civilizations or sports careers. In his book *The Ascent of Money,* for instance, historian Niall Ferguson explains the evolution of the financial system in the West in Darwinian terms.[18] He speaks of "mass extinction events," survival of the fittest banks, a "Cambrian Explosion" of new financial instruments, and so forth. This way of speaking can sometimes be illuminating, even if, at times, it's a stretch. Still, no one doubts that there are examples of the fittest surviving in biology and finance. We might have some sort of "evolution" here, but not in a theologically significant sense.

Finally, there's evolution in the sense of "progress" or "growth." Natural evolution has often been understood in this way, so that cosmic history is interpreted as a movement toward greater perfection, complexity, mind, or spirit. A pre-Darwinian understanding of "evolution" was the idea of a slow unfolding of something that existed in nascent form from the beginning, like an acorn eventually becoming a great oak tree. If anything, this sense of evolution tends toward theism rather than away from it, since it suggests a purposive plan. For that reason, many contemporary evolutionists (such as the late Stephen J. Gould) explicitly reject the idea that evolution is progressive, and argue instead that cosmic history is not going anywhere in particular.

Much more could be said, but it should now be clear that theism, properly understood, is compatible with many senses of evolution. For most of the senses of evolution we've considered, in fact, there's little appearance of contradiction. Of course, this is a logical point. It doesn't tell us what is true—only what could be true.

But there's one clear exception—the blind watchmaker thesis. Of all the senses of "evolution," this one seems, at least at first blush, to fit with theism like oil with water. It claims that all the apparent design in life is just that—apparent. That apparent design is really the result of natural selection working on *random* genetic mutations. (Darwin proposed "variation." Neo-Darwinism attributes new variations to genetic mutations. In the following chapters, we will follow convention and use "Darwinism" and "Neo-Darwinism" interchangeably, except where otherwise indicated.)

The word "random" in the blind watchmaker thesis carries a lot of metaphysical baggage. In Neo-Darwinian theory, "random" doesn't mean uncaused; it means that the changes aren't directed—they don't happen for any purpose. Moreover, they don't occur for the benefit of individual organisms, species, or eco-systems, even if, under the guidance of natural selection, an occasional mutation might ultimately redound to the benefit of a species.

Darwin, at least in his argument in *The Origin of Species*, assumed a form of radical deism in which God establishes general laws that govern matter, but then leaves the adaptation and complexity of life up to random variations and natural selection. (Note that Darwin's personal views are a separate matter from the structure and rhetoric of his argument in the *Origin*.[19]) Nowadays, though, most evolutionary biologists are more thorough-going materialists, as least when it comes to their science. So the blind watchmaker thesis is more or less the same as the mechanism of Neo-Darwinism as its leading advocates understand it.

The blind watchmaker thesis is usually wedded to some materialistic origin of life scenario, which isn't about biological evolution *per se*, though it is sometimes referred to as chemical evolution.

From the time of Darwin, who first proposed it, to the present, Darwinists have contrasted their idea with the claim that biological forms are designed. Here's how the late Darwinist Ernst Mayr put it:

> The real core of Darwinism, however, is the theory of natural selection. This theory is so important for the Darwinian because it permits the explanation of adaptation, the 'design' of the natural theologian, by natural means, instead of by divine intervention.[20]

Notice that he says "instead of." Darwinists almost always insist that their theory serves as a designer substitute. That's the whole point of the theory. This makes it different from other scientific theories, like Newton's law of gravity. Newton didn't formulate the law to get God out of the planet business (in fact, for Newton, God was involved in every aspect of the business.) And theories that invoke ordinary physical laws are determinate: they allow the scientist to make specific predictions about what will happen, all things being equal.

Darwin's theory isn't like that. It simply says that whatever has happened, and whatever will happen, the adaptive complexity we see in organisms is (primarily) the result of natural selection and random variation, not design. From the very beginning, the theory was intended to rule out teleological (purposive) explanations. As William Dembski once said: "The appeal of Darwinism was never, That's the way God did it. The appeal was always, That's the way nature did it without God."[21] That's why, even if not all agree with Richard Dawkins that Darwin "made it possible to be an intellectually fulfilled atheist,"[22] the vast majority of Darwinists claim that Darwin's mechanism makes God superfluous. It's their theory, so presumably they have a right to tell us what it means. Theists, in contrast to Darwinists, claim that the world, including the biological world, exists for a purpose, that it is, in some sense, designed. The blind watchmaker thesis denies this. So anyone wanting to reconcile strict Darwinian evolution with theism surely has a Grade A dilemma on his hands. It's akin to reconciling theistic evolution with anti-theistic evolution.

We noted above that the easy way to solve the problem of evil is to drop one of the offending premises. The same is true with theistic evolution: the easy way out is to drop or radically redefine the theistic part (dropping the Darwinian part is usually much riskier to one's career). Dissolving a dilemma, however, is not the same as resolving it. If the adjective "theistic" in "theistic evolution" is to be an accurate description, it should include a theistic view of God.

If you're unfamiliar with the debate over God and evolution, you might already be anticipating how to be a theistic evolutionist. A theistic evolutionist, as suggested above, would be someone who holds that God somehow sets up or guides nature so that it gives rise to everything from stars to starfish through a slowly developing process. Organisms share a common ancestor but reach their goal as intended by God. God works in nature, perhaps through cosmic initial conditions, secondary processes, discrete miracles, or some combination, to bring about his intended results, rather than creating everything from scratch. Or perhaps God created the universe as a whole primarily, but everything else he "delegates," as it were, to natural causes. But whatever the details, by definition the process of change and adaptation wouldn't be random or purposeless. It would implement a plan, and would have a purpose. So a theistic evolutionist, you might assume, would hold a *teleological* version of evolution, which includes cosmic evolution, the origin of life, and biological evolution, and would certain not endorse the Darwinian blind watchmaker thesis in biology.

But it's rarely so straightforward. Consider the view of Presbyterian pastor Timothy Keller. In his popular book, *The Reason for God,* he tells readers:

"For the record I think God guided some process of natural selection, and yet I reject the concept of evolution as an all-encompassing Theory."[23]

Earlier he says: "Evolutionary science assumes more complex life-forms evolved from less complex forms through a process of natural selection. Many Christians believe that God brought about life this way."[24] He also quotes approvingly from a Bible commentary, which affirms evolution as a mere "scientific biological hypothesis," but rejects it as a "world-view of the way things are." Thus partitioned, the reader is told, "there is little reason for conflict."[25] Elsewhere Keller observes that he has "seen intelligent, educated laypeople really struggle with the distinction.... Nevertheless, this is exactly the distinction they must make, or they will never grant the importance of" evolution as a biological process.[26]

But those "intelligent, educated, laypeople" struggle for a reason. What exactly is the distinction he is proffering, and what does it distinguish? Is he saying that while it's okay to speculate about various evolutionary hypotheses, we should not affirm any? Surely not, since he seems to affirm a broad, semi-Darwinian evolutionary hypothesis. So is he saying that Darwinian evolutionary theory explains hearts and arms and ears and bacterial flagella, but not our love of music and our moral intuition? And if so, on what basis is he maintaining the distinction? After all, it's not as if we have solid empirical evidence that natural selection acting on random genetic mutations can give rise to an avian lung but not to our belief in the Golden Rule. So at best, such a distinction would be *ad hoc*.

Or does Keller have something else in mind? He doesn't say. In any case, distinguishing evolution as a hypothesis from evolution as a "world-view of the way things are" doesn't offer much guidance one way or another. To be useful, he would need to specify exactly what he means by evolution, what he thinks it explains well, and what he thinks it leaves out that keeps it from constituting a "world view of the way things are." Instead, we get a vague distinction without a difference. It's no surprise that the laypeople to whom he's commended the distinction don't find it very illuminating.

Notice that he speaks of God "guiding some process of natural selection," but does not mention *random variation*, which is as much a part of Darwinian Theory as is natural selection. Perhaps that avoidance is intentional. But since he doesn't say outright that he rejects the idea that natural selection acts only on random genetic mutations, the careful reader is left guessing.

If we read him charitably, however, Keller seems to want to affirm that God guided the origin and development of life forms, all of which are linked by a chain of common ancestry, by coordinating his guidance with natural selection. So the outcome isn't really random. (Recall the generally accepted definition of "random" discussed above.)

At the same time, Keller explicitly rejects the blind watchmaker thesis. So he's apparently not an orthodox Darwinist. He doesn't quite realize that to hold this view consistently, however, he needs to embrace teleology and reject orthodox Neo-Darwinism and materialistic origin of life scenarios, and not merely reject "evolution as a worldview of the way things are," whatever that means.

I am not intending to pick on Keller, whose work I hold in high regard. I am using him to illustrate how confusing this issue can be, and how even smart, orthodox religious thinkers often get into a muddle when they try to wed their Christian beliefs with Darwinian evolutionary theory.

If we peel away these confusions and look for a straightforward, coherent position, however, we usually end up with the idea of God-guided common ancestry. This is probably what most people would think theistic evolution means. But they would be wrong, at least when it comes to describing the views of many who describe themselves as theistic evolutionists. These days most theistic evolutionists seek, somehow, to reconcile theism with *Darwinian* evolution. They may affirm design in some broad sense at the cosmic level, but things get patchy when it comes to biology. Though it's not always easy to understand what they're saying, many theistic evolutionists want to integrate the blind watchmaker thesis into their theology. Now that would be quite a trick to pull off. Is it possible? And even if it's possible, why believe it's true?

ORGANIZATION

THIS VOLUME IS ORGANIZED into four parts.

I. **Some Problems with "Theistic Evolution"** deals with broad thematic issues related to the God and evolution debate, such as the affinities between Gnosticism in early Christianity and the thought of certain theistic evolutionists; the unstable strategic alliance of theistic evolution and the "evolution lobby"; and the failure of theistic evolution to resolve the problem of evil. It engages the thought of a diverse group of theistic evolutionists including Karl Giberson, Kenneth Miller, and Stephen Barr.

II. **Protestants and Evolution** responds to several representative Protestant theistic evolutionists, including Francis Collins, Denis Lamoureux, and Howard Van Till.

III. **Catholics and Evolution** treats Catholic thought in particular, including the early Catholic responses to Darwinism from G. K. Chesterton, Hilaire Belloc, and St. George Jackson Mivart; the Catholic Church's response to Darwin's theory; and the often-perplexing response of certain Thomists to Darwinism and intelligent design.

IV. **Jews and Evolution** considers Darwinian evolution and intelligent design in light of the thought of ancient, medieval, and contemporary Judaism.

One might wonder why an ecumenical group of intelligent design proponents would write a book dealing with the subject of God and evolution. ID, after all, is not a religious program. Yet while intelligent design is based on non-sectarian, public evidence, and uses public modes of reasoning, ID, Darwinism, and other theories of origin obviously have theological *implications*. Individual ID proponents should be just as free to explore such implications as their critics, although such explorations should not be mistaken for ID *per se*.

Moreover, the resurgence of theistic evolution, or what might more accurately be called theistic Darwinism, has made theology a central concern in the current debate. Certain theistic evolutionists have argued, for instance, that Darwinian evolution is compatible with or even useful to theology. Others have claimed that ID is theologically suspect or even "blasphemous." This volume is, in large part, a response to these claims.

Finally, while it is often useful to distinguish natural science from, say, philosophy and theology, truth is ultimately a unity. While we may sometimes express a truth in different but complementary ways, a proposition cannot be true in theology but false in natural science. Sometimes, it is appropriate to explore the themes at the boundary of otherwise distinct academic disciplines. The chapters in this volume are intended to do just that.

Of course, the subject of God and evolution is far too big a subject for a single volume. In a sense, it touches all of the biggest questions we can ask about ourselves and the world we live in. If you find yourself perplexed by the current debate, however, the following contributions are intended to provide some relief.

Some Terms to Keep Straight

Young earth creationism: the view that God created the universe in six, roughly twenty-four hour days sometime in the last ten thousand years or so.

Old earth creationism: This view affirms that God created the universe from nothing, affirms the mainstream scientific views of the age of the Earth and the universe, and maintains that God worked directly in nature (as a "primary" or "efficient cause") to create some things.

Progressive creationism: those who describe themselves as progressive creationists are similar to old earth creationists. For instance, they usually hold to mainstream views of the age of the Earth and universe and also maintain that God acts directly at various points in cosmic history; but they tend to see more evolutionary development between the seams of God's specific acts than do old earth creationists.

Intelligent design (ID): ID proponents argue, on the basis of public evidence drawn from natural science, that nature, or certain aspects of nature, are best explained by intelligent agency. Most ID proponents are critics of Neo-Darwinism as an adequate explanation for the adaptive complexity of life, and of materialistic theories of the origin of life and biological information. Since ID is minimal, it is logically consistent with a variety of creationist and evolutionist views, but identical with none.

Theistic evolutionism: the view that theism and evolution are both true. The term is ambiguous because the word "evolution" means many different things. Logically, one could be an ID proponent and a theistic evolutionist, although the term is commonly used to describe individuals who affirm Darwinian evolution and are critical of contemporary intelligent design arguments in biology.

Theism: the view that a transcendent, all powerful, perfectly good, personal God created the world *ex nihilo*—from nothing—and continues to conserve and interact in and with it. Traditional Judaism, Christianity, and Islam are theistic.

Deism: the view that a transcendent God created the universe but does not interact with it or act directly in history.

Pantheism: the view that God and the universe are identical. Pantheists reject the concept of a transcendent God, and usually do not believe that God is fully personal.

Panentheism: the view that God has some transcendent qualities but is nevertheless in the world, or alternatively, that the world is "in" God. God and nature may be distinct but they're inseparable. Most panentheists believe that God evolves along with the world and did not create the world *ex nihilo*. Panentheism is somewhat popular among religious academics, especially those involved in the dialogue between faith and science; but it is extremely rare among ordinary religious believers.

Evolution: this word has many different meanings, not all compatible with each other, and only a few of which are theologically significant. Among the meanings of evolution are the following:

1. Change over time; history of nature; any sequence of events in nature.
2. Changes in the frequencies of alleles in the gene pool of a population.
3. Limited common descent: the idea that particular groups of organisms have descended from a common ancestor.
4. The mechanisms responsible for the change required to produce limited descent with modification, chiefly natural selection acting on random variations or mutations.
5. Universal common descent: the idea that all organisms have descended from a single common ancestor.
6. "Blind watchmaker" thesis: the idea that all organisms have descended from common ancestors solely through unguided, unintelligent, purposeless, material processes such as natural selection acting on random variations or mutations; that the mechanisms of natural selection, random variation and mutation, and perhaps other similarly

naturalistic mechanisms, are completely sufficient to account for the appearance of design in living organisms.[27]

7. A metaphor describing the rise, development, success, and collapse of sports careers, business enterprises, nations, and so forth, through a process of competition.
8. Progress or development through time of something that existed initially in a nascent form, such as a child emerging from an embryo or an oak tree from an acorn. This idea was common in pre-Darwinian views of biological evolution, which led to Darwin avoiding the word "evolution" in his *Origin of Species*. Contemporary Darwinists, following Darwin, generally reject this understanding of biological evolution, which suggests a purposeful or teleological process. Nevertheless, language that implies progress frequently appears even in the writings of those who officially reject it.

Meaning 6 (the blind watchmaker thesis) is the least compatible with theism. Meaning 8 is implicitly theistic or at least teleological.

Teleology: refers to a system, event, or process that is purposeful and goal-oriented. There are teleological and non-teleological versions of cosmic and biological evolution. A central purpose of Darwinian Theory is to explain the apparent teleology of life as merely apparent rather than real.

Darwinism: the theory that every form of life on Earth is descended from one or a few common ancestors, and that the adaptive complexity of life is largely the result of natural selection acting on random variations. Darwin proposed his theory as an alternative to the idea that species had been specially created, and most modern Darwinists have followed Darwin's lead. Strictly speaking, it is Darwin's mechanism of natural selection and random variation, and not common ancestry, that contradicts the intelligent design of life.

Neo-Darwinism: the modern version of Darwinism, according to which random variations are identified with random genetic mutations.

I.

Some Problems with "Theistic Evolution"

1. Nothing New Under the Sun

Theistic Evolution, the Early Church, and the Return of Gnosticism, Part 1

John G. West

I sing the goodness of the Lord, who filled the earth with food,
Who formed the creatures through the Word, and then pronounced them good.
Lord, how Thy wonders are displayed, where'er I turn my eye,
If I survey the ground I tread, or gaze upon the sky.

—Isaac Watts (1715)

From the words of English hymnist Isaac Watts to the music of Hadyn's *Creation*, from the ceiling of Michelangelo's Sistine Chapel to the pages of C. S. Lewis's novels *Perelandra* and *The Magician's Nephew*, the Christian doctrine of creation[1] has inspired countless poets, composers, authors, and artists to celebrate the beauty and artistry of God as Creator. Yet in his recent book *Saving Darwin: How to Be a Christian and Believe in Evolution* (2008), theistic evolutionist Karl Giberson writes dismissively of the Christian doctrine of creation, insisting that it is but "a secondary doctrine for Christians. The central idea in Christianity concerns Jesus Christ and the claim that he was the Son of God."[2] Giberson's point seems to be that so long as people accept the divinity of Jesus, their view of God as Creator is unimportant.

Early Christian thinkers would have disagreed vigorously. For example, when Irenaeus (c. 130–200 AD) began his refutation of Gnosticism in Book II of *Against Heresies*, he started not with the doctrine of Christ, but with what he called "the first and most important head," namely, the doctrine of "God the Creator, who made the heaven and the earth, and all things that are therein."[3] Similarly, the Nicene Creed, which reaches back nearly 1700 years and is accepted by all the major branches of Christianity as authoritative, begins by affirming "one God, the Father Almighty" who created "all things visible and invisible."[4] Many other affirmations of God as the Creator can be found in the early centuries of the church.[5] Thus, far from regarding the doctrine of creation as secondary, early Christians took it as the indispensable starting point for their theology.

Why were early Christians so insistent about the doctrine of creation? One obvious reason is that without God as Creator, the rest of the Christian story makes very little sense. Church historian Philip Schaff rightly observed that "without a correct doctrine of creation there can be no true doctrine of redemption."[6] According to the traditional Christian narrative, redemption is understood in light of the fall, and the fall is understood in light of a prior good creation. Thus, efforts to disassociate the doctrine of creation from the doctrines of redemption and the fall are likely to result in theological incoherence.

But there was another, more pressing reason why early Christians emphasized the doctrine of creation: They faced sharp opposition to the idea of God as Creator from the intellectual elites of their day. In many ways, that opposition foreshadowed debates over God and evolution in our own time. Perhaps there is no better way of gaining clarity about what is at stake theologically in today's debates over evolution than by understanding what was at stake in the conflicts over creation in the early church.

The Epicurean Materialists

DURING THE FIRST FEW centuries of the church, two distinct groups opposed the idea of God as Creator. Followers of the Greek atomists Democritus and Epicurus comprised the first group. They explicitly denied that the wonders of nature were produced by a designing intelligence, asserting that everything ultimately arose through a blind and impersonal material process involving the chance collisions of atoms.[7] In the century prior to the Christian era, the Roman poet Lucretius popularized Epicurean materialism in his epic poem *De Rerum Natura* ("On the Nature of Things"), where he proclaimed that "neither by design did the primal germs 'stablish themselves, as by keen act of mind." Instead, the colliding atoms continued "blow on blow, even from all time of old" until at last they combined fortuitously "into those great arrangements out of which this sum of things established is create[d]."[8]

Responding to the Epicureans' repudiation of design in nature, early Christians repeatedly argued that nature in fact provides compelling evidence that it was the product of a supreme intelligence. In their view, not only was design in nature real, it was plain and observable. Writing to Christians in Rome in the first century, the Apostle Paul argued that "since the creation of the world God's invisible qualities—his eternal power and divine nature—have been clearly seen, being understood from what has been made, so that men are without excuse" (Romans 1:20, NIV). Writing in the second century, Theophilus, Bishop of Antioch (c. 115–188 AD), likewise contended that "God cannot indeed be seen by human eyes, but is beheld and perceived through His providence and works."[9] What are these "works" through which we can see the intelligent activity of God? Theophilus went on to list the functional regularities of nature from astronomy, the plant world, the diverse species of animals, and the ecosystem. His conclusion? Just "as any person, when he sees a ship on the sea rigged and in sail, and making for the harbor, will no doubt infer that there is a pilot in her who is steering her; so we must perceive that God is the governor [pilot] of the whole universe."[10]

Writing in the third century, Dionysius, then Bishop of Alexandria (c. 200–265 AD), made similar arguments against those who claimed that the features of the universe were "only the works of common chance."[11] According to Dionysius, such persons fail to observe that "no object of any utility, fitted to be serviceable, is made without design or by mere chance, but is wrought by skill of hand, and is contrived so as to meet its proper use."[12]

Writing in the latter part of the third century and the early fourth century, Christian thinker Lactantius (c. 240–320 AD) likewise declared that "it is more credible that matter was made by God, because He is all-powerful, than that the world was not made by God, because nothing can be made without mind, intelligence, and design."[13] Lactantius went on to ask:

> If you had been brought up in a well-built and ornamented house, and had never seen a workshop, would you have supposed that that house was not built by man, because you did not know how it was built? You would assuredly ask the same question about the house which you now ask about the world—by what hands, with what implements, man had contrived such great works; and especially if you should see

> large stones, immense blocks, vast columns, the whole work lofty and elevated, would not these things appear to you to exceed the measure of human strength, because you would not know that these things were made not so much by strength as by skill and ingenuity?
>
> But if man, in whom nothing is perfect, nevertheless effects more by skill than his feeble strength would permit, what reason is there why it should appear to you incredible, when it is alleged that the world was made by God, in whom, since He is perfect, wisdom can have no limit, and strength no measure?

Such citations from the fathers of the early church could be multiplied.[14] Early Christians clearly and repeatedly taught that nature provides convincing evidence of God's design.

The debates between the early Christians and Epicurean materialists bear a striking resemblance to debates in our own day between theists and the so-called "new atheists" such as Richard Dawkins, Sam Harris, and Daniel Dennett.[15] Like the Epicurean materialists of old, today's "new atheists" repeatedly assert that nature is wholly the product of impersonal material forces rather than intelligent guidance. In the words of Dawkins, "[t]he universe we observe has precisely the properties we should expect if there is, at bottom, no design, no purpose, no evil and no good, nothing but blind pitiless indifference."[16] While materialists in the ancient world drew from the atomic theories of Democritus and Leucippus for support, the "new atheists" are inspired largely by the work of Charles Darwin, who is supposed to have shown how the apparent design observed throughout the biological world was actually produced by a blind and undirected process of natural selection acting on random variations. Much as early Christian thinkers responded to the Epicureans by pointing to the evidences of design in nature not easily explained as the result the chance collisions of atoms, today's intelligent design theorists have responded to the "new atheists" by pointing out evidences of design in physics, astronomy, cosmology, the origin of life, biochemistry and other fields that resist impersonal and purposeless explanations.[17] The scientific data being discussed today may be new, but the overall debate is not. The controversy over design in nature is one of the great and continuing debates in Western civilization.

Yet Epicurean materialists were not the only opponents of the idea of God as Creator in the early years of the church, and in many ways, they were not the opponents who most worried the early church fathers. That distinction fell to an eclectic group of writers and thinkers who became known collectively as the Gnostics. Unlike the materialists, the Gnostics considered themselves Christians. But that made matters worse in the eyes of the early church fathers, because it meant the Gnostics had a greater potential to confuse and mislead otherwise orthodox believers.

The Gnostic Heresy

What became known as the Gnostic heresy was complicated and esoteric, with many variations. Regardless of their differences, most Gnostics shared two key beliefs about God and the natural world: First, they denied that nature was created good—in their view, matter was evil and the material world was flawed from the start. Second, because the material world was evil, the Gnostics denied that God actually created it. Instead, they claimed that the world was created by another entity usually called the Demiurge (an idea borrowed from the ancient Greek philosopher Plato but modified in the intervening centuries). In creating the world, the Demiurge acted as if he were God, but in fact he operated blindly and ignorantly apart from God. According to Hippolytus:

> For the Demiurge, they say, knows nothing at all, but is, according to them, devoid of understanding, and silly, and is not conscious of what he is doing or working at ... he himself imagines that he evolves the creation of the world out of himself: whence he commenced, saying, "I am God, and beside me there is no other."[18]

The Gnostics' point was to disassociate God from any direct role in His creation, and thereby to deny that the world was the intentional and good result of God's specific design.

The leaders of the early church rejected the Gnostics' effort to distance God from His creation. Indeed, according to Irenaeus, the Gospel of John was written in part to counter these teachings of the Gnostics, especially of an early Gnostic known as Cerinthus: "Cerinthus ... taught that the world was not made by the primary God, but by a certain Power far separated from him, and at a distance from that Principality who is supreme over the universe, and ignorant of him who is above all."[19] It was to counter this claim that John 1:3 insists that "all things were made through" Christ who was God himself—not through a secondary entity like the Demiurge. Those today who intimate that Christians can dispense with the doctrine of God as Creator so long as they affirm Christ, run into a problem with John 1:3. The Apostle John couldn't be any clearer: If one denies that God was the direct agent of creation, one is also denying Christ. The same teaching is articulated by the Apostle Paul in Colossians (Col. 1:16) and by the writer of the Epistle to the Hebrews (Heb. 1:2).

Just as the debates between early Christians and the Epicureans resemble today's debates with the new atheists, the debates between early Christians and the Gnostics bear striking similarities to contemporary debates over theistic evolution. Indeed, in a certain sense, mainstream contemporary theistic evolution might well be regarded as a revival of Gnosticism.

Because the phrase "theistic evolution," like Gnosticism, can mean different things to different people, it would be helpful to define how the term is being employed here before going on. Broadly speaking, theistic evolution is an effort to reconcile Darwin's theory of unguided evolution with belief in God in general and Christian theology in particular.

Theistic evolution encompasses a wide array of approaches and views, which has generated considerable confusion about what its proponents actually believe. To a large extent, differences in opinion among theistic evolutionists are determined by how theistic evolutionists define both "theism" and "evolution." Does theism require a God who actively and intimately guides the development of life? Or does it allow a passive God who may not even know how the development of life ultimately will turn out? Alternatively, does evolutionary theory require an undirected process (as Darwin insisted)? Or can evolution include a process guided to specific ends by an intelligent cause? One's conception of theistic evolution will be markedly different depending on how one answers these questions.

In the initial decades after Darwin proposed his theory, theistic evolution typically was presented as a form of *guided* evolution. In this respect, most early forms of theistic evolution were incompatible with orthodox Darwinism. Theistic evolutionists of Darwin's era accepted the idea that there was a long history of life and that animals developed *via* descent with modification from a common ancestor. But they largely rejected Darwin's core contention that the development of life was a blind, undirected process dictated primarily by natural selection acting on random variations. As historian Peter Bowler points out, many of Darwin's contemporaries (including those in the scientific community) embraced the non-Darwinian idea "that evolution was an essentially purposeful process.... The human mind and moral values were seen as the intended outcome of a process that was built into the very fabric of nature and that could thus be interpreted as the Creator's plan."[20]

It is important to recognize just how strongly Darwin himself revolted against the view that evolution could have been guided toward particular goals. Darwin repeatedly made clear that natural selection neither required nor involved intelligent guidance:

> The term "natural selection" is in some respects a bad one, as it seems to imply conscious choice; *but this will be disregarded after a little familiarity*. No one objects to chemists speaking of "elective affinity"; and certainly an acid has no more choice in combining with a base, than the conditions of life have in determining whether or not a new form be selected or preserved.[21]

Indeed, according to Darwin, his law of natural selection provided a definitive refutation of the idea that the features of the natural world reflected a preconceived plan:

> The old argument of design in nature, as given by Paley, which formerly seemed to me so conclusive, fails, now that the law of natural selection has been discovered. We can no longer argue that, for instance, the beautiful hinge of a bivalve shell must have been made by an intelligent being, like the hinge of a door by man. There seems to be no more design in the variability of organic beings and in the action of natural selection, than in the course which the wind blows.[22]

Darwin was dismayed that the theistic supporters of evolution in his day rejected his formulation of evolution as an unguided process, and he was appalled when fellow naturalist Alfred Russel Wallace published an article advocating the idea of guided evolution.[23] Wallace shared credit with Darwin for developing the theory of evolution by natural selection, yet Wallace had concluded after looking at the evidence that human evolution had been guided by a "Higher Intelligence."[24] Darwin's most distinguished champion in the United States, Harvard botanist Asa Gray, espoused similar views. Not only did Gray believe that evolution was guided, in private he confessed that his teleological version of evolution was "very anti-Darwin."[25]

This widespread view of evolution as a purposeful process began to disintegrate early in the twentieth century after Darwinian natural selection underwent a resurgence due to work in experimental genetics.[26] Once Darwin's theory of undirected evolution became the consensus of the scientific community, the task for mainstream theistic evolution became considerably harder: Now one had to reconcile theism not only with the idea of universal common ancestry, but also with the idea that the development of life was an undirected process based on random genetic mistakes. How could God "direct" an "undirected" process? Modern theistic evolutionists have not offered clear or consistent answers. Instead, in recent years an increasing number of theistic evolutionists have explicitly advocated embracing evolution as an undirected process. Like the "new atheists" who draw on Darwin's theory in support of their views, these "new theistic evolutionists" appeal to Darwin's theory to justify a new theology that will fuse Christianity with unguided Darwinism.

But this new theology of theistic evolution turns out to be a repackaged version of the old theology of Gnosticism. Like the Gnostics of old, a growing number of theistic evolutionists explicitly deny that God actively guided the development of life, and they further deny that the material world was ever created originally "good."

Natural Selection as the New Demiurge

THE GNOSTICS OF OLD distanced God from his creation by assigning the creation of the world to a third party, the Demiurge. The Demiurge of the new theistic evolutionists is natural selection acting on random mutations. Just like the Gnostics' Demiurge, the Darwinian process acts ignorantly and blindly and apart from God's specific directions, allowing theistic evolutionists to disentangle God from the responsibility of a world they view as botched.

According to many new theistic evolutionists, God chose to "create" the world by setting up an undirected process over which he had no specific control and about which he did not even have foreknowledge of its particular outcomes. In a very real sense, God created a world that creates itself. In the words of Anglican theistic evolutionist John Polkinghorne: "an evolutionary universe is theologically understood as a creation allowed to make itself."[27] This view is hard to reconcile with traditional conceptions of God's foreknowledge and sovereignty, which becomes apparent when one reads the writings of leading theistic evolutionists. Former Vatican astronomer George Coyne claims that "not even God could know ... with certainty" that "human life would come to be."[28] Biologist Kenneth Miller of Brown University, author of the popular book *Finding Darwin's God* (used in many Christian colleges), flatly denies that God guided the evolutionary process to achieve any particular result—including the development of human beings. Insisting that "[e]volution is a natural process, and natural processes are undirected,"[29] Miller asserts that "mankind's appearance on this planet was not preordained, that we are here ... as an afterthought, a minor detail, a happenstance in a history that might just as well have left us out."[30]

No doubt recognizing that many of his fellow Christians would revolt at such a claim, Miller also states that "given evolution's ability to adapt, to innovate, to test, and to experiment, sooner or later it would have given the Creator exactly what He was looking for—a creature who, like us, could know Him and love Him."[31] But Miller is engaging in double-speak. He plainly does not believe that mankind (or any other outcome of evolution) represents an "exact" intention of God. In Miller's view, God apparently knew that the undirected process of evolution was so wonderful it would create *something* capable of praising Him eventually. But what that "something" would be was radically undetermined by God. Just how undetermined? At a 2007 conference, Miller admitted that if the history of evolution were to run again, the result "might be a big-brained dinosaur" or even "a mollusk with exceptional mental capabilities" rather than human beings.[32]

Modern theistic evolution is sometimes regarded as more "deistic" than theistic, but that comparison seems unfair to deism. Even deists typically believe that God originally designed and created the world before He left it to run itself. According to leading proponents of theistic evolution, however, God did not even do that much. God created an undirected process that then created the world, producing creatures He neither foresaw nor foreordained. Sometimes theistic evolutionists try to defend their view by claiming that just as God created human beings with free will, he bestows on nature the "freedom" to create itself without His direction.[33] While it is certainly a logical possibility that God could choose to create a world in this way, Christian theology provides scant support for the idea that this is how God actually acted in creating our world.

To the contrary, according to historic Christian teaching, God alone was Creator of our world. He did not delegate the task to an undirected third-party. No matter how metaphorically (or literally) one interprets Biblical discussions of creation, and no matter what the actual process God employed for bringing his creation into existence, the consistent claim of both the Bible and historic Christian theology is that the original creation embodied God's specific intentions. This does not mean that God could not have worked through secondary causes in creating the world. God very likely did, but they were not *undirected* secondary causes. That is the essential point. In historic Christian teaching, God—not a third party—is unquestionably the "master craftsman" as well as the artistic genius behind the exquisite order and beauty of the natural world.[34] By contrast, if one accepts the position of many theistic evolutionists, the only kind of artistry God seems to be permitted is that of the modern painter who creates a work by splattering paint haphazardly on empty canvass. God is more like Jackson Pollock than Michelangelo.

Some contemporary theistic evolutionists seem to recognize that their Darwinian conception of God creates tensions with traditional Christian theology. Perhaps that is why they tend to obfuscate just how much they adhere to Darwinism's conception of evolution as an undirected process. This seems to be the approach of those affiliated with the BioLogos Foundation started by Francis Collins to advance the cause of theistic evolution.

Although Collins and his foundation promote the books of champions of undirected evolution such as Kenneth Miller, its directors are more circumspect—or at least more muddled—about how undirected they think the evolutionary process might be.

Consider the writings of current BioLogos Vice President Karl Giberson (originally the Foundation's Co-President). On the one hand, Giberson says that "[a]s a believer in God I am convinced *in advance* that the world is not an accident and that, in some mysterious way, our existence is an 'expected' result."[35] On the other hand, Giberson states that he "side[s] with Darwin in rejecting the idea that God is responsible for the details."[36] Giberson glosses over what sorts of events in the history of life he considers "details." Are human beings a "detail" that God himself may not have intended? Giberson does not say in his book, and he did not provide much more clarity when asked publicly about his views at a forum at Biola University in 2009.[37] Just like Kenneth Miller, Giberson first stated: "I think there is reason to suppose that creatures like us were so predictable given the scenario that was provided that God anticipated that." But also just like Miller, Giberson indicated that "creatures like us" simply meant any "creatures that could have a special type of conscious awareness in relationship to the creator." He further admitted that "I think it's very hard to make a case that this particular form that we have is exactly what God intended." So would Giberson accept or reject a claim that human beings reflect God's specific intentions? "Well I'd want to know what 'specific' I guess meant there. I can't imagine that God just had to have creatures with five fingers and opposable thumbs and so on and until he got that he wasn't quite happy because that was his specific intention." When pressed on Miller's claim that Darwinian evolution was so undirected that it could have produced thinking clams or big-brained dinosaurs instead of us, Giberson responded:

> Well that's sort of an oxymoron. Because if you're thinking, you're not a clam, and if you're a clam you're not thinking. I mean he's making a point. We don't want to be a literalist. It's one thing to be a literalist with God's word; it's quite another to be a literalist with Ken Miller's word. I guess I would say that when Ken Miller is saying that he's saying that one could imagine a world where this kind of intelligence we have is embodied in a clam or in a dinosaur—that the image of God

and what God had in mind for our physical structure doesn't require the forms that we see before us.

We need to be clear that the issue here is not whether God *could* create rational beings other than human beings (of course he could), or even whether features like "five fingers and opposable thumbs" are necessary to being created in the image of God (of course they aren't). The issue is whether God was so uninvolved in the "details" of the evolutionary process that he did not determine whether evolution would produce thinking clams or thinking human beings. It is one thing to contend that Michelangelo could have created many other kinds of masterpieces than his famous statue of David. It is quite another thing to insist that he was so uninvolved in the sculpting process that he neither knew nor cared whether his sculpture would turn into the statue of a man or the statue of a clam—or, for that matter, a statue with twenty fingers instead of ten.

The trouble with trying to reconcile undirected Darwinian evolution with Christianity is not that God did not have the freedom to create the world differently, but that in the Darwinian scenario, strictly speaking, God is not the creator. At most, He set up a cosmic lottery to produce the forms of life and knew that with enough time He would eventually win some kind of jackpot. Whatever one thinks of this view of God, it is not the view offered by the Bible and the Christian intellectual tradition.

Like his colleague Giberson, Francis Collins is fuzzy about how far he is willing to go down the path of genuinely undirected evolution. On the one hand, Collins leaves open the possibility that God "could" have known and specified all of the outcomes of evolution (more about Collins's discussion of this possibility below). On the other hand, Collins delivered the keynote address to a 2008 conference on science and open theism (open theism is a position that explicitly denies God's exhaustive knowledge of future events).[38] Collins also sometimes writes as if he believes that significant parts of the history of life were in fact undirected by God, even when it comes to human beings. His discussion of so-called "junk DNA" is particularly instructive. He claims (wrongly, it turns out) that "roughly 45 percent of the human genome [is] made up of ... genetic flotsam and jetsam."[39] While conceding that "some might argue that these are actually functional elements placed there by the Creator for a good reason, and our discounting of them as 'junk DNA' just betrays our current level of ignorance," Collins ends up dismissing this explanation: "some small fraction of them may play important regulatory roles. But certain examples severely strain the credulity of that explanation." The clear implication of Collins's discussion of "junk DNA" is that he believes the human genome is "littered" with non-functional "junk" produced unintentionally during the undirected process of Darwinian evolution.

But Collins tries to shift the discussion from whether God knew and planned human beings to whether human beings really need their current set of physical features to be made in the image of God:

> How is this [idea that evolution "is full of chance and random outcomes"] consistent with the theological concept that humans are created 'in the image of God' (Genesis 1:27)? Well, perhaps one shouldn't get too hung up on the notion that this scripture is referring to physical anatomy—the image of God seems a lot more about mind than body. Does God have toe-nails? A belly button?"

This is a diversion. *Contra* Collins, the issue isn't whether God could choose to create a creature in his image without toe-nails. It is whether human beings as they do exist reflect God's specific choices and exquisite artistry or are the unintended by products of an undirected process.

To his credit, Collins seems to recognize there is a serious problem here for traditional Christian theists, and so after discounting the design of human beings in various ways, he offers an escape hatch for those who might be uncomfortable with his argument's trajectory. But the escape hatch has theological problems of its own.

God as the Cosmic Trickster

Collins suggests that God might have known and determined the outcomes of evolution from eternity but nevertheless created the world to *look* like it was produced by a random and undirected process:

> In that context, evolution could appear to us to be driven by chance, but from God's perspective the outcome would be entirely specified. Thus, God could be completely and intimately involved in the creation of all species, while from our perspective, limited as it is by the

tyranny of linear time, this would appear a random and undirected process."[40]

Collins's assertion that the biological world looks like the product of a "random and undirected process" places him in the peculiar position of being even less open to intelligent design in biology than Richard Dawkins, the world's foremost Darwinian atheist. Unlike Collins, Dawkins readily concedes that the biological world is rife with "complicated things that *give the appearance of having being designed for a purpose*."[41] In other words, Dawkins believes that things in biology look like they were designed; he simply thinks that Darwin's theory of unguided evolution provides sufficient reason for ignoring the clear appearance of design. Collins, by contrast, insists that the things in biology look "random and undirected," and only through the eyes of faith can we know that this appearance of non-design is deceiving. So Dawkins insists that things looked designed, but aren't, while Collins asserts that things don't look designed, but are.

Collins's escape hatch has the merit of being logically compatible with a more traditional understanding of God's sovereignty. His view would allow God to actively (if secretly) guide the development of His creation. But it still seriously conflicts with the Biblical understanding of God and His general revelation. Both the Old and New Testaments clearly teach that human beings can recognize God's handiwork in nature through their own observations rather than special divine revelation. From the psalmist who proclaimed that the "heavens declare the glory of God" (Psalm 19) to the Apostle Paul who argued in Romans 1:20 that "since the creation of the world His invisible attributes are clearly seen, being understood by the things that are made," the idea that we can see design in nature was clearly taught. Jesus himself pointed to the feeding of birds, the rain and the sun, and the exquisite design of the lilies of the field as observable evidence of God's active care towards the world and its inhabitants (Matthew 5:44–45, 48; 6:26–30). As discussed earlier, the observability of God's design in nature was a key theme in the writings of the early church fathers as well.[42] In his effort to head off a direct collision between undirected Darwinism and the doctrine of God's sovereignty, Collins seems to depict God as a cosmic trickster who misleads people into thinking that the process by which they were produced was blind and purposeless, even when it wasn't.

Whether Collins himself actually believes in the theological half-way house he has constructed to preserve God's sovereignty is unclear. As noted earlier, some of his other comments (e.g., his advocacy of the "junk DNA" argument) imply he harbors serious doubts about how much God actually directed the evolutionary process. Regardless of Collins's own view (and perhaps his own mind is ambivalent on the subject), many of his colleagues among the new theistic evolutionists forgo the half-way house and enthusiastically embrace the position that God neither knows nor intends the specific outcomes of evolution, which brings them to a position very similar to that of the early Gnostics. Just like ancient Gnosticism, much of modern theistic evolution replaces God as the active Creator of the world with a third party outside of God's specific direction and control.

Down playing God's active role as Creator is not the only way modern theistic evolution reinvents ancient Gnosticism. A second and equally striking parallel is the denial that the world was created originally good.

Denying the Fall

The Gnostics rejected orthodox Christian teaching that human beings were created good and then fell through a voluntary act of disobedience. In the Gnostics' view, the material world was never "good"; it was evil from the start. Leading theistic evolutionists today adopt a remarkably similar position. In their view, evolution makes unacceptable the idea that human beings were originally created good, which overturns the traditional Christian idea that human beings were created good and then fell into sin by a free choice. In 2001, Anglican Bishop John Shelby Spong explained how in his view Darwin made the traditional Christian account of redemption "nonsensical":

> I live on the other side of Charles Darwin. And Charles Darwin not only made us Christians face the fact that the literal creation story cannot be quite so literal, but he also destroyed the primary myth by which we had told the Jesus story for centuries. That myth suggested that there was a finished creation from which we human beings had fallen into sin, and therefore needed a rescuing divine presence to lift us back to what God had originally created us to be. But Charles

> Darwin says that there was no perfect creation because it is not yet finished. It is still unfolding. And there was no perfect human life which then corrupted itself and fell into sin, there was rather a single cell that emerged slowly over 4½ to 5 billion years, into increasing complexity, into increasing consciousness. And so the story of Jesus who comes to rescue us from the fall becomes a nonsensical story.[43]

Bishop Spong is widely known for his heterodox views on a variety of traditional Christian teachings, but one finds a similar denial of a real fall among some leading evangelical theistic evolutionists. Karl Giberson in *Saving Darwin* explicitly repudiates the idea that "sin originates in a free act of the first humans" and that "God gave humans free will and they used it to contaminate the entire creation."[44] In a section of his book subtitled "Dissolving the Fall," Giberson essentially argues that since human beings were created through Darwinian evolution, sin was there in human beings to begin with: "Selfishness ... drives the evolutionary process. Unselfish creatures died, and their unselfish genes perished with them. Selfish creatures, who attended to their own needs for food, power, and sex, flourished and passed on these genes to their offspring. After many generations selfishness was so fully programmed in our genomes that it was a significant part of what we now call human nature."[45] So in Giberson's view, human beings were sinful and flawed from their inception.

Giberson's repudiation of the traditional doctrine of the fall is obscured by his continuing usage of the term "fallen" in his book and public talks. Yet it is clear that for him the term "fallen" merely means that humans continue to be sinful, just like they were from the beginning. There was no actual "fall" in his view, as he frankly acknowledged during his appearance with me at Biola University in 2009:

> **John West:** Why do you continue to even use the word "fall"? ... Isn't your use of the word "fall" ... importing a theology that in fact you reject because there is no fall [in your view]? In your book you seem to say we're sinful to begin with: Selfishness drives evolutionary process so there wasn't a fall from anything—that's how we were originally developed, so the creation was flawed and sinful to begin with. Is that your view?

Karl Giberson: Yeah, no that's a fair description of my view. I was trying to be sort of consistent with the way theological language is used. There are a great many theologians—I remember reading essays by Karl Barth and Emil Brunner kind of arguing about original sin, and they talk about Adam and Eve and the fall in ways that sound almost fundamentalist, but neither of them accepted Adam and Eve as actual historical characters or the fall as an historical event. But that's theological language that has a particular meaning apart from what the English word itself entails. [46]

Francis Collins has been more circumspect on this topic in public than Giberson. In his 2006 book *The Language of God*, Collins gingerly skirts the issue of an historic fall, preferring to focus instead on whether Christians must believe in a literal Adam and Eve who were specially created by God with no biological antecedents. Collins appeals to a passage by C. S. Lewis in *The Problem of Pain* as his authority for believing that orthodox Christians can embrace the idea that the first human beings arose through a long period of animal evolution.[47] Although Collins fails to note Lewis's growing skepticism of Darwinism later in life,[48] he is right to point out that Lewis saw no theological objection to human beings sharing a common ancestor with the lower animals. The real problem with Collins's discussion of Adam and Eve is that it sidesteps the most serious challenge Darwinism poses for the Christian account of salvation by focusing on two lesser issues.

The literal existence of Adam and Eve and whether the first humans shared a common ancestor with other mammals are significant questions for many Christians. But there are two even more crucial ones: Did God originally create human beings morally good? And did the first humans become alienated from God by their own free choice? The historic Christian answer to both questions is an unequivocal yes, and C. S. Lewis himself forcefully articulated this historic Christian answer in the very passage Collins quotes, although Collins removes with ellipses some of Lewis's strongest comments on the subject.

Collins neglects to mention that his redacted quotation of Lewis comes from a chapter titled "The Fall of Man" in which Lewis explicitly defends an historic fall against those who contend that science has refuted it. According to Lewis, human beings really were morally good before the fall. "God came first" in man's "love and in his thought," and God received from man "obedient love and ecstatic adoration: and in this sense, though not in all, man was then truly the son of God, the prototype of Christ."[49] Lewis concludes his discussion by explaining that "the thesis of this chapter is simply that man, as a species, *spoiled himself*, and that good, to us in our present state, must therefore mean primarily remedial or corrective good."[50]

Yet Darwinism directly undermines this traditional Christian teaching. As Collins's friend and colleague Giberson points out, the Darwinian account of human evolution suggests that human beings were selfish and flawed from the very start. Hence, there can be no "fall" in the orthodox Darwinian view.

In *The Language of God*, Collins carefully refrained from either endorsing or rejecting the idea of an actual fall. But two years later he wrote a glowing foreword to Giberson's *Saving Darwin*, which contains Giberson's Darwinian repudiation of the idea. Giberson subsequently became a leader of Collins's pro-theistic evolution BioLogos Foundation. Given these facts, it might be reasonable to surmise that Collins agrees with Giberson's position. At the very least, he has no qualms about helping Giberson to promote his view.

Giberson rejects the idea that there is anything heterodox about his rejection of the fall, insisting that the fall "is inconsequential for our need for our salvation."[51] As he explains it, "I don't understand why we have to have been perfect and then fallen in order to be saved. I don't get this at all. It seems to me that no matter how we become sinful we can still be saved from that sin." Well, yes, if one is flexible enough, one can redefine the Christian concept of salvation to mean whatever one wishes. The point is that it will then be something very different from historic Christianity. In the traditional Christian view, salvation is the result of a perfect and holy God lovingly working to restore his creatures to a right relationship with him. In the new view of theistic evolution á la Giberson, salvation becomes the attempt of an absentee God to rescue his creatures from his own flawed creation in order to bring about a relationship that never existed.

Although Giberson may believe that his reformulation of the Christian narrative does no damage to the traditional concept of salvation, Darwinists who are former Christians might beg to differ. Noted historian of science Ron Numbers at the University of Wisconsin is a case in point. By his own admission, Numbers abandoned his belief in Christianity because of Darwinian evolution, and in an interview he outlined with stark clarity the theological implications that some theistic evolutionists do their best to evade. According to Christianity, said Numbers:

> We humans were perfect because we were created in the image of God. And then there was the fall. Death appears and the whole account [in the Bible] becomes one of deterioration and degeneration. So we then have Jesus in the New Testament, who promises redemption. Evolution completely flips that. With evolution, you don't start out with anything perfect.... There's no perfect state from which to fall. This makes the whole plan of salvation silly because there never was a fall.[52]

Contemporary theistic evolutionism—perhaps more accurately described as theistic *Darwinism*—is often presented to the public as a simple and common-sense solution to the conflict between Darwin and Christianity. But as the experience of Ron Numbers suggests, the problems raised by Darwinism for Christians are not so easily assuaged.

A New Theology?

Today's proponents of theistic evolution like to insist that the only tensions between Darwinian evolution and Christianity derive from narrow "fundamentalist" and "literalist" readings of the Bible. Given the negative connotations of the word "fundamentalist" in American culture, this charge is rhetorically clever. But it is also untrue. Darwin's conception of evolution as a blind and undirected process contradicts not just "fundamentalism" (whatever that is), but what C. S. Lewis liked to call "mere Christianity"—the core theological claims held in common by all major branches of Christianity for the past two millennia. Indeed, by rejecting directed evolution, the observability of design in biology, and an historic fall, mainstream theistic evolution proposes reconciling Christianity with Darwin by throwing overboard significant portions of historic Christian theology.

It is precisely because the disconnect between traditional Christian theology and mainstream evolutionary theory is so great that theistic evolutionists are now having to push for an overhaul of the traditional Christian message. Thus, Francis Collins has expressed a desire to bring together leading scientists, theologians, and pastors in order to "develop a new theology,"[53] while Karl Giberson has praised the "creative theology" of those seeking to reconcile Christianity with Darwinism.[54]

But as we have seen, the "new" theology of contemporary theistic evolution bears a striking resemblance to the old theology of Gnosticism that was repudiated by the early church. Before embracing this heterodox new theology, pastors and theologians need to be absolutely sure they have good reasons for what they are doing. Yet, as the next chapter will show, the reasons behind the resurgence of theistic evolution leave much to be desired.

2. Having a Real Debate

Theistic Evolution, the Early Church, and the Return of Gnosticism, Part 2

John G. West

One reason for retooling the Christian message to accommodate Darwinian Theory is presumably to make Christianity more attractive to the scientific community. However, the "new theology" of theistic evolution seems no more convincing to most Darwinists than the old theology of traditional Christianity. While the media often showcase high-profile theistic evolutionists such as Kenneth Miller and Francis Collins, such figures remain far from representative of modern evolutionary biology. The fact that this truth has been effectively obscured by the news media is largely a tribute to what Richard Dawkins drolly calls the "evolution defense lobby,"[1] which has mounted a concerted public relations offensive in recent years to put a faith-friendly face on the field of evolutionary biology.

Darwinian Evangelism

One of the leaders of this public relations effort has been Eugenie Scott, a self-described "evolution evangelist" and executive director of the National Center for Science Education (NCSE), the nation's preeminent pro-Darwin lobbying group.[2] Recognizing that most Americans believe in God, and that if evolution is seen as anti-religious it will continue to lag in popular support, Scott spearheads a variety of efforts to convince religious believers that Darwinian evolution and religion are compatible. On a federally funded website that the NCSE helped design, public school teachers and students are directed to a list of statements by religious groups endorsing evolution.[3] Scott, meanwhile, encourages biology teachers to spend class time having students read statements by religious leaders supporting evolution. She even suggests that students be assigned to interview local ministers about their views on evolution—but not if the community is "conservative Christian," because then the lesson that "Evolution is OK!" may not come through.[4] The NCSE's effort to inject religion into public school science classes in order to promote acceptance of Darwin's theory is a remarkable act of chutzpah for an organization that routinely chastises "antievolutionists" for supposedly trying to insert "religion" into science classes. Apparently, bringing religion into biology class is OK so long as it is used to endorse Darwin's theory.

The NCSE also advises evolution supporters to invite ministers to testify before school boards in favor of evolution,[5] and it has created a curriculum to promote evolution in churches.[6] The NCSE's former "Faith Network Director" assures people that "Darwin's theory of evolution ... has, for those open to the possibilities, expanded our notions of God."[7] Other evolutionists have collected signatures from clergy in support of evolution as part of "The Clergy Letter Project" and have urged churches to celebrate "Evolution Sunday" and "Evolution Weekend" on the weekend closest to Darwin's birthday.[8]

This attempt to reinvent the image of evolutionary biology as friendly to faith is an effort to deal with what might be called evolution's "Dawkins problem." Biologist Richard Dawkins is one of the world's best-known boosters of Darwinian evolution. Unfortunately for evolutionists, Dawkins zealously expounds the clear anti-religious implications of the theory and regularly denounces religion. One of his choicer comments is his description of religious faith as "one of the world's great evils, comparable to the smallpox virus but harder to eradicate."[9] Groups like the NCSE undoubtedly hope to reframe the faith and evolution debate by marginalizing Dawkins as a fringe figure whose views are not representative of Darwinists as a whole.

Ironically, many of the leaders of the "evolution defense lobby" who are so eager to spread the message that Darwinism is compatible with faith are themselves atheists, something Richard Dawkins relishes pointing out: "They are mostly atheists, but they are wanting to—desperately wanting—to be friendly to mainstream, sensible religious people. And the way you do that is to tell them that there's no incompatibility between science and religion."[10] The same Eugenie Scott who tells the public that evolution and faith are compatible also signed the "Humanist Manifesto III," which celebrates "the inevitability and finality of death" and proclaims that "humans are ... the result of unguided evolutionary change" as part of its case for "a progressive philosophy of life ... without supernaturalism."[11] The same *Scientific American* columnist Michael Shermer who insists that "evolution fits well with good theology"[12] is an avowed atheist who writes that "Science Is My Savior," because it helped free him from "the stultifying dogma of a 2,000 year old religion."[13]

Sometimes atheist defenders of evolution who are trying to build bridges to the faith community are more forthright in spelling out what they mean by the compatibility of evolution and faith. In his book *Living with Darwin* (2007), philosopher of science Phillip Kitcher of Columbia University criticizes the harshness of atheist Darwinists like Richard Dawkins because he thinks it would be more effective to defuse religious opposition to Darwin with honey rather than brimstone. Kitcher's proposed relief to the tension between Darwinism and Christianity is to convince Christians to "spiritualize" their religion. What does that mean? Christians must "abandon almost all the standard stories about the life of Jesus ... the extraordinary birth, the miracles, the literal resurrection" and even reject the "promise of eternal salvation" as "literally false."[14] As long as one does that, according to Kitcher, there need be no irreconcilable conflict between Darwin and Christianity.

Of course, in addition to the predominately atheist leaders of the "evolution defense lobby," there are sincere Christians in the scientific community who embrace theistic evolution like Francis Collins and those at his BioLogos Foundation. But the hard reality is that they represent a small eddy, not the mainstream of evolutionary biology. In fact, the closer one gets to fields dominated by Darwin's theory—and the higher one climbs on the ladder of scientific prestige—the fewer conventional theists one will find. Theistic evolution does not appear to be a live option for the majority of biologists. According to a national survey of faculty at both four-year and two-year colleges and universities published in 2007, sixty-one percent of college biology faculty consider themselves atheists or agnostics, a level of unbelief not only dramatically higher than in the general population but than in most other academic disciplines.[15] The anti-religious bent is even more pronounced among leading biologists. Nearly 95 percent of the biologists in the National Academy of Sciences (NAS) describe themselves as atheists or agnostics, again a percentage of unbelief far higher than any other scientific discipline represented by the NAS.[16] Similarly, according to a 2003 Cornell survey of leading scientists in the field of evolution, 87 percent deny the existence of God, 88 percent disbelieve in life after death, and 90 percent reject the idea that evolution is directed toward an "ultimate purpose."[17] Public relations efforts by the evolution defense industry notwithstanding, Richard Dawkins is far more representative of the beliefs of evolutionary biologists than Francis Collins.

This shouldn't be surprising. While Darwinian Theory does not necessitate atheism, its contention that all of the marvelous beauty and complexity of life are the products of a blind and undirected process certainly makes the case for traditional theism seem less plausible. Darwinian Theory has a similar corrosive impact on traditional conceptions of morality. If human behaviors and beliefs can be explained simply as byproducts of the struggle for survival, then it is hard to supply an objective standard for preferring traditionally moral behaviors over traditionally immoral ones. After all, according to Darwinism, the same struggle for survival that produced the maternal instinct also produced infanticide; the same process that produced kindness also bred cruelty. In Darwinian terms, all practices found in nature are equally natural and therefore equally normative. There is no Darwinian basis for distinguishing some natural behaviors as "good" and others as "evil." One can certainly do this if there exists an independent standard of right and wrong outside of nature, but that is precisely what orthodox Darwinism denies.

To their credit, many theistic evolutionists try to draw a line at Darwinian explanations of morality. Francis Collins, for example, has described the moral law as something unique to human beings that cannot be explained away by a Darwinian process. Influenced by C. S. Lewis, Collins cites the moral law as convincing evidence of a personal God who cares about the world:

> Where did that [moral law] come from? I reject the idea that that is an evolutionary consequence, because that moral law sometimes tells

> us that the right thing to do is very self-destructive. If I'm walking down the riverbank, and a man is drowning, even if I don't know how to swim very well, I feel this urge that the right thing to do is to try to save that person. Evolution would tell me exactly the opposite: preserve your DNA. Who cares about the guy who's drowning? He's one of the weaker ones, let him go. It's your DNA that needs to survive. And yet that's not what's written within me.[18]

Although Collins's effort to sustain an objective moral law is laudable, his line-in-the-sand seems rather *ad hoc* within his own framework. If one truly accepts Darwinian theory, there is no principled basis to exempt any capacity of human beings from the evolutionary process, and no orthodox Darwinist would ever accept the idea that Darwinian evolution does not fully explain man's capacities, including his mental and moral faculties. Darwin himself wrote an entire book—*The Descent of Man*—to demonstrate that all of man's intellectual, moral, and even spiritual faculties were produced by the same blind and undirected process that produced his bodily appendages.[19] When fellow scientists such as Alfred Wallace and St. George Jackson Mivart dared to disagree with Darwin about whether natural selection could explain such human faculties, Darwin became upset; by his lights, their views were heretical.[20]

Because Darwinian Theory purports to offer an all-encompassing explanation, the effort to contain its reach into human ethics and culture is hard to maintain. To be consistent, one either has to acknowledge that orthodox Darwinism is at least partially incorrect (and therefore it does not explain morality), or one eventually has to concede that Darwinism explains morality just as it explains every other feature in the biological world. Collins himself may be evolving toward the latter option. Prominent Darwinist Michael Shermer stated in a public debate in 2009 that Collins "conceded to me last week that he now largely accepts my own evolutionary theory on the origins of morality—that you don't even need an outside source for that, that God could have built into the system, through evolution, a development of a moral sense of right and wrong."[21]

Although one can try to ignore the theological and ethical implications of Darwinism by an act of will rather than logic, it is easy to see why atheist

Darwinists like Daniel Dennett view Darwinism as a "universal acid" that eats away traditional religion and morality. That is precisely how Darwinism has functioned in field after field over the past century.[22] Sadly, Darwinism seems to act just as much as a corrosive acid for some proponents of theistic evolution as it does for the champions of atheistic evolution.

Indeed, some contemporary theistic evolutionists have a hard time explaining exactly why they continue to believe in God at all. Consider Karl Giberson's rather depressing explanation near the end of *Saving Darwin*:

> As a purely *practical* matter, I have compelling reasons to believe in God. My parents are deeply committed Christians and would be devastated were I to reject my faith. My wife and children believe in God, and we attend church together regularly. Most of my friends are believers. I have a job I love at a Christian college that would be forced to dismiss me if I were to reject the faith that underpins the mission of the college. Abandoning belief in God would be disruptive, sending my life completely off the rails.[23]

Giberson's "compelling reasons" to believe in God are sociological; they are not about whether Christianity is actually *true*.

One wonders whether in ten years Giberson will go the route of former Calvin College professor Howard Van Till. In the early to mid 1990s, Van Till was the great champion of theistic evolution against Phillip Johnson, author of the book *Darwin on Trial* and an early figure in the modern revival of the idea of intelligent design. Van Till was supposed to show how an evangelical Christian could be wedded to Darwinism and still keep his faith. Van Till makes an appearance in Giberson's book as one of the "committed Christians" from "conservative evangelical traditions" who support theistic evolution. Unfortunately Giberson has not stayed up to date. Since retiring from his Christian college, Van Till has evolved well beyond Christianity and now considers himself a "freethinker."[24]

This critique will frustrate some theistic evolutionists who want badly to support Darwinian evolution and yet remain Christians. They don't want to think of themselves as double-minded. But the reality is that both orthodox Darwinism and orthodox Christianity point in different directions, and it is difficult to retain a permanent foothold when being pulled from opposite sides. Eventually the stronger force will prevail.

The Resurgence of Theistic Evolution

Although theistic evolution may have gained little traction with the atheists who dominate the field of evolutionary biology, it has become much more high profile in recent years in the faith community. This is due in part to extensive funding by the John Templeton Foundation, which has spent more than $20 million since 1996 on projects that promote theistic evolution as a major part of their mission, including start-up funding for two new groups to promote theistic evolution: the Faraday Institute in England, and the BioLogos Foundation in the United States.[25] The BioLogos Foundation explicitly includes as part of its mission "promoting dialog,"[26] which has been one of the key long-term goals of Francis Collins: "I think it's critical that we have a meaningful dialogue between people of faith and people involved in science."[27]

Christians should always be open to discovering fresh theological insights, and Collins is to be lauded for encouraging more dialogue on issues of faith and science. But the dialogue needs to be real, which means including voices that may not agree with each other. Unfortunately that does not appear to be what Collins had in mind. In late 2009 the BioLogos Foundation convened a closed gathering of evangelical Christian leaders ostensibly to promote dialogue on faith and science issues. But the range of scientific voices permitted at the meeting was sharply limited. Held in New York City, the gathering intentionally excluded any scientists or philosophers of science who were critical of Neo-Darwinism or supportive of intelligent design in biology. That was in keeping with the purpose of the grant from the Templeton Foundation, which was not directed at promoting an open exchange of views, but at convincing evangelical leaders to embrace evolutionary biology. In the words of the grant description on Templeton's website, the "invitation-only workshops" arranged by the BioLogos Foundation are designed to "bring scientists and evangelical leaders together to seek a theology more accepting of science, specifically evolutionary biology."[28]

At the same time Collins promoted one-sided "dialogue" at his own events, he refused to participate in public forums organized by others that include scholars who support intelligent design. Of course, Collins has the right to decline to participate in discussions with scholars who do not share his views. But no one should confuse the one-sided exchanges that result with genuine dialogue.

The theistic evolutionists' rationale for excluding scientific views critical of Darwinism from faith-evolution discussions seems to be that Darwinian Theory constitutes the consensus view of science, and questioning that consensus view supposedly demonstrates that one is "anti-science." But this way of thinking betrays a strikingly unsophisticated understanding of science. The more one knows about the history of the scientific enterprise, the more skeptical he or she is likely to be about equating the current consensus view of science with science itself.[29] Science is a wonderful human enterprise, but scientists can be as blinded by their prejudices as anyone else. In the last century, for example, the nation's leading evolutionary biologists at such institutions as Harvard, Princeton, Columbia, and Stanford enthusiastically promoted eugenics, the "science" of breeding better humans based on the principles of Darwinian biology. Eugenics was the "consensus" view of the scientific community for decades.[30] That did not make it good science. Eugenics provides a stark reminder about the susceptibility of even mainstream scientists to junk science.

Far from being anti-science, dissenting views in the scientific community help science thrive by counteracting groupthink and sparking debates that can lead to fresh discoveries. It is not too much to say that today's dissenting opinion in science may well turn into tomorrow's scientific consensus. So it is disheartening to come across otherwise thoughtful pastors or theologians who blindly accept claims that the consensus view of evolutionary theory is all that matters, or who have been lulled into believing that anyone who dissents from modern evolutionary orthodoxy is guilty of "opposing science."

In reality, a growing number of scientists and philosophers have offered sophisticated critiques of the power of undirected natural selection and random variation from within science itself.[31] These critiques are based on empirical investigations and standard modes of scientific reasoning, not on deductions from religious authority; and they are carried out by those with a love for the scientific enterprise. For example, research published by molecular biologist Douglas Axe in the *Journal of Molecular Biology* shows the astonishing rarity of certain working protein sequences, raising significant questions about how a blind process of natural selection acting on random mutations could generate them. In the words of Axe, the rarity of these working protein sequences among all the possible combinations is "less than one in a trillion trillion trillion trillion trillion trillion."[32] Before nature can build animals like mice, muskrats, or men, it first needs to build working protein sequences. If Darwinian evolution cannot account for the generation of such sequences, it can hardly account for the development of living organisms with even higher levels of complexity.

Similarly, multi-year experiments by University of Wisconsin-Superior biologist Ralph Seelke have demonstrated the sharp limits of what natural selection can do to evolve new functions in bacteria. Seelke explains that he has tested "the ability of evolution to produce a new function when two changes are both needed at effectively the same time ... in a population of trillions of bacteria and over thousands of generations."[33] The results? "A requirement for two changes effectively stops evolution."

Skepticism toward modern Darwinism's all-encompassing claims regarding natural selection and random mutation is far more widespread among scientists than most people realize. Despite the high cost of publicly airing one's doubts about Darwin, more than eight hundred doctoral scientists—from institutions such as Princeton, Ohio State, University of Michigan, and MIT—have signed a statement expressing their skepticism that the Darwinian mechanism is capable of explaining the complexity of life.[34] Geneticist Lynn Margulis at the University of Massachusetts—a critic of intelligent design—bluntly states that "new mutations don't create new species; they create offspring that are impaired."[35] National Academy of Sciences member Phillip Skell argues that "Darwinian evolution has functioned more as a philosophical belief system than as a testable scientific hypothesis. This quasi-religious function of the theory is ... why many scientists make public statements about the theory that they would not defend privately to other scientists."[36]

Before theologians and pastors decide to reinvent their theology in order to make it consistent with undirected Darwinism, they first ought to make sure that undirected Darwinism is actually true. But that cannot be done by listening only to the so-called consensus view of science; it also requires hearing the views of reasoned dissenters from the consensus.

Theistic evolutionists who want to exclude critics of undirected evolution from faith-science discussions betray an extraordinary level of insecurity. If the evidence for unguided Darwinism is really as overwhelming as they think, they should have nothing to fear from an open exchange of ideas. Unfortunately, the reticence of theistic evolution proponents to interact seriously with the scientific critics of Darwinism has bred in some of them a condescending attitude that is less than conducive to fruitful interchange. So certain they are right about the science, these theistic evolutionists do not really listen to what their critics are saying, let alone respond fairly to the issues they raise.

Francis Collins's superficial critique of intelligent design in *The Language of God* is a good example.[37] While his critique showed that he undoubtedly had read the critics' caricatures of intelligent design, it provided little indication that he had read what ID proponents have to say for themselves, including their responses to their critics. Superficial critiques of intelligent design are not limited to Collins, however. In 2010, the BioLogos Foundation commissioned a review of Stephen Meyer's book *Signature in the Cell: DNA and the Evidence for Intelligent Design* by evolutionary biologist Francisco Ayala.[38] But it was readily apparent from the review that Ayala had not read the book he critiqued (indeed, his review did not even get the title of Meyer's book right).[39] The same book Ayala apparently thought was too unimportant to treat seriously was endorsed by a member of the U.S. National Academy of Sciences, praised by one of Britain's top geneticists, and selected as a book of the year in the *Times Literary Supplement* by leading atheist philosopher Thomas Nagel.[40] Theistic evolutionists lose the opportunity to gain fresh insights as well as to refine their own views when they are so reflexively dismissive of the works of scholars with whom they disagree.

Even more troubling than straw-man critiques is the tendency of some theistic evolutionists to respond with thinly veiled disdain or contempt whenever their views are questioned. When Discovery Institute launched its Faith and Evolution website (www.faithandevolution.org) to explore both the scientific and cultural implications of the Darwin debate, Karl Giberson denounced the site as "slick, well-resourced, rhetorically clever, profoundly misleading, and almost completely devoid of any real science." Giberson's own broadside might have been charitably described as "almost completely devoid of any real substance," because it made patently false claims with no evidence to back them up.[41] Rather than refute the actual content of the Faith and Evolution website (which is extensive), Giberson's approach was to insinuate that intelligent design proponents are insincere or disingenuous (hence their "slick" and "profoundly misleading" website).

Sometimes the contempt expressed by theistic evolutionists is not veiled at all. In 2009, I engaged in an extensive online exchange with theistic evolutionist Stephen Barr. Barr took umbrage at a short article I wrote for the *Washington Post*'s "On Faith" website that briefly critiqued Francis Collins's view of theistic evolution.[42] As the exchange continued, Barr adopted an increasingly dismissive and exasperated tone, as if he found it beyond endurance that anyone would question or disagree with his views on evolution. By the end of the exchange, Barr was accusing me of questioning his integrity simply because I disagreed with some of his arguments.[43] Needless to say, this sort of approach is not conducive to open discussion. Superciliousness is not an argument, and if theistic evolutionists want genuine dialogue, they need to get beyond the attitude of dismissing anyone who disagrees with them as contemptible.

Not every theistic evolutionist adopts such an attitude; and not every critic of theistic evolution treats his opponents with the respect that they deserve. In the heat of the moment, it can be difficult to keep the focus on arguments rather than personalities. But the dismissive attitude among theistic evolutionists is sufficiently widespread that I think it needs to be called out.

Whether they realize it or not, theistic evolutionists' attempt to narrow the science-faith dialogue to the scientific status quo reinforces efforts by secular Darwinists to blacklist their opponents from the scientific community by firing, harassing, or denying tenure to scientists who are skeptical of unguided Darwinism or supportive of intelligent design. Italian geneticist Guiseppe Sermonti was right to call Darwinism "the 'politically correct' of science."[44] Those who do not embrace it are likely to find themselves excommunicated from the scientific community. When theistic evolution proponents seek to exclude scientists and scholars who criticize Darwinism from the dialogue about faith and evolution, they in effect legitimize the efforts of others to eject such scholars from the intellectual community as a whole.

Theistic evolutionists typically downplay the discrimination faced by Darwin dissenters, but those who are fair-minded need not look far for evidence of such persecution.[45] As a former tenured professor at an evangelical Christian university, I can attest to efforts by certain proponents of theistic evolution to blacklist supporters of intelligent design not only from the natural sciences but from the humanities and social sciences as well.

In their zeal for promoting Darwin, some theistic evolutionists exhibit an almost uncritical deference to the authority of scientists to set the agenda for the rest of the culture. They seem to assume that because the current majority of scientists embraces undirected Darwinism, that should be the end of the discussion for everyone else. Consciously or not, their reasoning reveals a tendency toward scientism, a belief in the primacy of science over every other kind of cultural authority. This underlying scientism undoubtedly impacts other areas besides the Darwin debate, making it less likely that theistic evolutionists will challenge an existing consensus of the scientific community on any issue of consequence. Perhaps this mindset explains in part why Francis Collins is just as unwilling to challenge the scientific majority on bioethical issues as he is on Darwinian evolution. Collins's support for embryonic stem cell research, his ambivalence toward the abortion of handicapped children, and his effort to raise doubts that human life begins at conception all mirror the conventional views of the current scientific elite.[46]

Theistic evolutionists who veer toward scientism are merely following the existing elite culture, where in recent years the authority of scientific experts has been employed increasingly as a trump card to shut down public discussion on issues ranging from embryonic stem cell research to sex education to climate change.[47] It is telling that since the 1990s there has been a dramatic increase in what some have called the "authoritarian tone" of science, exemplified by the growing use in science journalism of phrases such as "science requires" and "science tells us we should."[48]

Many decades ago, writers such as C. S. Lewis, G. K. Chesterton, and Richard Weaver predicted the danger to a free society of such an uncritical deference to those speaking in the name of science.[49] Pastors, priests and other religious leaders especially should be reluctant to relinquish their right to raise an independent voice in discussions of science and culture, and they should be especially protective of the right of others to participate in the conversation. For a Christian believer, "science tells us we should" can never be tantamount to "Thus saith the Lord," no matter what the prevailing culture might claim.

Modern theistic evolution raises serious challenges to all major branches of the Christian tradition, and those challenges require a full and open airing. Pastors, priests, and other church leaders need to make sure that they are not short-changed by a one-sided dialogue that prevents them from hearing all of the relevant facts and arguments. As Darwin himself acknowledged, "A fair result can be obtained only by fully stating and balancing the facts and arguments on both sides of each question."[50]

3. Smelling Blood in the Water

Why Theistic Evolution Won't Appease the Atheists

Casey Luskin

A RABBI ONCE TAUGHT THAT PEACE AND TRUTH ARE BOTH NOBLE goals, but in this imperfect world, it's not always possible to have both. Wisdom lies in knowing when to pursue peace, and when to stand one's ground and defend the truth.

This sage advice is relevant today. In fact, religious believers have argued incessantly about whether now is the time for peace and reconciliation with Darwinian evolution, or a time to divide over certain core principles.

On the one hand, atheist proponents of Darwinian Theory have disagreed over how to respond to perceived threats from religion: *Should atheists unleash full-scale war against all religion, or should they forge alliances with "moderate" religious communities, hoping thereby to more efficiently destroy those that refuse to compromise?*

On the other hand, religious communities have had comparable conversations about how to deal with Neo-Darwinian evolution: *Does Neo-Darwinism threaten fundamental orthodox religious beliefs, or is it an innocuous scientific theory that should be embraced, lest religion suffer embarrassment for inhibiting scientific progress?* Those who hold the latter view may even argue that opposing Darwinian Theory creates an unnecessary war between science and religion. By making peace with Neo-Darwinism, they believe, the atheist war against religion will dissipate.

Other chapters in this volume investigate the relationship between Neo-Darwinism and people of different faith traditions. In this chapter, I will examine the debates among Darwinian *atheists* about how to deal with religion. I hope to show that some Darwinian atheists feign friendship toward religion for strategic reasons, since they believe that religion will more easily yield to atheism in this way than it would if atheists attack religion outright.

More importantly, I intend to show that religious persons who embrace Darwinian evolution will never appease the atheist community as a whole. Many atheists believe that Neo-Darwinian science guts theism to its core, and they view religion as fundamentally inimical to scientific progress. So-called "new atheists" such as Richard Dawkins and Daniel Dennett go even further: In their view, full-scale war against religion is necessary until *all* religion—whether ostensibly Darwin-friendly or not—is destroyed. As a result, religious advocates for Darwin are, in many cases, allying themselves with people who favor the downfall of religion.

The Two Camps of Neo-Darwinism

When it comes to dealing with religion, proponents of neo-Darwinism fall roughly into two camps. First, there are those who view the modern theory of evolution as a vehicle for directly imposing atheism on society. Cornell's William Provine captures this position, proclaiming that "[e]volution is the greatest engine of atheism ever invented."[1] Likewise, philosopher Daniel Dennett describes Darwinism as a "universal acid," which "eats through just about every traditional concept,"[2] including religion. In recent years, a group calling itself the "new atheists" has taken up this banner.[3]

The second faction of modern evolutionists is a coalition of theistic evolutionists and non-theistic evolutionists styling themselves as kinder, gentler atheists. New atheist leader Richard Dawkins has collectively branded this group the "evolution defense lobby" (which I will shorten to just the "evolution lobby"):

> There's a kind of science defense lobby, or an evolution defense lobby in particular. They are mostly atheists. But they are desperately wanting to be friendly to mainstream sensible religious people. And the way you do that is to tell them that there isn't any incompatibility between science and religion.[4]

The evolution lobby's stated goal is to advance scientific literacy, public acceptance of Darwinian evolution, and supposedly nothing more. Eugenie Scott, executive director of the National Center for Science Education

(NCSE)—the cornerstone of the evolution lobby—explains that for her camp "the most important group we work with is members of the faith community."[5] Thus, though she herself is a signer of the Third Humanist Manifesto,[6] Scott counsels teachers to send students to interview pro-Darwin clergy in order to "stres[s] the compatibility of theology with the science of evolution."[7]

Where the two camps wholly agree is that the level of public acceptance of Neo-Darwinian evolution is depressingly low[8] and that they must take strong measures to reverse that trend.

Atheists in both camps yearn for the day when science is restored to its "rightful place"[9] in society. Both camps also believe that religion is the main cause of the public's rejection of Darwinism. What *is* in dispute between them is how to deal with religion.

The new atheists believe that religion is fundamentally evil and opposed to scientific progress and must therefore be greatly diminished, if not eliminated outright. Atheists in the evolution lobby might agree with that sentiment, but they prefer an incrementalist approach. They fear the atheist minority cannot defeat a religious majority if attacked frontally. Thus, "evolution lobby" atheists join hands with theistic evolutionists—for now at least—in order to overcome socially entrenched religious groups that challenge Darwinism.

These disagreements about how to deal with religion have led to debates between the new atheists and the evolution lobby atheists. Even evolution lobbyist Eugenie Scott admits that although she's "a philosophical materialist," she's taken "some lumps for being so conciliatory" toward religion.[10] One atheist scientist who has given the NCSE a few lumps is University of Chicago evolutionary biologist Jerry Coyne:

> The pro-religion stance of the NCSE is offensive and unnecessary—a form of misguided pragmatism.... The directors of the NCSE are smart people. They know perfectly well—as did Darwin himself—that evolutionary biology is and always has been a serious threat to faith. But try to find one acknowledgment of this incompatibility on their website. No, all you'll find there is sweetness and light. Indeed, far from being a threat to faith, evolution seems to reinforce it! Is it

> disingenuous to be a personal atheist, as some NCSE officials are, and yet tell others that their faith is compatible with science? I don't know. But the NCSE's pragmatism has taken it far outside its mandate. Their guiding strategy seems to be keep Darwin in the schools by all means necessary.[11]

Eugenie Scott's position makes sense, however, given her stated views. Scott describes herself as an "evolution evangelist"[12] whose organization's "goals are to help people understand evolution and hopefully accept it."[13] While Scott of course insists that NCSE's "goals are not to promote disbelief,"[14] she admits that for "someone who is serious about religion, Darwinian evolution needs to be coped with, and it may not be psychologically easy."[15] Nonetheless, in Scott's view "the job of a science teacher is to teach state-of-the-art science, and that means evolution," because "students who do not understand evolution cannot be said to be scientifically literate."[16] Yet she has acknowledged her view that evolution is "threatening" to those who take "a human exceptionalism kind of view, that humans are separate from nature and ... special to God as in some Christian traditions."[17]

Indeed, not all statements coming from the NCSE are religion-friendly. In early 2009, the NCSE released a set of "talking points" for activists testifying before the Texas State Board of Education, arguing, "Science posits that there are no forces outside of nature. Science cannot be neutral on this issue. The history of science is a long comment denying that forces outside of nature exist, and proving that this is the case again and again."[18] The talking points even encouraged activists to wax theological: "All educated people understand there are no forces outside of nature."[19]

This suggests the NCSE's posture to religious allies may be just that—a posture.

ENTERING THE CAMP OF THE PRAGMATIC EVOLUTION LOBBY

EVOLUTION LOBBYISTS COURT THE religious community as part of a long-standing public relations campaign to convince the largely skeptical and religious public that Darwinian evolution and religion are "fully compatible." This campaign has support from the highest echelons of the scientific community—including the U.S. National Academy of Sciences (NAS), which has been leading the charge for decades. Since 1981, the NAS has released many statements such as: "Religion and science are separate and mutually exclusive realms,"[20] and: "[T]he evidence for evolution can be *fully compatible* with religious faith."[21]

The NAS's official denial of any tension between evolution and religion might be easier to accept were it not that, as a survey in *Nature* reported, "rejection of the transcendent by NAS natural scientists" is "near universal."[22] What is more, atheism is highest among NAS biologists—where only 5.6 percent profess belief in God.[23] The authors of the study, historian Edward J. Larson and journalist Larry Witham, are skeptical of the official assurances from the NAS:

> The [NAS's *Science and Creationism*] booklet assures readers, "Whether God exists or not is a question about which science is neutral." NAS President Bruce Alberts said: "There are many outstanding members of this academy who are religious people who believe in evolution, many of them biologists." Our survey suggests otherwise.[24]

Of course these 94.4 percent of NAS biologists are not the only atheists who preach that evolution and religion are compatible while themselves rejecting belief in God. In his 2006 book *Why Darwin Matters: The Case Against Intelligent Design,* Michael Shermer presses his religious readers to accept Darwinism. He asks: "If you are a theist, what could possibly proclaim the greater glory of God's creation more than science, the instrument that has illuminated more than any other tool of human knowledge the grandeur in this evolutionary view of life?"[25]

Yet this same Shermer is the founder of The Skeptics Society. He says he "was once a born-again evangelical Christian," but now is an "atheist" or "skeptic."[26] He has proclaimed that "[t]here is no God, intelligent designer, or anything resembling the divinity as proffered by the world's religions."[27] Shermer once wrote: "Science is my Savior"[28] and even argues, "[W]ere we to take a strictly scientific approach to the God question, we would have to reject the God hypothesis."[29]

Shermer may claim that "science knows no religious ... boundaries,"[30] but his own view suggests he doesn't really believe that. University of Toronto philosopher James Robert Brown charges that Shermer merely "wants to pacify Christians, especially the fence-sitters, hoping to keep them from falling into the antievolution camp" and thus "tells tales to make believers think they can swallow the Darwin pill with no side effects."[31] Brown observes, "I understand this strategy, but I doubt it will work. Better to be straightforward and to point out that there are serious tensions."[32]

There are other notable academics who lost their faith at least in part due to Darwinism, and yet nonetheless preach to the public that Darwinian evolution and religion are compatible. Some are more forthright than Shermer, even admitting they take this stance due to a "strategy" (as Brown puts it) related to public opinion and education policy—not because they actually believe that Darwinism is metaphysically neutral.

In 2006, University of Wisconsin-Madison professor Ronald Numbers, a renowned historian of creationism, co-published an article in the *Journal of Clinical Investigation* in which he claimed: "A majority of people ... like a large block of religious scientists, do not see any conflict between religious belief and evolutionary theory." Numbers accused "creationists" of fostering "a false duality between science and religion."[33] Yet he admits that evolution played a role in his own drift away from a religious upbringing to a state of unbelief, even though he says he "really wanted to have religious beliefs for a long time."[34]

Unlike Shermer, Numbers admits that some advocate peace between evolution and religion for politically motivated reasons. He decries Dawkins's attitudes, claiming that they do "a terrible disservice to public policy in the United States,"[35] and fears that if evolution becomes too closely linked with atheism, then people like Dawkins could unwittingly "undercut the ability of American schools to teach evolution."[36]

Philosopher of biology Michael Ruse is another self-described "ardent evolutionist" and "ex-Christian"[37] who once said: "I find it a great relief no longer to believe in God."[38] Yet he wrote a book asking, "Can a Darwinian be a Christian?" and answered "Absolutely!"[39] In 2006, Ruse leaked private e-mails he had written to Dawkins and Daniel Dennett, which unwittingly confirm that the push to promote compatibility between evolution and religion stems from political and legal concerns:

> I think that you [Daniel Dennett] and Richard [Dawkins] are absolute disasters in the fight against intelligent design—we are losing this battle, not the least of which is the two new supreme court justices who are certainly going to vote to let it into classrooms—what we need is not knee-jerk atheism but serious grappling with the issues—neither of you are willing to study Christianity seriously and to engage with the ideas—it is just plain silly and grotesquely immoral to claim that Christianity is simply a force for evil, as Richard claims—more than this, we are in a fight, and we need to make allies in the fight, not simply alienate everyone of good will.[40]

Similarly, Francisco J. Ayala is a leading scholar who disavows any challenge to religion from Darwinism, yet was driven toward Darwinism by the very natural evil that drove him away from his Catholic faith. Called "the Renaissance man of evolutionary biology"[41] by the *New York Times*, Ayala served as president of the American Association for the Advancement of Science (AAAS) and was the founding chair of the AAAS Dialogues on Science, Ethics, and Religion.[42] From that platform, he's become one of the most outspoken defenders of the compatibility between God and Darwinian evolution.

A favorite subject of the *New York Times* and recent recipient of the Templeton Prize, since 1999 Ayala has argued in various interviews that (Darwinian) evolution "is more consistent with belief in a personal god than intelligent design,"[43] and that evolution "is not only NOT anti-Christian, but the idea of special design ... might be ... blasphemous."[44] His stature within the evolution lobby is seen in his co-authorship of the preface to the aforementioned NAS booklet which asserts that God and evolution are "fully compatible."[45] Ayala's 2007 book *Darwin's Gift to Science and Religion* likewise declares that "Christians need not see evolution as a threat to their beliefs."[46]

But is Ayala even a Christian? When speaking at the "Beyond Belief" atheism conference in 2006, Ayala told the audience, "I know how Richard [Dawkins] feels on matters of religion. *I could agree with you on many things which I will not make explicit here.*"[47] The following year Ayala was deposed as an expert witness in a lawsuit over the use of creationist textbooks in private schools where he affirmed it is "fairly-accurate" to say he "spent five years in the priesthood until he said his intellectual side could no longer rationalize evil and human tragedy under the auspices of a supposedly loving God. As a result he not only left the priesthood, he left the Roman Catholic Church never to return."[48] When questioned further, Ayala refused to answer specific questions about his personal religious views. This is not unusual for Ayala. A 2008 *New York Times* interview reported he "will not say whether he remains a religious believer," because, in Ayala's words, "I don't want to be tagged ... by one side or the other."[49]

Ayala represents perhaps the most eminent proponent of the view that Darwinian evolution poses no threat whatsoever to religion, and yet he categorically refuses to publicly answer simple questions about whether he is religious. This posture may be rhetorically useful, but it is hardly forthright.

A final example of an atheist who publicly has denied—at least lately—any tension between God and Darwinian evolution is science journalist Chris Mooney. In his 2009 book with Sheril Kirshenbaum, *Unscientific America: How Science Illiteracy Threatens our Future,* Mooney argues that "atheism is not the logically inevitable outcome" of Darwinism, because "[a] great many scientists believe in God with no sense of internal contradiction, just as many religious believers accept evolution as the correct theory to explain the development, diversity, and interrelatedness of life on Earth."[50]

But this wasn't always Mooney's line. In a 2001 article in *Slate* titled "Darwin's Sanitized Idea," Mooney laments that the PBS *Evolution* miniseries' "attempt to divorce Darwinian science from atheism, though well intentioned, is finally naïve."[51] According to Mooney in 2001, "Darwinism presents an explanation for life's origins that lacks any supernatural element and emphasizes a cruel and violent process of natural selection that is tough to square with the notion of a benevolent God."[52] He even rebukes Darwin for inserting the word "Creator" into later editions of *The Origin of Species* as a form of "appeasement."[53]

So which is the real Mooney? Mooney's own personal views and history of atheist activism may help resolve these apparent contradictions.

Before graduating from Yale in 1999, Mooney co-founded the Yale College Society for Humanists, Atheists and Agnostics.[54] During this time he interned with the Campus Freethought Alliance (CFA) where he helped draft a "Bill of Rights for Unbelievers" and even spoke on CFA's behalf at the National Press Club.[55] Mooney also wrote for the *Yale Herald*, railing against conservative religious viewpoints, claiming there is a "plague of religious homophobes" and asserting that "[p]egging all Christians as anti-gay would be premature—but not by much."[56] In another article he used labels like "unreason" and "dangerous, bigoted irrationalism" to describe the views of *First Things* founder, the late theologian Father Richard John Neuhaus.[57] After graduation, Mooney launched his professional career working for *The Skeptical Inquirer*, writing its Doubt and About column from 2002–2006,[58] and serving as contributing editor at least until 2008.[59]

In an apparent effort to rebrand himself, by early 2009 none of the information in the above paragraph appeared in Mooney's bio posted with a group calling itself "Science Progress."[60] But the old Mooney isn't completely gone. He continues to say, "I am as much an atheist as I have ever been."[61] He's just adopted new lingo.

While the old Mooney used Darwinism to attack religion, the new Mooney claims a different cause—increasing "scientific literacy." *Unscientific America* hopes for an America that is "as scientifically literate as anyone could reasonably hope for"[62]—which requires greater public acceptance of evolution. Mooney charges that we can only "think about taking a rest after the percentage of Americans" who have certain non-evolutionary views "drops below 20%."[63] "But until then," he counsels, "we must be constantly vigilant."[64]

The new Mooney acknowledges that his endgame is for science to "become the common culture,"[65] fulfilling his "goal ... to create an America more friendly toward science and reason."[66]

There's nothing wrong with creating a society that supports science, but what does Mooney believe will happen to religion in this society? His vision becomes clear in a 2009 blog post titled, "Why We Celebrate Darwin." According to Mooney, "Darwin means far, far more to us than anything his science, alone, can convey." That's because Darwin "epitomizes ... [a] secular worldview, and ... [a] science-focused life" where "religion is not particularly important."[67]

The new Mooney's direction appears to be a mere tack, not a full change of course. Though labeled a campaign for "scientific literacy," he seems embarked on a pragmatic strategy for diminishing religion and advancing atheism. Indeed, when commenting on his aforementioned 2001 *Slate* article, the new Mooney defends his prior argument. "I don't entirely disagree with this—especially the observation that 'Many students who study evolution will find themselves questioning the religions they have grown up with,'" he wrote in a 2009 blog post. *"I think that's very true, and all to the good."*[68]

In the same post, Mooney attempts to explain why he now sings a more conciliatory tune. "I would not have written the *Slate* piece today" because "I've changed my emphasis," wrote Mooney. "A lot more reading in philosophy and history has moved me toward a more accomodationist (sic) position. So has simple pragmatism."[69]

One of Mooney's pragmatic concerns is likely that overtly linking Darwinian evolution with atheism could create constitutional problems for teaching evolution in public schools. Expressing fears much like those of Numbers and Ruse, Mooney describes the dangers posed by the new atheists:

> What if Coyne and the New Atheists are right, and evolution (or science itself) isn't actually neutral? What if there really is a fundamental conflict between science and religion?... I fear that were the New Atheists to somehow prevail on this point, the anti-evolutionists might wreak some serious havoc in the courtroom in a later case. This is one reason to be concerned about the New Atheist position.[70]

So it isn't so difficult to reconcile the new Mooney with the old one. It's not that the new Mooney actually believes God and evolution are compatible. On the contrary, he still believes it's "all to the good" when Darwinism erodes faith. He simply has realized that linking atheism with Darwin could be dangerous to public opinion and public evolution education.

In summary, many atheists in the evolution lobby believe that simply by advocating Darwinism under the guise of *science literacy*, they can naturally diminish religion without expressing any outright hostility toward it. They publicly assert there are no tensions between God and Darwin, but privately believe otherwise. Like any good political strategist, atheists in the NCSE-backed evolution lobby court the middle-ground—in this case religious moderates—in hopes of increasing their base of support.[71]

But since this is a temporary marriage of political convenience, once they no longer need religious moderates to provide cover for Darwinism, atheists in the evolution lobby will probably end up sounding a lot like the new atheists.

Entering the "Big Debate" of the New Atheist Camp

Chris Mooney is not the only atheist to describe some evolution defenders as engaging in "appeasement." Richard Dawkins, called the "pope of atheism"[72] by his followers, used the same provocative metaphor in 2006:

> Scientists divide into two schools of thought over the best tactics with which to face the threat [of religion]. The Neville Chamberlain 'appeasement' school focuses on the battle for evolution. Consequently, its members identify fundamentalism as the enemy, and they bend over backwards to appease 'moderate' or 'sensible' religion.... Scientists of the Winston Churchill school, by contrast, see the fight for evolution as only one battle in a larger war: a looming war between supernaturalism on the one side and rationality on the other. For them, bishops and theologians belong with creationists in the supernatural camp, and are not to be appeased.[73]

According to Dawkins, "If you are concerned with the stupendous scientific question of whether the universe was created by a supernatural intelligence or not," then "fundamentalists are united with 'moderate' religion on one side," and of course Dawkins finds himself "on the other."[74]

An NCSE spokesman once quipped that "if Dawkins didn't exist, [ID proponents] would invent him."[75] But Dawkins is no anomaly. Many other scientists in Dawkins's "Winston Churchill school" will never be placated by theistic evolution. They have made it clear that they will not stop until religion is *completely* destroyed. One outspoken scientist who concurs with Dawkins is University of Chicago evolutionary biologist Jerry Coyne.

"The Big Debate," wrote Coyne during an August 2009 blog war with evolution lobbyists, "continues about whether faith and science are compatible and whether scientists should criticize those religious people who agree with them about matters like evolution."[76]

The big debate was sparked in February 2009 when Coyne published an article in *The New Republic* charging that theistic evolutionists "fail to achieve their longed-for union between faith and evolution" because "a true harmony between science and religion requires either doing away with most people's religion and replacing it with a watered-down deism, or polluting science with unnecessary, untestable, and unreasonable spiritual claims."[77] In Coyne's view, "Attempts to reconcile God and evolution keep rolling off the intellectual assembly line. It never stops, because the reconciliation never works."[78]

This debate got especially heated in April 2009 after Coyne published a provocative piece titled, "Truckling to the Faithful: A Spoonful of Jesus Helps Darwin Go Down." Levying criticisms mentioned earlier, Coyne wrote:

> Professional societies like the National Academy of Sciences—the most elite organization of American scientists—have concluded that to make evolution palatable to Americans, you must show that it is not only consistent with religion, but also no threat to it.... Given that many members of such organizations are atheists, their stance of accommodationism appears to be a pragmatic one.[79]

For Coyne, accommodating religion is unconscionable because "by consorting with scientists and philosophers who incorporate supernaturalism into their view of evolution, they erode the naturalism that underpins modern evolutionary theory."[80] Moreover, Coyne charges that the strategy of evolution lobbyists is ineffectual: "We've been making nice with religion for decades" but "America remains as 'unscientific' as ever."[81] He puts it bluntly to the NCSE: "The strategy you suggest has not worked."[82]

For new atheists like Coyne, the issue isn't just a lack of public acceptance of evolution, because, as he says: "We don't just perceive religion as the root of the problem, it IS the root of the problem."[83] "The scientific way of looking at the world," he declares, is "fundamentally incompatible" with religion.[84] Coyne stakes out the position of the new atheists clearly: **"The 'new atheists' are against religion because it is inimical to rational thought."**[85]

Coyne's "A Spoonful of Jesus Helps Darwin Go Down" post attracted widespread support from like-minded scientists across the Internet.

University of Toronto biochemist and leading textbook author Larry Moran chimed in. "I don't like the fact that NCSE cozies up to theistic evolutionists like Ken Miller and Francis Collins while, at the same time, actively distancing itself from vocal atheist scientists like Richard Dawkins," he complained. "NCSE shouldn't ... promote the idea that science and religion are compatible."[86]

In a post titled "Science and Religion are Not Compatible," CalTech physicist Sean M. Carroll endorsed Coyne's view, arguing that the "accomodationism" [sic] approach is a "rhetorical strategy on the part of some pro-science people and organizations to paper over conflicts between science and religion so that religious believers can be more comfortable accepting the truth of evolution and other scientific ideas."[87]

Richard Dawkins not only sympathized with Coyne for exposing the NCSE's strategy as "purely political,"[88] but also outlined his way forward. He suggested that "ridicule, of a humorous nature, is likely to be more effective than the sort of snuggling-up and head-patting that Jerry is attacking." But Dawkins implores his fellow atheists to "go beyond humorous ridicule, sharpen our barbs to a point where they really hurt," to target "the fence-sitters,"[89] and to levy an intense barrage of ridicule at religious moderates to bully them into surrendering to atheism. "I think that [fence-sitters] are likely to be swayed by a display of naked contempt," writes Dawkins. "Nobody likes to be laughed at. Nobody wants to be the butt of contempt."[90]

Biologist and popular blogger P. Z. Myers has wholeheartedly embraced Dawkins's strategy. "I'm on the radical side of things," he confesses. "And I think really the only way we can resolve [the debate between science and religion] is for some day religion to be reduced to little more than a hobby, or a little eccentricity that some people practice."[91] According to Myers, "We need widespread social stigmatization of religion to eradicate religion."[92] And he's more than happy to put that strategy into practice.

For Myers, "science is a threat to religion" and "if you're doing religion, you're not thinking scientifically."[93] But religion is not just anti-scientific, according to Myers, for "[i]f they believe in Jesus, they've agreed to accept an absurdity."[94] Myers pines for the day "[w]hen we achieve post-theism" and "the question of what god is will be regarded as ... a string of nonsense syllables."[95]

In an interview by RavingAtheists.com, Myers was asked whether "religion should be respected." He replied, "Am I allowed to use profanity here? OK, simply put, ***NO!*** You don't get to declare that an idea is to be respected by default, especially not when that idea is patent bullshit."[96] When asked about his take on federally paid U.S. military chaplains, Myers called their employment, "Welfare for the intellectually deficient."[97]

On his popular blog *Pharyngula*, Myers gave "ten enthusiastic thumbs up" to Coyne's arguments, agreeing that the NCSE employs "a failed strategy that is leading us down a dangerous path."[98] For Myers, the central issue is preserving a worldview devoid of anything supernatural. Darwin once said, "I would give absolutely nothing for the theory of Natural Selection, if it requires miraculous additions at any one stage of descent." In interpreting Darwin, Myers says that "what Darwin is rejecting in that statement is what we now call theistic evolution."[99] In Myers's view, "by favoring theism as much as they have, [the NCSE has] broken away from the spirit" of Darwin's science.[100]

Myers is unmistakably clear that he and other new atheists will not stop until religion is destroyed. "[O]ne reason some of us 'New Atheists' are not compromising our attack on religion," he writes, is that they "aim for a post-theistic world in which the religious rationale is recognized as a toxic pathology that diminishes the legitimacy of an argument, and that includes the humble homilies of the Christian moderate."[101]

Another science blogger of like sympathies is James Madison University mathematics professor Jason Rosenhouse. "Many others look at the savagery and unpredictability of evolution and see threats both to God's goodness and to human specialness," wrote Rosenhouse. "They may not be logically forced to abandon their faith, but they are not being unreasonable in seeing a fundamental conflict between evolution and their faith."[102]

According to Rosenhouse, "attempted reconciliations seem terribly implausible"[103] and thus few will "see the arguments of [Ken] Miller and [Michael] Ruse (and many others) to be credible."[104] Such "a strategy of trying to convince people to move to a more moderate sort of religion is doomed to failure,"[105] he argues. "So long as traditional religion remains a dominant force in our society we are going to have this problem."[106] Akin to Dawkins and Myers, Rosenhouse says: "The only hope for a long-term solution is to marginalize religion in public discourse. I don't know if we can accomplish that, but I do know it won't happen without a whole lot of screaming and yelling."[107]

Coyne's provocative *New Republic* article also critiqued the NAS's 2008 *Science, Evolution, and Creationism* booklet.

> True, there are religious scientists and Darwinian churchgoers. But this does not mean that faith and science are compatible, except in the trivial sense that both attitudes can be simultaneously embraced by a single human mind. It is like saying that marriage and adultery are compatible because some married people are adulterers.[108]

Coyne castigated the NAS for only offering statements on the relationship of evolution and religion from theistic evolutionists, but giving none "by anyone who sees faith and science as in conflict."[109] Instead, Coyne suggests that "there are plenty of scientists and philosophers, including myself, Richard Dawkins, Sam Harris, Steven Pinker, P. Z. Myers, Dan Dennett, A.C. Grayling, and Peter Atkins, who feel strongly that science and religion are incompatible ways of viewing the world."[110]

In a similar vein, Daniel Dennett quips that Brown University biologist Ken Miller is "disingenuous" to argue that God and evolution don't conflict because "[t]he theory of evolution demolishes the best reason anyone has ever suggested for believing in a divine creator."[111] Dennett contends that "those evolutionists who see no conflict between evolution and their religious beliefs have been careful not to look as closely as we have been looking, or else hold a religious view that gives God what we might call a merely ceremonial role to play."[112]

Oxford chemist and noted author Peter Atkins contends religion is "evil"[113] and will not stop until religion loses out in society. According to Atkins, "Darwin effectively swept purpose aside in the living world," and "[a]ll reimpositions of purpose are artifices of the religious to feed their faith."[114] Atkins writes that "[s]cience is almost totally incompatible with religion,"[115] for science offers "a far more honest approach" than the "empty gulping and the verbal flatulence that passes for theistic exposition."[116] Atkins anticipates that science will show "that the creator had absolutely no job at all to do, and so might as well not have existed."[117]

Neuroscientist Sam Harris, one of the most influential new atheists, also argues that "Science Must Destroy Religion."[118] In a 2007 editorial in *Nature*, Harris offered a rallying cry: "Scientists should unite against [the] threat from religion." He charged Francis Collins with "high-minded squeamishness" for asserting that religion and science are compatible.[119] Collins's book *The Language of God*, according to Harris, "should have sparked gasping outrage from the editors at *Nature*."[120]

Harvard ecologist and acclaimed author E. O. Wilson also envisions a showdown between religion and science where the "eventual result of the competition between the two world views ... will be the secularization of the human epic and of religion itself."[121] In Wilson's view, evolutionary biology is the focal point of this clash. "It's only when you get into evolutionary biology and the origin of human nature and consciousness," writes Wilson, that we "begin to see conflict" between science and religion.[122] In his Pulitzer-prize winning book *On Human Nature*, Wilson elaborates: "If humankind evolved by Darwinian natural selection, genetic chance and environmental necessity, not God, made the species."[123]

Jerry Coyne also makes clear that the foe is not just "creationism" (however defined):

> It's not just about evolution versus creationism. To scientists like Dawkins and Wilson, the *real* war is between rationalism and superstition. Science is but one form of rationalism, while religion is the most common form of superstition. Creationism is just a symptom of what they see as the greater enemy: religion.[124]

During his "big debate" with atheists in the evolution lobby, Jerry Coyne insists that "the real problem with evolution in this country is not creationists, but religion." He predicts that the new atheists "will only win this war by either vanquishing religion or waiting for it to disappear in the US, as it has in Europe."[125] But Coyne is hopeful. "I am absolutely confident that some time in the distant future," he writes, "we will put away our childish things and religion will disappear in America."[126]

It should be clear that the new atheists will never be pacified by theistic evolutionism or even the defeat of intelligent design. For them, an alliance with theistic evolutionists is a deal with the devil. Their goals will only be achieved when religion is completely eradicated.

The False Refuge of the NOMA Model

The debate between new atheists and evolution lobbyists presents a problem for theistic evolutionists. As we've seen, (1) out of political expediency, many atheist members of the evolution lobby promote the view that God and evolution are compatible out of pragmatic political convenience, while rejecting the very reconciliation they purport to advocate, and (2) the vocal and determined new atheist community will not stop until all religion is vanquished.

Some theistic evolutionists respond to the new atheists by claiming that science and religion can't conflict because they speak entirely different languages. So long as religious folks don't challenge Darwinism, they believe religion will be immune to attacks from the new atheists. Theistic evolutionist Francis Collins seems to promote this idea in a passage from his book *The Language of God* endorsed by the NAS: "Science's domain is to explore nature. God's domain is in the spiritual world, a realm not possible to explore with the tools and language of science."[127]

In this posture, Collins harks back to the compromise proposed by the late Harvard paleontologist Stephen J. Gould. Using a term of Catholic origin, Gould argued that science and religion are "non-overlapping magisteria" (NOMA). Gould's proposal is a sucker's bargain, however, since Gould effectively proposed that science gets the domain of reality, whereas religion gets "values."

Yet, for new atheists, even Gould's NOMA bargain concedes too much to religion. As Richard Dawkins explains:

> Chamberlainites are apt to quote the late Stephen Jay Gould's 'NOMA'—'non-overlapping magisteria'. Gould claimed that science and true religion never come into conflict because they exist in completely

separate dimensions of discourse.... This sounds terrific, right up until you give it a moment's thought.[128]

Unsurprisingly, many new atheists have opposed the NOMA model because they believe that evolutionary science falsifies or contradicts religious claims of any kind. "Although Gould's desire for peace between these often warring perspectives was laudable," writes Daniel Dennett, "his proposal found little favor on either side, since in the minds of the religious it proposed abandoning all religious claims to factual truth and understanding of the natural world (including the claims that God created the universe, or performs miracles, or listens to prayers), whereas in the minds of the secularists it granted too much authority to religion in matters of ethics and meaning."[129]

Evolutionary biologist Massimo Pigliucci likewise argues that NOMA will appeal to few, because "NOMA applies to the very special concept of God that a deist would feel comfortable with, not to what most people think of as 'God.'"[130] E. O. Wilson contends: "Science has always defeated religious dogma point by point when the two have conflicted ... the two beliefs are not factually compatible."[131] In Wilson's view, "those who hunger for both intellectual and religious truth will never acquire both in full measure."[132]

According to biologist and philosopher John Dupré, NOMA "does nothing to save the theological claims—centrally the existence of a God—from the threat that they encounter from the growth of science."[133] Because theistic religion postulates a God, Dupré argues that "Darwinism undermines the only remotely plausible reason for believing in the existence of God" because, "without the argument from design there is nothing very credible left of theism generally, and Christianity in particular.[134] (Whether Dupré is right that the design argument is the only plausible reason for believing in God is, of course, beyond the scope of this chapter.)

In *The God Delusion*, Richard Dawkins claims that NOMA is "a peculiarly American political agenda, provoked by the threat of populist creationism"[135] promoted by the NCSE and other groups who "bend over backwards" for religious moderates.[136] These charges that NOMA is politically motivated seem confirmed by Bora Zivkovic, a biology lecturer (and Darwinist) at Wesleyan College, who admits that, "NOMA is wrong, but is a good first tool for gaining trust ... to help [religious students] accept evolution."[137]

Additional examples of atheists attacking NOMA could be given,[138] but atheists are not the only critics. Legal scholar (and ID-critic) Kent Greenawalt observes: "Few religious believers will be found within the category of people who suppose that the domains have no overlap" because "many religious believers think that a persuasive religious account of ultimate reality bears on subjects to which natural science speaks."[139]

From the ID side, Stephen Meyer argues, "Christianity in particular does not simply address questions of morality and meaning as Gould's NOMA principle asserts, but it also makes factual claims about history, human nature and, it would seem, the origin of the natural world." In Meyer's view, the problem is that "[t]he NOMA principle consistently applied, therefore, requires subtracting content not just from science or literalistic fundamentalism, but from basic orthodox Christianity."[140]

Meyer argues that science also suffers under NOMA because it is prevented from speaking to subjects covered by religion. Thus, if there is any area where the inadequacy of the NOMA model becomes clearest it may be found in Darwinian attempts to explain the origin of morality and religion. As philosopher David Hull wrote in *Nature*, the "compromise" that "science and religion ... cannot conflict" fails when "[a]dvocates of evolutionary ethics claim to provide totally naturalistic explanations of ethics."[141]

The tradition of explaining religion or morality in evolutionary terms traces back at least to Darwin. In *The Descent of Man*, Darwin argued that "[t]here is no evidence that man was aboriginally endowed with the ennobling belief in the existence of an Omnipotent God."[142] Under Darwin's evolutionary scheme, "dreams may have first given rise to the notion of spirits" and subsequently, "[t]he belief in spiritual agencies would easily pass into the existence of one or more gods."[143] Seven decades later, *Time* magazine quoted Julian Huxley proclaiming: "For a justification of our moral code we [need] no theological revelation.... Freud in combination with Darwin suffice."[144]

Historian of evolutionary theory and AAAS fellow Peter J. Bowler explains the importance of such claims, asking: "If evolutionism implied that humans were merely improved apes, and nature only a senseless round of struggle and death, where was the divine source of moral values?"[145] Michael Shermer shows no hesitation in answering Bowler's question, asserting, "Morality is the natural outcome of evolutionary and historical forces, not divine command."[146] Such views are not uncommon in the scientific community: a 2007 survey found that 72% of responding evolutionary biologists viewed religion merely as an "adaptation, a part of evolution."[147]

Indeed, many have elaborated on the implications of Darwinian explanations for the origin of religion and morality. According to Harvard evolutionary psychologist Marc D. Hauser, "Either a divine power created our universal moral sense or evolution did."[148] His conclusion is that the explanation lies in "some source other than the divine," for "[b]iology would be the logical candidate."[149] Michael Ruse states that "the recognition of morality as merely a biological adaptation" means that "the evolutionist and the Christian part company" because an objective morality is "expressly denied by the Darwinian evolutionist."[150] Writing with E.O. Wilson, Ruse asks: "If God does not stand behind the Sermon on the Mount, then what does?"[151] Their answer is clear: "[T]he basis of ethics does not lie in God's will" because in the end, "ethics as we understand it is an illusion fobbed off on us by our genes to get us to cooperate. It is without external grounding."[152]

In his 2006 book *Breaking the Spell: Religion as a Natural Phenomenon*, Daniel Dennett contends that "[r]eligious practices can be accounted for in the austere terms of evolutionary biology"[153] and compares faith in God to believing in "Santa Claus or Wonder Woman" or an "imaginary friend."[154] He explains why such arguments have serious implications for religion: "[M]any readers ... will see me as just another liberal professor trying to cajole them out of some of their convictions. And they are dead right about that—that's what I am, and that's exactly what I'm trying to do."[155]

As noted, Francis Collins has encouraged religious believers to adopt a kind of NOMA, proposing that "science is the way to understand the natural world" and "faith is the way to understand questions that science can't answer."[156] But when Darwinism answers questions traditionally within the religious domain—such as the origin of religion or morality—guess which side has to fall silent? As ID-proponent Phillip Johnson has insightfully commented, under NOMA science and religion are "'separate but equal' of the *apartheid* variety."[157]

While not all these arguments are equally strong, clearly NOMA dies the death of a thousand counterexamples. Various religions—including Collins's Christianity— simply do make all sorts of claims that fall within the purview of natural science. And Darwinian scientists make all sorts of claims of interest to religious believers. To be sure, the relationship between science and religion is complicated, and probably differs for different sciences and different religions. But the new atheists, whatever their excesses, are right to reject the simplistic NOMA strategy as an accurate description of the relationship of real science and real religion. At the very least, new atheists believe NOMA is false. So if theistic evolutionists think they can find refuge in the NOMA model, they will discover that it is barely a speed bump for atheists seeking to explain all of biology—including the origin of religion and morality—in Darwinian terms. As *The Economist* astutely predicted, "Evolutionary biologists tend to be atheists" and thus "most would be surprised if the scientific investigation of religion did not end up supporting their point of view."[158]

Unconditional Surrender

Upon realizing that religion and science do not always deal with neatly quarantined subjects, some religious persons might be tempted to capitulate completely to Neo-Darwinism in the hope of immunizing their privatized and carefully constrained religious views from further assault. The problem is that the Darwinism of the new atheists is the Darwinism of mainstream science. And as the new atheists observe, even if one can find ways to reconcile it with God, Neo-Darwinism is not theologically inert.

In 1998, after the National Association of Biology Teachers (NABT) removed the words "unsupervised" and "impersonal" from its official description of evolution, over seventy Darwinian biologists, including notable scientists such as Richard Lewontin, John Lynch, and Niall Shanks, wrote a letter of protest to the NABT arguing that "evolution indeed is, to the best of our knowledge, an impersonal and unsupervised process."[159] Also attacking theistic evolutionists, the letter claimed that the notion that an intelligence is "supervising evolution in a way to perfectly mimic an unsupervised, impersonal process" is a viewpoint "that has been repeatedly invalidated on philosophical grounds ever since David Hume and well before Darwin."[160] They harshly criticized the NABT's removal of the "unsupervised" descriptor for Darwinian evolution:

> The NABT leaves open the possibility that evolution is in fact supervised in a personal manner. **This is a prospect that every evolutionary biologist should vigorously and positively deny.**[161]

William Corben observes in the journal *Science and Education* that the NABT's move provided an empty remedy because "[t]he problem is that 'unsupervised and impersonal' describes what many evolutionary biologists believe about the universe and they take this as a granted part of science."[162] Corben's contention seems valid: in 2005, thirty-nine Nobel laureates wrote the Kansas State Board of Education to inform them that "evolution is understood to be the result of an unguided, unplanned process of random variation and natural selection."[163] Indeed, a wide variety of mainstream biology textbooks have used identical language to describe and define Darwinian evolution.

Theistic evolutionist and biologist Ken Miller estimated that 35 percent of high school students use his biology textbooks, as well as over two hundred colleges.[164] Yet previous editions of one of Miller's most popular textbooks—the "elephant" edition of *Biology*—describe evolution as follows: "[E]volution works without either plan or purpose.... **Evolution is random and undirected.**"[165] Indeed, both the 1991 and 1994 editions of Miller and Levine's *Biology: The Living Science* left readers with a stark description of the implications of Darwinian evolution:

> Darwin knew that accepting his theory required believing in philosophical materialism, the conviction that matter is the stuff of all existence and that all mental and spiritual phenomena are its byproducts. Darwinian evolution was not only purposeless but also heartless—a process in which the rigors of nature ruthlessly eliminate the unfit. Suddenly, humanity was reduced to just one more species in a world that cared nothing for us. The great human mind was

no more than a mass of evolving neurons. Worst of all, there was no divine plan to guide us.[166]

Presaging the Miller-Levine textbook, Stephen Jay Gould's *A View of Life* (co-authored with Salvador Luria and Sam Singer), teaches that "Darwin held a strong allegiance to philosophical materialism—the notion that matter is the ground of all existence and that 'spirit' and 'mind' are the products or inventions of a material brain."[167] Gould, who proposed the NOMA bargain, and his co-authors suggest that "biology demonstrated that we were not created in the image of an all-powerful God but had evolved from monkeys by the same process that regulates the history of all organisms."[168] Likewise Guttman's *Biology* teaches that a species survives "if the cosmic dice continue to roll in its favor" and therefore "just by chance, a wonderful diversity of life has developed during the billions of years in which organisms have been evolving on earth."[169]

Douglas Futuyma's popular college-level textbook *Evolutionary Biology* candidly argues: "By coupling undirected, purposeless variation to the blind, uncaring process of natural selection, Darwin made theological or spiritual explanations of life superfluous."[170] According to Futuyma's text, "[t]he profound, and deeply unsettling, implication of this purely mechanical, material explanation for the existence and characteristics of diverse organisms is that *we need not invoke, nor can we find any evidence for, any design, goal, or purpose anywhere in the world*,"[171] and thus, "Darwin's theory of evolution ... provided a crucial plank to the platform of mechanism and materialism."[172]

Strickberger's *Evolution* describes Darwinism as "an attempt to displace God": "To the question, 'Is there a divine purpose for the creation of humans?' evolution answers no." To the question 'Is there a divine purpose for the creation of any living species?' evolution answers no."[173] In fact, if we take mainstream biology textbooks at their word, then Darwinian evolution is a "blind,"[174] "uncaring,"[175] "heartless,"[176] "undirected,"[177] "purposeless,"[178] and "chance"[179] process that involves "extreme randomness,"[180] that acts "without either plan or purpose,"[181] and implies "materialism,"[182] because we are "not created for any special purpose or as part of any universal design."[183] We are informed that "a god of design and purpose is not necessary,"[184] as "evolutionary randomness and uncertainty replaced a deity having conscious, purposeful, human characteristics."[185]

The purpose of this recitation is not to show that biology textbooks intentionally seek to oppose theism (though perhaps some do), nor to claim that there are no ways to reconcile God and evolution. Rather, it is to show how *standard Neo-Darwinian theory* is described by mainstream biology textbooks. As a result, religious persons who capitulate to the Neo-Darwinian theory of evolution should not reasonably expect to convince the new atheists—or even significant portions of the public—that Darwinian Theory and traditional theistic religion are compatible. They may deceive themselves. They may get modest tactical support when they attack intelligent design. But they occupy a very small world of their own imagining.

The Neo-Darwinian claim that life is the result of an unplanned and undirected processes is fundamental to the worldview promoted by the new atheists. William Provine puts it thus: "One can have a religious view that is compatible with [Darwinian] evolution only if the religious view is indistinguishable from atheism."[186] New atheists see no good reason to accept the claims of theistic evolutionists. For them, theistic (Darwinian) evolution is an oxymoron.

Conclusion

IF THE NEW ATHEISTS are right, then alliances between atheists in the evolution lobby and the religious community are only temporary and cynical. Someday—perhaps soon, these atheists believe—Darwin's universal acid will complete its task and there won't be anything left for religious persons to defend. Surely the more compelling response to the new atheists' war against religion—if it is supported by the data—is a sound, civil, and compelling demonstration that their scientific arguments are wrong. But one thing is clear: Baptizing their scientific views with theology will do nothing to stop the covert or overt efforts of many Darwinian atheists against religion.

4. Death and the Fall

Why Theistic Evolution Does Nothing to Mitigate the Problem of Evil

William A. Dembski

For young-earth and old-earth creationists, humans bearing the divine image were created from scratch.[1] In other words, God did something radically new when he created us—we didn't emerge from preexisting organisms. On this view, fully functioning hominids having fully human bodies but lacking the divine image never existed. For most theistic evolutionists, by contrast, primate ancestors evolved over several million years into hominids with fully human bodies. What happened next? Physician Paul Brand finds a possible answer in Genesis 2:7, which reads, "And the Lord God formed man from the dust of the ground and breathed into his nostrils the breath of life, and man became a living being." In reflecting on this passage, Brand writes:

> When I heard that verse as a child, I imagined Adam lying on the ground, perfectly formed but not yet alive, with God leaning over him and performing a sort of mouth-to-mouth resuscitation. Now I picture that scene differently. I assume that Adam was already biologically alive—the other animals needed no special puff of oxygen, nitrogen, and carbon dioxide to start them breathing, so why should man? The breath of God now symbolizes for me a spiritual reality. I see Adam as alive, but possessing only an animal vitality. Then God breathes into him a new spirit, and infills him with His own image. Adam becomes a living soul, not just a living body. God's image is not an arrangement of skin cells or a physical shape, but rather an inbreathed spirit.[2]

Accordingly, hominids that evolved from primate ancestors initially lacked whatever spiritual reality was required—presumably including cognitive and moral capacities—to bear the image of God.[3] Then, at some particular moment, they received God's image and became fully human.

On the assumption that humans evolved under divine guidance, God must at some point have transformed them so that they became rational moral agents made in God's image. Yet, is such a transformation of being and consciousness compatible with biological evolution as it is generally understood by most scientists? Yes and no. Evolutionary geneticist Jerry Coyne defines biological evolution as follows:

> There is only one going theory of evolution, and it is this: organisms evolved gradually over time and split into different species, and the main engine of evolutionary change was natural selection. Sure, some details of these processes are unsettled, but there is no argument among biologists about the main claims.[4]

This is the standard definition of biological evolution—it is textbook orthodoxy—though Coyne fails to mention random variation. According to the orthodox view, biological evolution rests on two pillars: universal *common descent*, the historical claim that all organisms trace their lineage back to a common ancestor; and the *Darwinian mechanism*, the theoretical claim that natural selection acting on random variations accounts for biological diversification.

Coyne's definition is quite general, and leaves "some details ... unsettled," so perhaps it's possible that God could use evolution as characterized by Coyne to endow hominids with the cognitive, moral and other capacities required to bear the image of God. And perhaps God could transform human consciousness, moral capacity, and the like, without performing a miracle in the traditional sense. For instance, the image of God might be an emergent property of human brains that attain a certain level of complexity. This assumes a monism in which the human is viewed as essentially material (though also as a creature of God).[5] Alternatively, perhaps a dualism in which the human is viewed as essentially a union of matter and spirit could be reconciled with Coyne's characterization of biological evolution.[6] For instance, perhaps God could "add" spirit to the human body to impress on it the image of God. Nancey Murphy is a proponent of the monistic view;[7] Francis Collins is a proponent of the dualistic view.[8] Each affirms both evolution and Christian faith.

This is all quite vague and speculative, but perhaps Coyne has still given God enough room to maneuver in endowing humans with the divine image. Nonetheless, these moves contradict the views of most Darwinists, who deny any fundamental, qualitative discontinuity of consciousness and capacities between humans and the presumed primate ancestors. And their denial has a fine pedigree, since Charles Darwin himself rejected a fundamental divide between human cognitive and moral capacities and those of the rest of the animal world. In *The Descent of Man*, he wrote:

> The difference in mind between man and the higher animals, great as it is, certainly is one of degree and not of kind. We have seen that the senses and intuitions, the various emotions and faculties, such as love, memory, attention, curiosity, imitation, reason, etc., of which man boasts, may be found in an incipient, or even sometimes in a well-developed condition, in the lower animals.[9]

Moreover, this isn't merely an atheist gloss on the theory. On the contrary, many *theistic* evolutionists now agree with Darwin that humans are in every way continuous with the rest of the animal world. For instance, in his book *Saving Darwin*, Karl Giberson states:

> Once we accept the full evolutionary picture of human origins, we face the problem of human uniqueness. The picture of natural history disclosed by modern science reveals human beings evolving slowly and imperceptibly from earlier, simpler creatures. None of our attributes—intelligence, upright posture, moral sense, opposable thumbs, language capacity—emerged suddenly. Every one of our remarkable capacities must have appeared gradually and been present in some partial, anticipatory way in our primate ancestors. This provocatively suggests that animals, especially the higher primates, ought to possess an identifiable moral sense that is only *quantitatively* different from that of humans. Not surprisingly, current research supports this notion.[10]

Unfortunately, as the recent Climategate scandal underscores,[11] scientific research can often be manipulated to support just about any conclusion. Giberson goes on to make a case against human exceptionalism based on primate research. In *The Design of Life*, Jonathan Wells and I make a case for human exceptionalism based on linguistics, mathematics, and cognitive psychology.[12] Let the reader decide who has the better argument.[13]

Human exceptionalism acknowledges that important similarities exist between humans and primates. But it insists that far-reaching differences also exist, especially differences in cognitive and moral capacities, and that these represent a difference in kind and not, as Darwin and many contemporary Darwinists hold, merely a difference in degree.[14]

Giberson's hardcore Darwinism leads him to reject that at some particular point in time evolving hominids might have been divinely transformed into the divine image: "One could believe, for example, that at some point in evolutionary history God 'chose' two people from a group of evolving 'humans,' gave them his image, and then put them in Eden, which they promptly corrupted by sinning. But this solution is unsatisfactory, artificial, and certainly not what the writer of Genesis intended."[15] If our concern is with what the writer of Genesis intended, however, then we probably shouldn't be trying to graft a theory of evolution onto it. When the writer of Genesis 1:21 and 1:25 stated that organisms were created "after their kind," it's hard to imagine that he intended the fluidity of all species as required by evolution. Nonetheless, as soon as one decides to read Genesis from an evolutionary perspective, a distinct transformation of being or consciousness into the divine image becomes mandatory—provided, that is, one is serious about preserving the Fall. After all, there must be some state from which the original human beings *fell*.

Giberson, unfortunately, is not serious about preserving the Fall. The very passage just quoted from *Saving Darwin* appears in a section titled, "Dissolving the Fall." Giberson rejects any traditional conception of the Fall.[16] Indeed, his understanding of sin is simplistic and heterodox. He sees the essence of sin as selfishness. And coincidentally, "selfishness," for Giberson, "drives the evolutionary process."[17] Simply put, we are selfish because evolution is selfish, and we are a product of evolution. Salvation for him, then, is transcending our evolutionary past. By contrast, in a traditional view of the Fall, we are saved not from what evolution has produced in us but from what we have done to ourselves in willfully sinning against a holy God. In the traditional view, the evil humanity experiences is the evil it has brought on itself.

Solving the Problem of Evil?

Many theistic evolutionists, however, commend this heterodox element of their view because they think it answers the problem of evil better than traditional orthodoxy. In fact, most theistic evolutionists increasingly argue that Christian theism benefits from the idea that God created indirectly by Darwinian means rather than directly (as in special creation). Theistic evolutionists worry that a God who creates directly renders the problem of evil insoluble. Such a God would be responsible for all the botched and malevolent designs we find in nature. By letting Darwinian natural selection serve as a designer substitute, theistic evolutionists can refer all those botched and malevolent designs to evolution. This, in their view, is supposed to resolve the problem of natural evil and thereby help validate Christian theism.[18]

Well-known Darwinian and former Catholic priest Francisco Ayala makes precisely such an argument: "A major burden was removed from the shoulders of believers when convincing evidence was advanced that the design of organisms need not be attributed to the immediate agency of the Creator, but rather is an outcome of natural processes." According to Ayala, "if we claim that organisms and their parts have been specifically designed by God, we have to account for the incompetent design of the human jaw, the narrowness of the birth canal, and our poorly designed backbone, less than fittingly suited for walking upright."

In Ayala's view, right-thinking Christians need to "acknowledge Darwin's revolution and accept natural selection as the process that accounts for the design of organisms, as well as for the dysfunctions, oddities, cruelties, and sadism that pervade the world of life. Attributing these to specific agency by the Creator amounts to blasphemy." Charging Christian opponents of Darwin's theory with blasphemy may seem unduly harsh. Ayala therefore attempts to soften this charge by granting that those who oppose evolution and support special creation "are surely well-meaning people who do not intend such blasphemy." Ayala's concession (and condescension) here is to the intellectual feebleness, as he sees it, of those who cling to the old naïve creationist outlook and have yet to wrap their minds around the stark truth of evolution.

In any case, he doesn't retract the charge of blasphemy: "This is how matters appear to a biologist concerned that God not be slandered with the imputation of incompetent design."[19]

In turning the table on special creation, however, Ayala has in fact turned it 360 degrees. The table is back to where it was originally, and the problem he meant to shift to special creation confronts him still. Ayala worries that a God who creates by direct intervention must be held accountable for all the bad designs in the world. Ayala's proposed solution is therefore to have God set up a world in which evolution (by natural selection and random variation) brings about bad designs. But how does this address the underlying difficulty, which is that a creator God has set up the conditions under which bad designs emerge? In the one case, God acts directly; in the other, indirectly. But a Creator God, as the all-powerful source of all being, is as responsible in the one case as in the other.

We never accept such shifting of responsibility in any other important matter, so why here? What difference does it make if a mugger brutalizes someone with his own hands (that is, uses direct means) or employs a vicious dog on a leash (that is, uses indirect means) to do the same? The mugger is equally responsible in both cases. The same holds for a creator God who creates directly or indirectly by evolution. Creation entails responsibility. The buck always stops with the Creator. That's why so much of contemporary theology has a problem not just with God "intervening" in nature but also with the traditional doctrine of creation *ex nihilo,* which makes God the source of nature.

The rage in theology these days is to diminish the power and ultimacy of God so that God is fundamentally constrained by the world and thus cannot be held responsible for the world's evil. Process theology, which sees God as evolving with the world and the world as having an autonomy beyond God's reach (thereby enabling God to shed responsibility for evil), is a case in point. Process theologian Robert Mesle elaborates:

> [S]ince God cannot control the evolutionary process, there is no reason even to assume that God was aiming that process specifically at us. The history of evolution has been filled with more crucial events than we can dream of, and God could not control them. God and

> the world have been involved in a continuous dance in which God must continually take the decisions of the creatures and work with them—whatever they may be. For better or worse, each decision of each creature plays some role in the world's process of becoming. And God works to create something good out of what the world makes possible. Evolution, then, is an ongoing adventure for God, as it is for the world.[20]

In my view, process theology unleashes a raging pack of new problems and thus is not a suitable replacement for the traditional doctrines of God and creation. But let's grant, for the sake of argument, that it resolves the problem of dysteleology (bad designs) resulting from natural selection. The problem is that Ayala and prominent theistic evolutionists are not arguing for process theology (or some other diminished deity) but for the compatibility of fully Darwinian evolution with classical Christian theism. We are told, "Embrace Darwinian evolution and you can still be a good Christian."[21]

Ayala is therefore in no position to require that Christian believers revise their doctrine of God in light of Darwinian Theory. In particular, he cannot require that believers in divine omnipotence and creation *ex nihilo* revise these beliefs to suit a more evolution-friendly theology. Christians who hold to a traditional doctrine of creation and accept the mechanism of selection and random variation as God's method of creating organisms therefore confront the problem of evil with the same force as believers who hold the identical doctrine of God but reject Darwinism and accept special creation. Indeed, for the Christian it does nothing to resolve the problem of evil by passing the buck to a naturalistic evolutionary process (a process, in that case, created by God). This is filling one hole by digging another.

Theistic evolutionists deny this conclusion. Responding to my mugger/attack-dog analogy, Giberson remarks:

> The contribution evolution makes to this discussion ... is the remarkable discovery that nature has built-in creative powers. As Christians we affirm that these powers—which include the power to create both wonderful and terrible things—come from God, but they are wielded by nature. This is a traditional theological concept that understands that God works through *secondary* as well as *primary* causes.... [T]he gift of creativity that God bestowed on the creation is theologically

> analogous to the gift of freedom God bestowed on us. Both we *and* the creation have freedom. Our freedom comes with a moral responsibility to use it properly. But that does not prevent us from doing terrible things. The freedom God gave humans was exercised most effectively in the construction of gas chambers at Auschwitz and Dachau. But, because humans have freedom, we do not say that God created those gas chambers. God is, so to speak, off the hook for that evil. In exactly the same way, less the moral dimension, when nature's freedom leads to the evolution of a pernicious killing machine, God is "off the hook." Unless God micromanages nature so as to destroy its autonomy, such things occur. Likewise, unless God coercively micromanages human decision making, we will often abuse our freedom.... When God grants freedom to creatures this means, in ways often difficult to understand, that those creatures can act *independently* of God, to not be robotic automatons or trained attack dogs. In the case of the holocaust, we always do exactly what Dembski says we never do: we shift the responsibility from God to the Nazis. Such reflections have long characterized Christian thinking about the problem of evil.[22]

Giberson suggests that by looking to evolution and the freedom it confers on nature, it is possible to dissolve long-standing problems concerning evil (especially natural evil) that confront classical theism with its teachings about special creation. Giberson's argument, however, turns on an equivocation. It is one thing to ascribe freedom to humans created in God's image. Our freedom carries with it moral responsibility and the ability to make freely willed choices. These would be impossible if our choices were determined in every way. So, one could argue, even an omnipotent God would have to choose between creating free and responsible agents and strongly determining what they will choose.

But what moral responsibility attaches to nature's freedom? What choices does a star or a planet make? Does it make any sense to say that nature *does* this but *ought to* do that? Giberson admits that nature's freedom operates "less a moral dimension." Such an admission, however, defeats the whole enterprise of using evolution to mitigate the problem of evil. Nature simply does what it does and cannot embark on a course of self-improvement to do what it ought to do. Moreover, what nature does results entirely from the capacities that God has given it. Giberson conflates two radically different forms of freedom. As Clive Harden notes in a reply to Giberson:

> Why would God be off the hook for creating a mechanism (evolution) that kills and destroys the way it does? For in Giberson's theodicy, not only did God make the process of evolution, He set it in place and started it. This would be like me letting a bunch of mice, some infected with a plague, loose into a town. The mice have their own freedom to do whatever they want and go wherever they want, and do it all without a "moral dimension." This does nothing to get me "off the hook" for whoever as a result dies.[23]

Invoking the freedom of nature to mitigate the problem of evil is a longstanding theme among theistic evolutionists. John Polkinghorne, for instance, sees a certain inevitability to sin and evil, regarding them as a necessary cost of God bestowing freedom on nature. Thus, in coming to terms with natural evil, Polkinghorne will recount the following anecdote:

> Austin Farrer once asked himself what was God's will in the Lisbon earthquake (that terrible disaster of 1755, when 50,000 people were killed in one day). Farrer's answer was this—and it's a hard answer, but I think a true answer—that God's will was that the elements of the earth's crust should behave in accordance with their nature. God has given them freedom to be, just as he has given us freedom to be.[24]

Farrer would presumably have given the same rationalization for the Asian tsunami of 2004 and the Haitian earthquake of 2010.

In any case, invoking the freedom of nature does little to answer the worries raised by such evils. We can imagine a world far more violent than ours in which many more people die annually of natural disasters. Alternatively, we can imagine a world far more halcyon than ours in which no one dies of natural disasters because the whole world is a serene tropical paradise. Ascribing natural evil to the freedom of nature does little to address the amount of natural evil in the world or whether the freedom of nature could have taken a different form so that there would be less of it (or perhaps none at all).

Was the Lisbon or Haitian earthquake really nothing more than a consequence of the freedom of the earth's crust? How does such an answer comfort the victims and survivors? Why, as just remarked, didn't God simply place us on a less dangerous planet where earthquakes don't ravage human life? Or was this not an option for the Creator, and if not, why not? What are we to make of divine providence in a world with the freedom to crush us? Why, in most classical liturgies of the Christian churches, do we pray for favorable seasons and good crops if the freedom of nature means that the land is going to do whatever it will regardless of our prayers? Or does God constrain the freedom of nature? But, if so, why doesn't God place tighter constraints on this freedom in relation to evil?

An irony gets lost in many of these discussions about the world's freedom: How can the freedom of creation, which results from a freely acting God who freely bestows freedom on creation, *force* the world to be a dangerous place full of evil? Shouldn't the freedom of creation rather give us freedom *not* to sin? And shouldn't it be possible for God to create a world whose freedom is not destructive and does not entail evil? Ironically, invoking nature's freedom in an attempt to avert the problem of (natural) evil requires at crucial points sacrificing God's freedom in creation.

Although many theistic evolutionists see Darwin's theory as helping to resolve the problem of evil, many agnostic and atheistic evolutionists see it as doing just the opposite. What sort of God, they ask, would create life by so brutal and wasteful a process? Darwin himself continually referred to the history of life as "the great battle of life" and "the war of nature."[25] He elaborated: "From the war of nature, from famine and death, the most exalted object which we are capable of conceiving, namely, the production of the higher animals, directly follows."[26] The struggle for life is absolutely central to his theory—herein lies its creative potential.

Darwin's most famous work is his *Origin of Species*, often simply called the *Origin*. Few remember the full title: *On the Origin of Species by Means of Natural Selection or the Preservation of Favoured Races in the Struggle for Life*. Favored races? Doesn't that mean superior races? And don't superior races imply inferior races? Within Darwinian Theory, it is the destiny of inferior races to be rooted out and destroyed. Thus, in *The Descent of Man*, the sequel to the *Origin*, Darwin noted:

> At some future period, not very distant as measured by centuries, the civilised races of man will almost certainly exterminate, and replace, the savage races throughout the world.... The break between man and

> his nearest allies will then be wider, for it will intervene between man in a more civilised state, as we may hope, even than the Caucasian, and some ape as low as a baboon, instead of as now between the negro or Australian and the gorilla.[27]

Just so there is no doubt, Darwin saw, as a consequence of his theory, that whites (whom he regarded as "superior") would exterminate blacks (whom he regarded as "inferior"). He didn't celebrate this, of course, but he did see it as a natural consequence of his view.

To soften evolution's harshness, theistic evolutionist Denis Alexander asks us to imagine that God, when creating by Darwinian means, is a "great artist in the studio, with energy, creativity and paint flying in all directions, out of which process emerges the richness and diversity of the created order."[28] It hardly follows from this analogy to an artist's studio that creativity must, as a matter of course, be messy and wasteful. Having at one point in my career been an art dealer (the family business consisted in buying and selling oil paintings), I can testify that the studios of artists can be reasonably neat and that neatness seems not to have impaired their creativity. Nor does history bear out that great artists have tended to be slobs. Even so, it's not clear what waste or mess would mean to an omnipotent God of unlimited resources. Still, let's grant, for the sake of argument, that Alexander is correct in not faulting God for any presumed waste or mess in the evolutionary process.

The charge that evolution is inherently cruel now poses a more difficult problem for reconciling theism and evolution. Some theorists have tried to soft-pedal the cruelty inherent in evolution by suggesting that cooperation plays as important a role in it as competition.[29] But cooperation, far from eliminating or mitigating evolution's cruelty, is merely an outgrowth of it. That's because, in a Darwinian context, organisms tend to cooperate when other organisms are competing against and trying to destroy them—in other words, they cooperate when a competitor is being cruel to them. Cruelty from an out-group is, we might say, Darwinian evolution's way of making us nice to our in-group.[30]

The preferred way that theistic evolutionists deal with nature's cruelty, however, is denial and rationalization. Sure, natural selection involves pain, but, as Darwin stressed, the pain is worth it: "As natural selection works solely by and for the good of each being, all corporeal and mental endowments will tend to progress towards perfection."[31] Thus Darwin sanctified evolution and deified natural selection. It's just too bad that natural history had to be littered with casualties "from the war of nature, from famine and death."[32] But to make an omelet, you have to break a few eggs, and evolution certainly knows how to break eggs.

Besides rationalization, there's denial. Thus we are told that cruelty is not really cruel unless conscious moral agents (like us) are suffering it: "Whilst cruel rats and malevolent weasels might exercise such wicked designs in the pages of children's books," writes Denis Alexander, "to the best of our knowledge the real animal world is amoral and has no ethics."[33] But Alexander here fails to distinguish between cruelty as a conscious motivation (which is culpable in us but lacking in other animals) and cruelty as we experience it (such cruelty comes against us as much from nature as from the malevolent intentions of fellow humans). We rightly see the Darwinian interpretation of the fossil record—as a history of predation, parasitism, disease, death, and extinction as the creative engines of life—as cruel even if the animals in it cannot properly be said to have cruel motivations.

There are subtleties here, to be sure. Something might seem cruel from our perspective but seem quite different from the perspective of a rat. But the Darwinian picture transforms massive death into a sort of sub-creator or demiurge. It takes the problem of death and evil and raises it exponentially.[34]

Bottom line: the problem of evil is a difficult issue, but Darwinian evolution, with or without God, does little to mitigate it.

II.

Protestants and Evolution

5. Random Acts of Design

The Inconsistency of Francis Collins

Jonathan Witt

Evangelical Christian Francis Collins was appointed by President Obama to direct the National Institutes of Health and was formerly the head of the Human Genome Project, the monumental and successful effort to map the 3.1 billion letters of the human genetic code. In his bestseller, *The Language of God: A Scientist Presents Evidence for Belief* (2006), he draws upon his training as a geneticist and physician to argue for both Darwinian evolution and a transcendent Creator.

The evolution he argues for involves no direct intelligent input after the origin of the universe until the origin of humans, and yet he also makes a case for Christian theism, arguing not only for a Creator but also for the possibility of miracles, the deity of Christ, and a literal resurrection. He insists that a scientist can believe these articles of Christian doctrine without checking his brain at the door.

The mainstream media might have focused on the fact that such a well-regarded scientist would defend these robust Christian doctrines, all of which are an affront to our elite and overwhelmingly materialist culture. Instead, they have mostly emphasized two other aspects of the book: Its insistence that Darwinism is no threat to Christianity, and its argument that Darwinism better explains a range of physical evidence than either creationism or intelligent design. What has gone begging for ink, however, is a feature of the book hidden in plain sight: Francis Collins makes a scientific case for intelligent design.

According to intelligent design, which extends from the origin of matter to the origin of mind, *an intelligent cause is the best explanation for certain features of the natural world*. In chapter nine Collins argues against intelligent design in biology, and this is what the mainstream media underscored, assisted no doubt by Collins' own subsequent activism against intelligent design (including his founding of the BioLogos Foundation to advocate for Darwin's theory in the Christian community). But notwithstanding Collins's support for Darwinian Theory or his high-profile attacks on intelligent design, in the book's third chapter, "The Origins of the Universe" he argues that *an intelligent cause is the best explanation for certain features of the natural world*—in this case, features that existed before the origin of life.

Collins's Case for Intelligent Design

Collins begins this portion of his book by reviewing twentieth-century discoveries in physics and cosmology, many of which reinforce Christian teaching. For example, whereas scientists of the nineteenth century generally believed that the universe was eternal, a growing body of evidence in the twentieth century convinced them that the universe began about 14 billion years ago, a theory, Collins notes, nicely in harmony with the biblical doctrine of creation *ex nihilo*, that is, creation out of nothing.

Next, he summarizes the fine-tuning problem, the growing body of evidence suggesting that the physical constants of nature (gravity, electromagnetism, and the mass of the universe, among many others) are exquisitely calibrated to allow for complex and even advanced life. A very tiny difference in *any* of these and life as we know it would be impossible.

Collins then describes the three live explanations for fine tuning among the international community of physicists, chemists, astronomers, and cosmologists: (1) There are many universes in addition to our own, perhaps an infinite number, and at least one was bound to have the right physical constants for advanced life; (2) we're just incredibly lucky; and (3) the physical constants look fine-tuned because they were fine-tuned. That is, they were designed.

He does not wrap up the chapter by saying, "I prefer option 3 because it confirms my prior religious commitments." Instead, he appeals to physical evidence and standard methods of reasoning to argue that the design hypothesis best explains the physical evidence in question.

His conclusion is clear, though his language guarded. He says of the two non-design options: "On the basis of probability, option 2 is the least plausible. That then leaves us with option 1 and option 3. The first is logically defensible, but this near-infinite number of unobservable universes strains credulity. It certainly fails Occam's Razor."

He then responds—again with considerable delicacy—to the objection that introducing a supernatural designer itself violates Occam's Razor: "It could be argued, however, that the Big Bang *itself* seems to point strongly toward a Creator."

Is Collins merely changing the subject? No. One properly invokes Occam's Razor (that is, the principle of parsimony, which predates Occam and is present in the work of Scotus, Aquinas and Aristotle among others) only when competing explanations have equal explanatory power. In such situations, the principle urges us to prefer the simplest of these explanations. But if one of the explanations has much greater explanatory power, the explanatory power trumps the principle of parsimony.

Without going into all this, Collins deftly reminds readers that we are not dealing with explanations that explain equally well: the Creator hypothesis has substantially greater explanatory power for the origin of the universe and so it doesn't violate Occam's Razor.

Collins's appeal to the Big Bang and the fine-tuned cosmos form two of his key design arguments. (The third is discussed below.) That he made them publicly is, by itself, noteworthy. In our present intellectual climate, scientists have been harassed and even fired for advocating intelligent design, while the idea is routinely attacked in news stories and the popular books of writers like Richard Dawkins and Daniel Dennett. The fact that such a prominent scientist has made a case for a designing intelligence based on physical evidence should not go unacknowledged.

Collins's Case *Against* Intelligent Design

The reason that Collins's case for intelligent design *has* been largely overlooked was alluded to earlier: He not only explicitly attacked intelligent design in biology in his book, but he later publicly crusaded against it. Indeed, by 2008, Collins was darkly warning that intelligent design "is not only bad science but is potentially threatening in other deeper ways to America's future."[1]

This subsequent emphasis is unfortunate because Collins's critique of intelligent design in biology is not nearly as nuanced as his support for intelligent design in cosmology. For one thing, Collins accepts and repeats in chapter nine of his book a misleading definition of intelligent design, one regularly employed by its critics. According to them, intelligent design, or ID, is a purely negative argument against Darwinism coupled with a disreputable God of the gaps theology.

Having set up an ID straw man, they claim that design theorists merely poke holes in Darwinism and then insist that the holes prove that God designed life. More broadly, they claim that ID proponents supposedly argue *from* our present ignorance of any adequate material cause for certain natural phenomena directly *to* intelligent design.

But this is not the case. Design theorists in biology do offer an extensive critique of Darwinian Theory, but they also offer *positive* evidence for intelligent design. They argue from our growing knowledge of the natural world, including the cellular realm, and from our knowledge of the only kind of cause ever shown to produce information or irreducibly complex machines (both found at the cellular level): namely, intelligent agents.

Take three examples from chapter 9 of Collins's book—examples Collins uses to argue for Neo-Darwinism. First, he refers to the "backward wiring" of the vertebrate eye—an apparently inefficient structure that forces light to pass through the nerves and blood vessels on its way to the eye's light sensors—and argues that this is evidence for Neo-Darwinism and against the idea that a wise designer played a direct role in the evolution of this organism. "The design of the eye does not appear on close inspection to be completely ideal," he writes, and its imperfection seems "to many anatomists to defy the existence of truly intelligent planning of the human form."

Arch-Darwinist and atheist Richard Dawkins has made this argument, and Darwinists love to recycle it. Now even if this argument showed that the vertebrate eye failed to meet our standards of perfection, it wouldn't follow that design had nothing to do with it. However, geneticist and physician Michael Denton has demonstrated that the "suboptimality" argument is misguided. The "backward" wiring actually improves oxygen flow, an important advantage not achievable by the tidier approach demanded by Darwinists.[2] Design theorists have called attention to this point repeatedly, but Collins shows no evidence that he is aware of it. (Perhaps that is because Dawkins and other Darwinists studiously avoid mentioning it.)

Second, Collins points to so-called "junk DNA" as evidence that a hit-and-miss tinkerer in the form of Darwinian evolution, rather than an all-wise designer, laid down the millions of pages of genetic information essential for life. But as William Dembski and I explain in chapter 6 of *Intelligent Design Uncensored*, the genetic regions that Darwinists assumed were junk are proving to have several important functions previously overlooked. They regulate the timing of DNA replication and transcription; they tag sites that need to have their genetic material rearranged; they guide RNA splicing and editing; they help chromosomes fold properly; and they regulate embryo development. This is just a partial list. Far from an example of bad design, "junk DNA" has turned out to be one more demonstration that life's information-processing systems make the best computers humans have managed to devise look like child's play. (For more on this, see chapter 6 in this volume by Jonathan Wells. [3])

A third instance where Collins betrays his lack of familiarity with the work of leading design theorists is his handling of the scientific controversy surrounding a microscopic rotary engine called the bacterial flagellum. The flagellum is a favorite of design theorists because they are convinced that attempts to explain its origin apart from design are manifestly inadequate. Besides, it makes a great ID mascot, because images of the flagellum practically scream design.

In his book *Darwin's Black Box*, Lehigh University biochemist Michael Behe made this sophisticated molecular machine famous by arguing that it was "irreducibly complex" and therefore evidence of design. He used the simple mechanism of a mousetrap as an example of irreducible complexity. If any part of the mousetrap is missing (the base, spring, hammer, holding bar, or catch), the trap cannot work. Even with four of the five parts in place, it is utterly useless. The mousetrap, then, is irreducibly complex. It is either complete, or it is not a mousetrap.

In the same way, the bacterial flagellum, composed of more than 40 distinct kinds of protein machinery, needs every one in place to function. If it has only 39 proteins, it will not work.

What does irreducible complexity have to do with Darwinian evolution? A conscious designer can pull together several dysfunctional parts and assemble them into a functional whole, but Darwinian evolution—which denies the possibility of intelligent guidance or foresight—must progress by one slight, functional mutational improvement at a time. So how can the Darwinian mechanism build an irreducibly complex motor one part at a time, if the motor cannot propel at all until all of its parts are in place?

Summarizing the arguments of leading Darwin defender Kenneth Miller and others, Collins argues that nature could have co-opted simpler molecular machines to create the bacterial flagellum, and points to the syringe-like "type three secretory apparatus" as evidence of such an indirect pathway. But design theorists have noted three crucial problems with this explanation.

One, the micro-syringe at best accounts for only ten proteins, leaving thirty or more unaccounted for, and these other thirty proteins are not found in any other living system. Second, as a wider body of literature suggests, the system probably developed after the more complicated flagellum, not the other way around.

Finally, even if nature had on hand all the right protein parts to make a bacterial flagellum, something would still need to assemble them in precise temporal order, the way cars are assembled in factories. How is such a task presently accomplished? As biologist Scott Minnich and philosopher Stephen Meyer explain, "To choreograph the assembly of the parts of the flagellar motor, present-day bacteria need an elaborate system of genetic instructions as well as many other protein machines to time the expression of those assembly instructions."[4]

Collins never mentions any of this. In these and other instances—the question of testability, for example, and the claim that intelligent design makes no predictions—he comes across as one who, while a fine experimental biologist and science administrator, simply has never engaged the best arguments for intelligent design in biology. So far as one can tell, he's never even seen these arguments.

Collins the Naturalist

Also in chapter 9 of his book, he implicitly invokes a rule known as methodological materialism (also called methodological naturalism) to argue that biologists should not give up looking for a material cause (in this case, a Darwinian cause) for particular biological structures just because scientists have yet to discover it.

This forms part of his argument against intelligent design. As he puts it, "ID is a 'God of the gaps' theory, inserting a supposition of the need for supernatural intervention in places that its proponents claim science cannot explain," and its "proponents have made the mistake of confusing the unknown with the unknowable, or the unsolved with the unsolvable."

The suggestion here is that design theorists are hobbled by a failure of the imagination, an inability to imagine how the Darwinian mechanism could have achieved anything as sophisticated as the flagellar motor. But it is the Darwinists who have been unable to imagine, much less demonstrate in the laboratory, a credible Darwinian pathway to the flagellum.

The situation suggests two possibilities: Either (1) there is an unguided evolutionary pathway and scientists will eventually discover it; or (2) there is no viable evolutionary pathway apart from one guided by intelligence. By refusing to consider the second option, Collins commits the fallacy of begging the question.

Imagine a boy who tells a girl he could climb to Jupiter because a natural ladder stretches from one planet to the other. The girl points out that nobody on earth has ever found such a ladder and there is reason to believe it doesn't exist—because of the constantly changing distance between the planets, the sun getting between them, etc. The boy shakes his head at her and patiently explains, "We haven't found it yet? That's an argument from ignorance. Scientists are finding all sorts of new things in our solar system all the time. Look at the moon. And the asteroid belt between Mars and Jupiter. We used to not know anything about that. Already we have several steps along the way. You see, everything is falling into place."

Collins's suggestion that we are sure to find a Darwinian pathway for the bacterial flagellum isn't this outlandish, of course, but it employs the same reasoning. He assumes the Darwinian pathway certainly exists. He insists that any scientist skeptical of us ever finding the pathway is simply giving up. And he ignores the positive arguments for design, proceeding as if the design argument is entirely a negative critique of a scientific project still in progress.

Following Behe

But here is the oddest thing, the thing that makes *The Language of God* such an interesting study of how certain theistic evolutionists treat intelligent design. As seen earlier, Collins does not always commit this error. For instance, in his arguments for design from the origin and fine-tuning of the universe, Collins makes the same kind of argument for design that Behe makes in biology, inferring design as the best explanation of the current evidence. In each case a critic could note that Collins has violated the rule of methodological materialism that he himself invoked against intelligent design theory.

This criticism could be leveled against his first two design arguments, noted above, and also against a third design argument, in which he appeals to the moral law in the human heart as evidence of design. Collins critiques the other leading explanation for the moral law—that the "moral law" is only a bundle of survival instincts instilled by Darwinian evolution. And he argues that a better explanation is that we are not just matter but also spirit.

To this argument, the thoroughly consistent methodological materialist could respond, "But Dr. Collins, just because we are ignorant of a detailed Darwinian pathway to things like human altruism doesn't mean we won't ever find the pathway. You're arguing from ignorance to design, and you can't do that."

Collins was right not to let himself be governed by this line of thinking when he inferred design from the origin of the universe, cosmic fine-tuning, and the moral law within. The objection that he argued to design from mere ignorance of an adequate material cause assumes ahead of the evidence that such a cause actually exists, and that the evidence has no tell-tale fingerprints of intelligence. The truly reasonable approach is to do what historical scientists routinely do: compare the available evidence, make an inference to the best explanation, and then see how that inference holds up in light of subsequent discoveries.

By insisting on that right in the realms of cosmology and human experience, one of the world's leading geneticists unintentionally may have nudged the scientific community a step closer to the day when such an approach will be tolerated in the life sciences as well.

Collins's Theology

Collins does offer a theological argument for his selective application of methodological materialism and his belief that Darwinism is no threat to Christianity. He suggests that God fine-tuned the initial conditions of the universe so perfectly that no further intervention was needed until he was ready to raise up one form, *hominid*, by investing it with an immortal soul that evolution could not instill. Collins contends that "humans are also unique in ways that defy evolutionary explanation and point to our spiritual nature."

On this view, God acted directly in the origin of the universe and in the origin and history of humanity, but his perfect wisdom meant that nature required no additional guidance or direction (or design) during the intervening 14 billion years. Collins suggests that anything less than such a "fully gifted creation" (I am borrowing physicist Howard Van Till's term) is unworthy of a God who is both omnipotent and omniscient.

As Collins puts it:

> ID portrays the Almighty as a clumsy Creator, having to intervene at regular intervals to fix the inadequacies of His own initial plan for generating the complexity of life. For a believer who stands in awe of the almost unimaginable intelligence and creative genius of God, this is a very unsatisfactory image.

Thus, between the origin of matter and man, he suggests, we have a good theological reason to consistently apply the principle of methodological materialism.

But in making this argument, Collins treats God's relationship to time in a manner inconsistent with his treatment of this subject in chapter 3 of his book. There he notes that the God of Christianity invented and transcends time, both past, present, and future. He makes this point to explain how God could exist "before" the Big Bang and how he could know that his finely tuned new universe would one day lead to the evolution of planet Earth and human beings.

But this theological subtlety has an implication he overlooks when he criticizes intelligent design theory for (according to him) positing a God who can't get the design right the first time (that is, at the origin of a "fully gifted" universe). If God is outside of time, if he stands over past, present, and future, then such interventions would occur in the eternal present of the "I Am"[5] whether they occurred "all at once" 14 billion years ago or at different points throughout the history of the universe. If Collins can use this reasoning to explain design at the level of physics and cosmology, then he has no basis to criticize an ID theorist who (Collins thinks) wants to posit God's activity within the cosmic order. This isn't really an accurate account of the ID claim, which simply claims that design is detectable but makes no claims about the discrete locations of design activity. But that's not important here; for even on his own terms, Collins's argument doesn't hold up.

Also notice how blithely Collins equates God's ongoing involvement in creation with incompetence. (Ken Miller, whose book *Finding Darwin's God* he recommends, told the *Philadelphia Inquirer* that the God of intelligent design theorists "is like a kid who is not a very good mechanic and has to keep lifting the hood and tinkering with the engine."[6])

Why? What if the creator likes to stay involved? What if he doesn't want to wind up the watch of the cosmos and simply leave it to crank out everything from supernovas to sunflowers? What if his relationship to the cosmos is also like a gardener to his garden? What if he wants to get his hands dirty? A Gnostic God wouldn't be caught dead with earthly soil on him, but ours is no such God, as the incarnation, crucifixion and bodily resurrection together attest. Whatever the subtleties of ID theory, there's no basis in Christian theology for rejecting the possibility of God acting directly within the created order.

God's Chances

Collins's synthesis possesses another crucial shortcoming. It undercuts either God's sovereignty or the random element at the heart of Darwinian Theory. The relevant passage is in chapter 10, in which he asks, "How could God take such chances? If evolution is random, how could He really be in charge, and how could He be certain of an outcome that included intelligent beings at all?" The answer, he continues,

> is actually readily at hand, once one ceases to apply human limitations to God. If God is outside of nature, then He is outside of space and time. In that context, God could in the moment of creation of the universe also know every detail of the future. That could include the formation of the stars, planets, and galaxies, all of the chemistry, physics, geology, and biology that led to the formation of life on earth, and the evolution of humans, right to the moment of your reading this book—and beyond.

This being the case, we who are "limited ... by the tyranny of linear time" would think evolution "driven by chance, but from God's perspective the outcome would be entirely specified."

If God merely knew about future events like the origin of humans, while granting an element of random play to the unfolding of the universe, Darwinian randomness might be preserved. But then God would not have specified the various outcomes as Collins suggests. If, on the other hand, God did not grant the evolutionary process an element of random play, then we are no longer talking about Darwinian evolution, and Collins's admission that the outcome was entirely specified by God is as good as saying that it was intelligently designed by God, albeit through the use of secondary causes, a mode perfectly at home in design theory.

In an earlier chapter Collins blamed Darwinian evolution for supposed bad design (like the backward wiring of the eye), but if every physical event unfolded according to a plan hard-wired into the universe from the beginning, then God is every bit as responsible for the backward wiring of the eye as if he had designed it directly. Christian theologians through the ages have defended a similarly strong role for Providence, but Collins cannot invoke a meticulously detailed Providence to explain the evolution of life and humanity while at the same time suggesting that a random process rather than God was responsible for supposed evolutionary failures.

THE UNFULFILLED PROMISE OF FRANCIS COLLINS

IN *THE LANGUAGE OF God*, Collins makes a sincere but unsuccessful effort to synthesize Darwinism and Christianity. The book succeeds where he perhaps least expects it: In some cases, happily, he violates the rules of methodological materialism by allowing himself to consider design as the best explanation for the origin of the universe, the fine-tuning of the physical constants, and the moral law within the human heart.

In granting himself this freedom, Collins was returning to the origins of the scientific revolution. Modern science was born of the twin convictions that the universe was the rational product of a rational mind, and that this maker was not bound at every turn by the deductive syllogisms of an earlier age, meaning that the best way for a scientist to determine how the Creator did things is to turn to nature and carefully scrutinize it.

Alas, in the years following the publication of *The Language of God*, Collins did not live up to the promise of the best parts of his book. Instead of becoming more open to the evidence for intelligent causes throughout nature, he seems to have become more close-minded. There even have been reports that he has abandoned his belief that Darwinian Theory cannot account for human morality.[7]

Fortunately for the future of science, many other scientists and students are refusing to be bound by the question-begging restrictions of methodological materialism and are following the evidence where it leads.

6. Darwin of the Gaps

Francis Collins's Premature Surrender

Jonathan Wells

On June 26, 2000, President Bill Clinton announced the completion of the first survey of the human genome—the DNA sequence of a human cell. "Today," he said, "we are learning the language in which God created life." At the President's side was Francis S. Collins, director of the Human Genome Project, who had helped to write Clinton's speech. "It is humbling and awe-inspiring," Collins said, "to realize that we have caught the first glimpse of our own instruction book, previously known only to God."[1]

The complete sequence was announced three years later,[2] and in 2006 Collins published a book titled *The Language of God: A Scientist Presents Evidence for Belief.*[3] Ironically, however, Collins's "evidence for belief" did not include DNA, which he claimed provides compelling evidence for Darwin's theory of evolution instead.

According to Collins, the data from DNA sequencing provide "powerful support for Darwin's theory of evolution, that is, descent from a common ancestor with natural selection operating on randomly occurring variations." Darwinian evolution is "unquestionably correct"—indeed, as a mechanism it "can and must be true"—and people who argue that it has failed to explain the apparent design in living things are clinging to a "God of the gaps" position that is doomed to collapse with further advances in science. Collins argues that the only "scientifically consistent and spiritually satisfying" way to be a believer—one that "will not go out of style or be disproven by future scientific discoveries"—is "theistic evolution, or BioLogos."[4]

Yet Collins ignores evidence that has come out of the very project he administered. Instead of supporting Darwinian evolution, the new DNA evidence actually undercuts it. Indeed, the more we learn about our genome, the less tenable Darwin's theory becomes. Collins is clinging to a "Darwin of the gaps" position that becomes more precarious with each new discovery.

God of the Gaps?

In *The Language of God*, Collins bases his religious belief primarily on the "Moral Law," the concept of right and wrong that "appears to be universal among all members of the human species." He also sees grounds for belief in the Big Bang, the orderliness of natural laws, and the so-called Anthropic Principle. "The fact that the universe had a beginning, that it obeys orderly laws that can be expressed precisely with mathematics, and the existence of a remarkable series of 'coincidences' that allow the laws of nature to support life ... point toward an intelligent mind." But Collins warns against "God of the gaps" thinking. "Faith that places God in the gaps of current understanding about the natural world may be headed for crisis if advances in science subsequently fill those gaps."[5]

A skeptic might object that Collins himself relies on a God of the gaps. "In my view," Collins writes, "DNA sequence alone, even if accompanied by a vast trove of data on biological function, will never explain certain human attributes, such as the knowledge of the Moral Law." And "the Big Bang cries out for a divine explanation," because "I cannot see how nature could have created itself."[6] Yet some evolutionary psychologists think they have already explained the Moral Law as a product of Darwinian processes, and some cosmologists think they have already explained the Big Bang as a result of cyclical expansions and contractions of an eternal universe—or as the beginning of just one of countless members of the multiverse.

No matter. Collins ignores these considerations and focuses instead on what he considers to be the most egregious God of the gaps position: intelligent design in biology. According to Collins, intelligent design (ID) rests upon three propositions: (1) "evolution promotes an atheistic worldview;" (2) "evolution is fundamentally flawed, since it cannot account for the intricate complexity of nature;" and (3) "if evolution cannot explain irreducible complexity, then there must have been an intelligent designer."[7]

None of these represents ID accurately. According to the Discovery Institute's Center for Science and Culture, "Intelligent design holds that certain features of the universe and of living things are best explained by an intelligent cause, not an undirected process such as natural selection."[8]

So, contrary to Collins's first proposition, ID does not claim that "evolution promotes an atheistic worldview." "Evolution" can mean simply change over time, which carries no religious or anti-religious implications. Or it can mean minor changes within existing species, an uncontroversial phenomenon that is also religiously neutral.

It is not evolution in general, but Darwinism's exclusion of design, that ID proponents reject. Darwin wrote that he saw "no more design in the variability of organic beings, and in the action of natural selection, than in the course which the wind blows."[9] So he was "inclined to look at everything as resulting from designed laws, with the details, whether good or bad, left to the working out of what we may call chance."[10] Because Darwinism excludes design in the details, many people (including many Darwinists) have argued that it "promotes an atheistic worldview." But the conflict between Darwinism and ID concerns only the question of design.

Collins's second and third propositions falsely imply that ID consists only of a negative argument—as though the absence of evidence for Darwinism automatically justifies an inference to design. Yet (as we shall see below) a design inference is warranted only when effects resemble those that we know from experience require an intelligent cause. The fact that Darwinian Theory cannot explain something may be a necessary condition for inferring intelligent design in biology, but it is not sufficient.

Collins specifically criticizes biochemist Michael Behe's argument that some features of cells are irreducibly complex. "By *irreducibly complex*," Behe wrote in 1996, "I mean a single system composed of several well-matched, interacting parts that contribute to the basic function, wherein the removal of any one of the parts causes the system to effectively cease functioning."[11] Behe argued that whenever we encounter irreducible complexity in our daily lives we quite reasonably attribute it to an intelligent agent, since that is the only cause that we know can produce it.

Behe described several features of living organisms that are irreducibly complex. One of these is the human blood-clotting cascade, a system of interacting proteins that ensures that clots will form only when and where they are needed. But Collins asserts that "the well-established phenomenon of gene duplication" shows that the component parts of the blood-clotting cascade "reflect ancient gene duplications that then allowed the new copy ... to gradually evolve to take on new a function, driven by the force of natural selection."[12] Behe wrote a decisive response to this criticism way back in 2000,[13] but Collins ignores it.

The most "damaging crack in the foundation of intelligent design" for Collins, however, is recent research that allegedly undercuts Behe's argument for the irreducible complexity of the bacterial flagellum, a propulsion system driven by a microscopic high-speed motor. Collins first re-casts Behe's view of irreducible complexity as a claim that "the individual subunits of the flagellum could have had no prior useful function of some other sort," then he proceeds to argue against that straw-man claim.[14]

As evidence, Collins cites the type III secretory system (TTSS), "an entirely different apparatus used by certain bacteria to inject toxins into other bacteria that they are attacking" that is composed of proteins similar to some of those in the bacterial flagellum motor. "Presumably," Collins writes, the elements of the TTSS:

> were duplicated hundreds of millions of years ago, and then recruited for a new use; by combining this with other proteins that had previously been carrying out simpler functions, the entire motor was ultimately generated. Granted, the type III secretory apparatus is just one piece of the flagellum's puzzle, and we are far from filling in the whole picture (if we ever can). But each such new puzzle piece provides a natural explanation for a step that ID had relegated to supernatural forces, and leaves its proponents with smaller and smaller territory to stand upon.[15]

As in the case of the blood-clotting cascade, Collins ignores Behe's counterargument—in this case, that "there's no reason that parts or subassemblies of irreducibly complex systems can't have one or more other functions."[16] The fact that a fuel pump can be used for other purposes doesn't mean that the automobile engine of which it is a part is undesigned. Furthermore, evidence suggests that the bacterial flagellum is *older* than the TTSS. If anything, the latter probably *de*-volved from the former.[17]

Collins's main complaint against ID, however, is that it:

> fails in a way that should be more of a concern to the believer than to the hard-nosed scientist. ID is a "God of the gaps" theory, inserting a supposition of the need for supernatural intervention in places that its proponents claim science cannot explain. Various cultures have traditionally tried to ascribe to God various natural phenomena that the science of the day had been unable to sort out—whether a solar eclipse or the beauty of a flower. But those theories have a dismal history. Advances in science ultimately fill in those gaps, to the dismay of those who had attached their faith to them. Ultimately a "God of the gaps" religion runs a huge risk of simply discrediting faith. We must not repeat this mistake in the current era. Intelligent design fits into this discouraging tradition, and faces the same ultimate demise.[18]

But Collins's "God of the gaps" description of the history of science is inaccurate, except perhaps as an account of the demise of animism. As physicist David Snoke has written, "Did anyone ever argue for the existence of God because we did not understand magnets or the orbits of the planets? Perhaps some pagan shaman somewhere has argued that way, but I see no evidence for any serious Christian argument along these lines."[19] And Collins's suggestion that scientific advances have eliminated the possibility of intelligent design is exaggerated, to say the least. After all, for the blood-clotting cascade and the bacterial flagellum the only thing Collins offers is speculation about ancient gene duplications.

In any case, ID is not a "God of the gaps" argument.

First, ID is not inserting supernatural intervention, unless the effects of intelligence themselves are defined as supernatural. ID makes only the minimal claim that it is possible to infer from the evidence of nature that some features or patterns in nature are explained better by an intelligent cause than by undirected processes. True, one can then ask about how the design is implemented, or about the nature of the intelligence. To the latter, a reasonable answer would be God. But ID alone does not take us that far; it is not natural theology.

Second, and more importantly, design inferences are not arguments from ignorance. No sane person argues, "I don't know what caused X, therefore it must be designed." We infer design in our daily lives when X resembles things that we know are produced by intelligence and could not plausibly have been produced without it. Irreducible complexity is one hallmark of designed things; the "specified complexity" of William Dembski is another.[20] In either case, we infer design most reliably when we have more evidence, not less.

Darwin and his contemporaries thought living cells were blobs of protoplasm; it was easy for them to assume that such blobs were undesigned. But as modern biologists learn more and more about the irreducibly complex biochemical cascades and molecular machines needed for life, it becomes less and less plausible to dismiss cells as accidental by-products of unguided natural forces. If anything is having to retreat in the face of scientific advances, it is Darwinism.

Collins doesn't see it that way.

Overwhelming Evidence?

Collins claims that the evidence provides "powerful support for Darwin's theory of evolution, that is, descent from a common ancestor with natural selection operating on randomly occurring variations." Regarding the second part of the theory (natural selection acting on random variations), Collins writes: "Darwin could hardly have imagined a more compelling digital demonstration of his theory than what we find by studying the DNA of multiple organisms. In the mid-nineteenth century, Darwin had no way of knowing what the mechanism of evolution by natural selection might be. We can now see that the variation he postulated is supported by naturally occurring mutations in DNA." Although most mutations are neutral or harmful,

> on rare occasions, a mutation will arise by chance that offers a slight degree of selective advantage. That new DNA "spelling" will have a slightly higher likelihood of being passed on to future offspring. Over the course of a very long period of time, such favorable rare events can become widespread in all members of the species, ultimately resulting in major changes in biological function.[21]

Of course, no one doubts that mutations can lead to minor changes within existing species ("microevolution"). The best-known example is antibiotic resistance in bacteria. But can similar changes lead to new species, organs and body plans ("macroevolution")? In 1937, evolutionary biologist

Theodosius Dobzhansky noted that "there is no way toward an understanding of the mechanisms of macroevolutionary changes, which require time on a geological scale, other than through a full comprehension of the microevolutionary processes observable within the span of a human lifetime." He concluded: "For this reason we are compelled at the present level of knowledge reluctantly to put a sign of equality between the mechanisms of macro- and microevolution, and proceeding on this assumption, to push our investigations as far ahead as this working hypothesis will permit."[22]

An assumption is one thing; evidence is another. Collins writes, "Some critics of Darwinism like to argue that there is no evidence of 'macroevolution' (that is, major change in species) in the fossil record, only of 'microevolution' (incremental change within a species)." But he writes that "this distinction is increasingly seen to be artificial."[23] To prove his point, Collins cites a Stanford University study of stickleback fish.[24] Marine sticklebacks typically have armor plates extending from head to tail, but many freshwater sticklebacks lack such plates, and biologists have found a correlation between this difference and variations in the gene for Ectodysplasin (EDA), a molecule involved in the formation of the plates. Collins concludes:

> It is not hard to see how the difference between freshwater and saltwater sticklebacks could be extended to generate all kinds of fish. The distinction between macroevolution and microevolution is therefore seen to be rather arbitrary; larger changes that result in new species are a result of a succession of smaller incremental steps.[25]

But marine and freshwater sticklebacks are merely varieties of the same species, *Gasterosteus aculeatus*. There is no evidence here that variations in their gene for EDA would (or even *could*) lead to the origin of a new species, much less to the major changes in anatomy needed for macroevolution. The same can be said for the only other example cited by Collins, minor variations in disease-causing viruses, bacteria, and parasites.

So, like Dobzhansky, Collins merely *assumes* that microevolution can be extrapolated to macroevolution. Despite seventy years of genetic research the extrapolation remains an assumption, and the distinction between micro- and macroevolution is no more "arbitrary" now than it was then.

Regarding the first part of Darwin's theory (descent from a common ancestor), Collins writes that "the study of multiple genomes" enables evolutionary biologists "to do detailed comparisons of our own DNA sequence with that of other organisms." Since DNA mutations accumulate over time, organisms with a recent common ancestor would be expected to show fewer differences in their DNA than organisms that diverged in the distant past. "At the level of the genome as a whole," Collins writes, "a computer can construct a tree of life based solely on the similarities of DNA sequences," and he includes an evolutionary tree ("phylogeny") of mammals constructed in this way. Collins concludes: "This analysis does not utilize any information from the fossil record, or from anatomical observations of current life forms. Yet its similarity to conclusions drawn from studies of comparative anatomy, both of existent organisms and of fossilized remains, is striking."[26]

This claim is simply false. First, DNA data often lead to conflicting phylogenies. For example, Collins's evolutionary tree in *The Language of God* shows flying lemurs related to tree shrews, and rabbits and monkeys on more distant branches. But a phylogeny published in 2002 in the *Proceedings of the National Academy of Sciences USA* shows flying lemurs related to monkeys and tree shrews related to rabbits.[27] Conflicts among different DNA-based trees are a major headache for evolutionary biologists, some of whom spend their entire careers attempting to resolve them.

Second, phylogenies constructed with DNA often conflict with phylogenies based on anatomy. Take whales, for example—fossils of which Collins asserts are "consistent with the concept of a tree of life of related organisms." On anatomical grounds, evolutionary biologist Leigh Van Valen proposed in the 1960s that modern whales are descended from an extinct group of hyena-like animals.[28] Then, in the 1990s, molecular comparisons suggested that whales are more closely related to hippopotamuses,[29] though evolutionary biologist Kenneth D. Rose reported in 2001 that "substantial discrepancies remain" between the anatomical and molecular evidence.[30] In 2007, J. G. M. Thewissen and his colleagues pointed out that since whales appear in the fossil record 35 million years before hippopotamuses "it is unlikely that the two groups are closely related," and they concluded from anatomical comparisons that whales are descended from a raccoon-like animal instead.[31]

The conflict between anatomical (morphological) and molecular phylogenies continues, and the problem is bigger than whales. In 2007, British scientists analyzed 181 molecular and 49 morphological trees and observed that "molecular and morphological phylogenies often seem to be at odds with each other."[32]

Junk DNA?

Even more surprising than Collins's misplaced reliance on DNA phylogenies is his reliance on so-called "junk DNA." According to Darwinian Theory, natural selection would tend to eliminate DNA changes that are harmful, but not changes that have no effect on function. When molecular biologists discovered in the 1970s that the vast majority of the mammalian genome consists of DNA that does not code for proteins, some assumed that this was simply molecular garbage that had accumulated over the course of evolutionary history. Collins cites certain segments of this "junk DNA" known as "ancient repetitive elements (AREs)," which he attributes to transposable elements ("jumping genes"). Collins writes that almost half of the human genome is "made up of such genetic flotsam and jetsam." Remarkably, when one compares AREs in mice and humans "many of them remain in a position that is most consistent with their having arrived in the genome of a common mammalian ancestor, and having been carried along ever since."[33]

The key assumption underlying Collins's argument is that the AREs are functionless. In a revealing passage Collins writes:

> Some might argue that these are actually functional elements placed there by a Creator for a good reason, and our discounting of them as "junk DNA" just betrays our current level of ignorance. And indeed, some small fraction of them may play important regulatory roles. But certain examples severely strain the credulity of that explanation. The process of transposition often damages the jumping gene. There are AREs throughout the human and mouse genomes that were truncated when they landed, removing any possibility of their functioning. In many instances, one can identify a decapitated and utterly defunct ARE in parallel positions in the human and the

> mouse genome. Unless one is willing to take the position that God has placed these decapitated AREs in these precise positions to confuse and mislead us, the conclusion of a common ancestor for humans and mice is virtually inescapable.[34]

Here (and elsewhere in his book) Collins is using a peculiarly Darwinian form of argument. In *The Origin of Species,* Darwin repeatedly argued that his theory must be true because divine creation is false.[35] This is an odd way to defend a scientific theory, yet it is common in the Darwinian literature. For example, in a section on "Evidence for Evolution" in the 2005 college textbook *Evolution,* Douglas J. Futuyma wrote: "There are many examples, such as the eyes of vertebrates and cephalopod molluscs, in which functionally similar features actually differ profoundly in structure. Such differences are expected if structures are modified from features that differ in different ancestors, but are inconsistent with the notion that an omnipotent Creator, who should be able to adhere to an optimal design, provided them."[36]

How do Futuyma and Collins know what a Creator would or would not do? Where else in science are statements about a Creator used to support a theory? Obviously, there's something very strange about Darwinian "science."

Stripped of its dubious theological content, Collins's argument reduces to this: Darwin's theory predicts the accumulation of functionless DNA differences, and that is what we find. But recent genome research provides growing evidence that much so-called "junk DNA" is not functionless at all. For example, in 2006 Japanese and American researchers discovered that "a large number of nonprotein-coding genomic regions are under strong selective constraint"—meaning that they have functions, otherwise selection would not affect them. The researchers wrote:

> Transposable elements are usually regarded as genomic parasites, with their fixed, often inactivated copies considered to be 'junk DNA' ... [but many such] sequences have been under purifying selection and have a significant function that contributes to host viability.[37]

In other words, the very "decapitated and utterly defunct" transposable elements that Collins considers his best evidence are turning out not to be functionless after all.

A similar result was reported by California scientists in 2007, who surveyed 10,402 noncoding elements in the human genome and found that a surprisingly high percentage functioned in gene regulation. They concluded that "mobile elements may have played a larger role than previously recognized."[38] The same year, Australian molecular biologists reported: "While less than 1.5% of the mammalian genome encodes proteins, it is now evident that the vast majority is transcribed, mainly into non-protein-coding RNA ... [of which] increasing numbers are being shown to be functional." The Australians concluded that the percentage of the genome that encodes functional information "may be considerably higher than previously thought."[39] And in 2008, American researchers demonstrated an important function for noncoding RNAs transcribed from segments of repetitive DNA that had previously been considered junk.[40] It seems that with each scientific advance Collins's "inescapable" evidence for common ancestry shrinks.

In addition to junk DNA, Collins also cites "silent mutations" in protein-coding segments of DNA. Since three letters of the DNA code are needed to specify one amino acid, and since there are sixty-four possible arrangements of such letters but only twenty amino acids, most amino acids can be specified by more than one three-letter "word." This means that even in the protein-coding segments of DNA some mutations do not change the resulting amino acid sequence. These are sometimes called "silent mutations."

In the course of evolution, natural selection would tend to eliminate proteins that have been damaged by changes in their amino acid sequences, so according to Darwinian Theory organisms are more likely to carry DNA mutations that do not produce such changes than those that do. According to Collins, when we compare DNA sequences of related species,

> silent differences are much more common in the coding regions than those that alter an amino acid. That is exactly what Darwin's theory would predict. If, as some might argue, these genomes were created by individual acts of special creation, why would this particular feature appear?[41]

There he goes again, invoking a theological argument to support what is supposedly a scientific theory. Theology aside, his argument (as above) relies on the assumption that "silent" mutations are functionless. In 2002, however, Uruguayan scientists discovered that a "silent" mutation in a gene in bacteria decreased the solubility of the resulting protein, even though it did not change the amino acid sequence.[42] And in 2007, scientists at the U.S. National Cancer Institute found that a "silent" mutation in mammalian cells significantly altered the functional properties of a multi-drug resistance protein while leaving its amino acid sequence unchanged.[43] If "silent" mutations are not silent after all, but have a discernible function, then Collins's argument falls apart.[44]

DARWIN OF THE GAPS

COLLINS CLAIMS HE'S BASING his case for Darwinism on new knowledge from genome sequencing, but he's actually basing it on falsehoods: namely, that mutations and natural selection have explained macroevolution, or that DNA phylogenies are consistent and reliable, or that most of our DNA is junk. In the case of junk DNA, Collins is arguing from gaps in our knowledge—gaps that are now being filled. The more molecular biologists learn about DNA, the more functions they discover in what were previously thought to be functionless segments.

How ironic.

Collins's own words may come back to haunt him. Faith that places Darwinian evolution in the gaps of our current understanding about the natural world—then asserts dogmatically that it "can and must be true"—is headed for a serious crisis as science advances. Like many other ideas in the past, theistic evolution has hitched its wagon to a falling star. Theistic Darwinism—or what Collins calls BioLogos—fits into this discouraging tradition, and faces the same ultimate demise.

7. Making a Virtue of Necessity

Howard Van Till's "Robust Formational Economy Principle"

Jay W. Richards

For over a decade, physicist Howard Van Till defended an (alleged) solution to a misguided and protracted battle between "creationists" and "evolutionists." As he saw it, the so-called "creation-evolution" debate fostered unnecessary conflict between natural science and religion, often leading to a loss of faith among Christian scientists and intellectuals. His view, he thought, provided a way to integrate theology and the natural sciences that avoided such conflicts and preserved faith.[1]

Ironically, since his retirement as a professor from Calvin College, Van Till has left the Christian faith and become a "free thinker" and perhaps an agnostic (I will discuss this more below). Nevertheless, a form of his argument is practically universal among theistic evolutionists, and some Christian theistic evolutionists still cite his work as reflecting an orthodox Christian view. So despite the evolution of Van Till's personal views, his argument is still relevant.

In this chapter, I will focus on what Van Till has called the "Robust Formational Economy Principle,"[2] which he proposed as an alternative to both intelligent design and to various forms of creationism. ("Robust Formational Economy Principle" is obviously a cumbersome phrase. So, to reduce the number of syllables in this chapter, I will refer to the idea as the "Principle" in what follows.) Van Till's arguments for this Principle are empirical, philosophical, and theological. I will focus primarily on his theological arguments.

Once we've worked our way through his arguments, it becomes clear that the Principle is on very shaky ground theologically. The best reason for affirming the Principle, in fact, is not theological but pragmatic: it prevents conflict with those who insist on "methodological naturalism" in science. But we still have to ask, Is there any good reason to believe that it's true?

Van Till's Argument Briefly Described

WHILE VAN TILL ORIGINALLY set his sights on "young earth creationism," in the 1990s he took up the case against "intelligent design theorists" as well. He refers to both views as "special creationism." For Van Till, the most objectionable aspect of so-called "special creationists" is their denial that the natural world has a "robust formational economy." Van Till directs his argument against the ID argument that intelligent design is empirically detectable[3] in the natural world and that methodological naturalism (or materialism) in science is misguided.[4]

Van Till calls his theistic evolutionary view the *fully gifted creation perspective*.[5] At times he has framed his basic idea in the form of a question: "Does the universe have the 'right stuff' for making full evolutionary development possible, or are certain key capacities lacking?"[6] Other times he expands the question:

> Is the Creation's formational economy sufficiently robust (that is to say, is it equipped with all the necessary capabilities) to make it possible for the creation to organize and transform itself from elementary forms of matter into the full array of physical structures and life-forms that have existed in the course of time?[7]

According to Van Till, all "special creationists" answer "no" to this question. Van Till and presumably everyone else say "yes."

As the question suggests, Van Till argues that there is no simple split between naturalists and theists when it comes to judging the Principle. A theist could answer yes to the question. So would the Principle count in favor of metaphysical naturalism or in favor of theism, which claims that nature is a creation? While many naturalists and theists assume that the truth of the Principle would count for naturalism and against theism, Van Till argues that this is a mistake. Why? Because theism could still be the most reasonable answer to questions such as:

1. Why is there something rather than nothing?

2. How did the universe come to have (or exhibit) a Robust Formational Economy Principle?
3. How did there come to be a universe with rational creatures, who have such things as moral and aesthetic sensibilities?[8]

Combining (1) through (3) with our moral and aesthetic experiences might still provide justification for being a theist and a Christian. More strongly, Van Till argues, a universe expressing the Principle may be better explained by theism than by naturalism, especially when all the facts of experience are added to the mix. Thus it's a mistake to assume that the existence of a universe, which does not require any "extra-natural, form-imposing agent" would be evidence for naturalism over theism.

The Problems with the Principle (and One Advantage)

The Advantage of the Principle

Van Till's Robust Formational Economy Principle has one obvious advantage. It allows theists to adhere to methodological naturalism. That means there is unlikely to be any conflict, or potential for conflict, between theological and scientific beliefs (at least in the natural sciences). In other words, the Principle protects the Christian scientist from having to object to methodological naturalism in some or another scientific discipline. And all things being equal, Christian scientists obviously prefer not to have conflicts between their religious and scientific beliefs.

Van Till shows that there is no obvious *logical* conflict between a world that exhibits the Principle and the truth of theism or Christianity. The Principle is something that, for the Christian theist, might be true. That is, God could create some world with the capacities to generate all the things that are in that world. Let us call such a world, a "Principle-conforming World." Surely, since God is omnipotent, God could create a Principle-conforming World.

So Van Till's argument has an understandable pragmatic motivation and it makes a valid logical point. It's possible for a Principle-conforming World to exist, and the Principle helps Christians avoid arguments with their colleagues who adhere to methodological naturalism.

The Central Question: Why Think the Principle Is True?

But how good is this argument as a rejoinder to contemporary arguments for intelligent design? Not very good, since few of those who doubt the Principle, including most proponents of intelligent design, would claim that God could not create a Principle-conforming World. In fact, so far as I know, no one has ever argued that. The issues at stake are more specific.

For instance, you might ask: Is it possible for there to be a world, containing all of the things that *this* world contains, that is a Principle-conforming World? *Did* God create such a world? And: Is our world, with its actual physical and biological entities, one that conforms to the Principle?

There are other needed distinctions in the neighborhood. For even if a Principle-conforming World is possible, there is still the question of whether it's possible for the actual world, with its particular physical and chemical laws, to generate all the physical and biological structures it actually has. To determine this, we would need either the right kind of in-principle argument or the right kind of empirical evidence, or probably both. It's not self-evident or true *a priori* that, say, the known physical and chemical laws are sufficient to explain all the biological and physical structures we find in this world.[9] That's something we would need to discover, not just assume.[10]

These are simple logical points. But there are more. Some ID theorists, such as Stephen Meyer[11] and William Dembski,[12] have argued that it's probably not possible for the known, repetitive, physical laws, or those laws plus chance, to generate large amounts of complex, specified biological information. If this is correct, and the actual world has such information, then the actual world does not conform to the Principle.

But let's bypass these arguments for now, and suppose that the Principle really holds in this world. That is, let's assume that the actual world has the formational capacities to give rise to all the physical and biological structures it contains. Still, this would not establish whether those capacities, even if sufficient, actually generated all such structures. It also would not establish the *likelihood* that those formational capacities gave rise to all the physical and biological structures in the actual world. For establishing such a thing would require looking at the empirical evidence, where possible, and weighing alternative explanations for past events, such as the origin of life and the universe. It's doubtful that we have the relevant evidence to conclude such a thing with any confidence.

So even if we concede for the sake of argument that the actual world has enough formational capacities to generate all the physical, biological and mental things there are, we would still have a problem knowing if those capacities actually did so. How would one decide whether the world's formational capacities are exclusively responsible for all the (natural) structures in the world?

It is here that we see that, whatever may be logically possible, the Principle has a much closer affinity with naturalism than with theism because, for the metaphysical naturalist, the Principle has to be true. If one assumes that there is nothing other than the physical universe, then obviously whatever formational capacities the universe has must be adequate to produce whatever there is. Argument and evidence are irrelevant, since the naturalist can deduce the (presumed) truth of the Principle from naturalistic first principles. So while there may be no obvious logical conflict between the Principle and theism, naturalism requires the Principle whereas theism does not. In that sense, naturalism has a clear *prima facie* attraction to the Principle. For the theist, the Principle may be logically possible, but is neither required nor preferable *a priori*.

If you're a theist, then, how should you decide whether the Principle applies to the actual world? Van Till sometimes offers an inductive historical argument, which points to the continued success of scientists in finding natural explanations for various natural phenomena. Given such a track record, he argues, it's reasonable to assume that there will be natural explanations (that is, explanations that appeal to unintelligent or material causes) for those phenomena still unexplained.

Van Till also offers a few pragmatic concerns. One is that the natural sciences, or more precisely, most natural scientists, assume that there are "natural" (that is, non-teleological) explanations for all phenomena; so it would be unwise for Christians and other theists to challenge this assumption.[13] The other related argument is really apologetic advice.[14] Van Till urges Christians not to appeal to God's action every time some sort of "gap" appears in our scientific knowledge. For instance, just because scientists do not yet know how life originated on earth, Christians should not seek to fill this gap by claiming that God did it. Why not? Well, (1) it's not a scientific explanation (where scientific explanation equals a non-purposive explanation), and (2) when Christians have attempted this in the past, they have been forced to retreat before the advance of scientific knowledge, and brought disrepute to the faith. This "God of the gaps" strategy, according to Van Till, makes it look like divine agency is only possible where there are "gaps" in the natural economy. This tactic discredits Christian belief over time.[15]

These are all important though flawed arguments in their own right. Since others have answered them adequately, however, I will not belabor them here.[16] Instead, I would like to consider the *theological* basis on which, so far as I can tell, Van Till concludes that ours is a Principle-conforming World.

The Principle of Divine Propriety: God Wouldn't Do It That Way

Despite a number of secondary arguments, Van Till ultimately seems to ground his case for the Principle in a certain theological aesthetic. He finds it distasteful to depict God as acting directly in the course of nature. Let's call this the *principle of divine propriety*. This principle-behind-the-Principle informs Van Till's conception of God as well as his understanding of God's relation to the world.

The gist of Van Till's theological argument for the Robust Formational Economy Principle is that an omnipotent God would not create the world with natural laws and entities that have to be supplemented from time to time.[17] The all-powerful Christian God would have the foresight to "get it right the first time," so to speak. Van Till's choice of words betrays this aesthetic conviction. He always uses pejorative terms to refer to those who think that God acts directly in nature. So he speaks of "interventionism," the "god-of-the-gaps" fallacy, "special creationism," "episodic naturalism," "episodic creationism" and so on.

To justify this aesthetic preference, Van Till argues that there is a disanalogy between divine and human creative activity. God does not "design" nature as an artificer designs a clock or an artist paints a portrait. This image of an artificer seems more fitting for a Platonic Demiurge working with an intransigent and pre-existing material substratum than for the transcendent God of Christianity.[18] The Christian God, unlike a mere artificer, gives nature and its constituents their very being.[19]

Arguments along these lines are standard fare among theistic evolutionists. Despite their popularity, however, they're demonstrably invalid. Of course the creative capacity of God is much greater than that of a human artificer, qualitatively greater. But surely God's capacity is *at least as great* as a human artificer. Just because the notion of design in the sense of "artificer" does not exhaust God's creative capacity does not mean that it doesn't apply at all.

The former argument is also problematic, since it assumes the virtue of efficiency, which is irrelevant to an omnipotent God, with, well, unlimited resources. What seems to lie behind these complaints is a particular image, namely, the metaphor of the creation as a self-sustaining and even self-generating machine, like a watch.[20] Or, if one finds the notion of a machine off-putting,[21] think of "a self-generating organism" or just a "self-generating entity." If this is one's ideal, then it would seem inappropriate for God to create a watch that needed intermittent winding and tinkering. This is true enough; but there is little reason to allow this single, rather clunky metaphor to control the discussion. There are many other ways to understand God's relation to his creation.

This leads to what is perhaps the most obvious objection to Van Till's theological aesthetic, at least for Christians: it contradicts the biblical and Christian description of God's character and actions. It looks like a deduction from an abstract and debatable definition of omnipotence, rather than an implication of the Christian view of God. It does not comport well with a God who creates a world from nothing, courts a stubborn and rebellious people, makes covenants with individuals and nations, performs miracles, becomes incarnate as Jesus, and then raises Jesus from the dead. It's also difficult to square with the Christian life, which assumes that God answers prayers and works in the lives of believers through justification, sanctification, and so on.[22]

I can think of one way that Van Till could respond to objections such as these. He could distinguish between "natural history" and "salvation history." In this way, he could affirm that God acts in extraordinary and direct ways in salvation history, while insisting that God acts only through secondary causes in natural history.[23] Thus Van Till essentially would subsume natural history under the traditional category of *secondary causality* while allowing the category of salvation history to fall under both primary and secondary causality. This way of dividing up God's actions looks somewhat *ad hoc*;[24] but I will not tarry on this point, since it does not quite get to the core of the problem with the Principle.

The Central Theological Problem

The fundamental theological question for justifying one's belief that the Principle holds in our world, it seems to me, is this: Is it possible that God could create a world that does not conform to the Principle? Or, to put it differently, Is it possible for God to have a reason for creating such a world? More to the point, is it possible for God to create such a world, which has all the physical and biological structures that we find in our world? And is it possible that the Christian God could ever have good reasons to create such a world? Finally, is it possible for us to have good reasons to believe that God has created just such a world?

Perhaps to ask these questions is to answer them. We could spend all day constructing scenarios by which God might create such a world. Maybe he wants to be able to interact with his creation or with his creatures in a variety of ways, both directly and indirectly. Or maybe he wants his creation, or certain pockets of it, to reflect the fact that he has not only given it being and created it, but that he has designed both it and its constituents. Maybe he prefers a world that has a certain "artifactuality" to it, and that is clearly not self-explanatory, at least for the open-minded. Maybe he fancies a world where natural scientists, in searching the starry heavens, would get a glimpse of the King of Heaven.

Maybe God is like a novelist who includes himself as one of the characters in his novel. Maybe he has fashioned a world that bears empirical marks of his creative activity. Perhaps he desires a world that is more like a violin than a self-winding watch, an instrument he can play, or an arena in which he can act in different ways, sometimes directly, sometimes indirectly. Maybe he wants a world that exhibits a certain predictable regularity, but is by no means closed to his direct influence and agency. Maybe he seeks, as Dallas Willard puts it, to create a "God-bathed world." Perhaps he prefers a world in which, to quote Paul, "what can be known about God is plain to [human beings]," and in which God's "eternal power and divine nature" can be "understood and seen through the things he has made" (Romans 1:19–20, NRSV). Maybe God is like a hobbyist, who enjoys having a "work in progress." Perhaps God is like a gardener, who gives his garden ample opportunity to develop "on its own," while nevertheless giving it being, tending, plowing, weeding, and watering from time to time. Or maybe he just wants to annoy methodological naturalists and all those intellectuals who presume to set restrictions on his activity.

Now if it is possible that God could create a non-Principle-conforming-World, then it's possible for there to be evidence of the same. In such circumstances, it would surely be improper to say, as Van Till often does, that God has "withheld" certain capacities, or failed to bless it sufficiently, or that God "overpowers" and "coerce[s] material into assuming forms that it was insufficiently equipped to actualize with its God-given capabilities."[25] These assertions would imply that God somehow acted stingily or inappropriately in creating the world with the capacities it has. The capacities for a created thing are determined by the purposes for which it was created. I shouldn't complain that God "withheld" a set of wings from me or deprived me of the ability to fly unprotected in deep space or to move objects with my mind. There is no principle of divine propriety that would justify my complaint that human beings were created wingless. It would be obnoxious for me to complain that I lack these capacities, especially since I have the capacities God intended for me, and most human beings, to have.

In the same way, the propriety of nature having the ability to generate all its own complexity will depend on whether God intended for it to do so. If God means for nature to conform to Van Till's Principle, then it's appropriate that it do so. If he does not, then it is not. There's no *a priori* theological rule that determines that God should have done so, let alone that he has done so.

So it's possible that God would create a world in which the Principle doesn't apply. And this fact leads back to the central point at issue: how do we know that the actual world is not such a world? The issue here isn't what is possible, but what is true. How do we know, or what good reason do we have to think that our world is not one such world? There may be other reasons for thinking that the world conforms to the Robust Formational Economy Principle, but the principle of divine propriety is not one of them.

Making a Virtue of Necessity

Although it might seem improper to speculate on Van Till's motivations for defending the Principle, the issues mentioned above lead me to doubt that the reason for affirming the Principle is primarily theological. I suspect that it's actually an example of making a theological virtue of necessity. This is a risky charge to make, of course, since it requires going behind an explicit argument and looking for a subtext, by considering things like the argument's function. Moreover, this sort of thing can occur unconsciously, so one must clearly distinguish between making a virtue of necessity and a conscious intention to deceive. Nevertheless, anyone who does much reading in contemporary theology will recognize that many arguments that appear theological on the surface actually derive from quite different sources.

The unstated reasoning could go something like this. Perhaps some Christian academic concludes, for a variety of reasons, that nature probably does have all the resources necessary to generate all the entities it contains. He's convinced that under the right conditions, chemistry can give rise to life, and that the Darwinian mechanism of natural selection and random variation can generate the complexity and adaptations in living creatures.

Moreover, perhaps he thinks that, regardless of the truth of the matter, it's not career-enhancing for Christians to challenge this assumption.

At the same time, he realizes that some people, Christians and naturalists alike, think that the fact that the world can generate everything it contains is evidence for naturalism. So he decides the best way to block that inference is to construct a theological argument for the Robust Formational Economy Principle, give it a fancy theological name like the *fully gifted creation perspective,* and wax indignant about the alternatives, which, from that perspective, look theologically deficient or even heretical.

After all, a primary, although unstated, function of the Principle is to immunize and compartmentalize Christian belief from scientific challenges. Whether or not this is why Van Till adopted it, it's obvious that the Principle looks tailor-made to prevent any conflict between Christian belief and methodological naturalism.[26] Armed with the Principle, Van Till can argue that natural scientists are generally on a path toward truth when they assume that there are "natural explanations" (that is, impersonal explanations) for all natural phenomena.

Ordinarily, this argument would be implausible, since it's surely unlikely that God's actions and attributes would conform so nicely to a contentious rule like methodological naturalism. But the Principle turns the necessity for maintaining methodological naturalism into a theological virtue.

Rather than concluding that Christian theology implies a view of God in keeping with the canons of methodological naturalism, then, it seems more likely that Van Till, perhaps unwittingly, has made a theological virtue out of a perceived necessity. If so, then the Principle reduces to little more than a theological attempt to accommodate methodological naturalism. As such, it's powerless to counter design theorists who challenge methodological naturalism itself.

Conclusion

The theological justification for assuming that the world conforms to the Robust Formational Economy Principle is flimsy. It provides little guidance in deciding whether nature has generated all of the entities it contains, and whether nature points beyond itself or exhibits evidence of intelligent design. Whatever case can be made for Principle, it's not a theological one.

So what other reason could we have to affirm it? Well, perhaps we could discover it, or least find lots of evidence for it, through various experiments in the natural sciences and inferences in the historical sciences, aided by disciplines such as information, confirmation and probability theory.[27] Perhaps we would be justified in believing it after following a long inferential chain, comprised of many empirical links. But if so, then we should be open to the possibility that the Principle does not obtain in the world. Being open to both these possibilities seems to be the proper course. Therefore, the debate returns to the issues to which design theorists have appealed all along, namely, whether there is empirical evidence of intelligent design in nature.

Post-Script: Responding to Van Till

Until his retirement in 2003, Howard Van Till was a professor at Calvin College in Grand Rapids, MI, a school affiliated with the Christian Reformed Church. After his retirement, Van Till abandoned the Christian beliefs that he had affirmed as a professor. He now considers himself a "free thinker."

The previous year, I had published a version of my critique of his views in the journal *Philosophia Christi*. Van Till wrote a response, which was published, along with my article, and with a final response by me.[28]

In his response, Professor Van Till made some surprising concessions, indicating that he was already beginning to abandon Christian theology. He also confirmed many of my original critiques. I think Van Till's responses, and subsequent developments, should serve as a cautionary tale for those who seek to incorporate naturalism, methodological or otherwise, into Christian theology.

The Principle as an A Priori Commitment

In his response, Van Till confirmed one of my central claims, namely, that the Robust Formational Economy Principle functions for him as an intellectual *a priori*. As a result, it prevents him from considering evidence or arguments against the Principle or for intelligent design, on their own terms. Whether one adopts it for philosophical, scientific, or theological reasons, the Principle bars entry to any contrary evidence. Of course Van Till admits that the Principle might be wrong, and he recognizes that we cannot know its truth with absolute certainty; nevertheless, he says forthrightly, "*[T]hat's the horse I'm betting on.*"[29] But why bet on one horse? Why not spread your bets and be open to other possibilities?

After all, I can consider what sorts of things would count in favor of the Principle. Similarly, why could Van Till not consider evidence and arguments against it? In particular, why would he not engage the arguments of intelligent design theorists, some of which contradict the Principle? I argued that the theological and empirical reasons for adopting the Principle do not nearly justify adopting it in advance of unfettered investigation. He gave no reasons in his response to counter that argument.

In fact, Professor Van Till's response substantiated my charge that the Principle imposes procrustean restrictions on his thinking. It's not just that he had concluded that the Principle is true; rather, he did not consider the disputed issues without already presupposing it. The way in which he frames virtually every question assumes the Principle, even the title of his response: "Is the Creation's Formational Economy Incomplete?" This is a contentious way of putting it, since no design theorist argues that the creation's formational economy is *incomplete*. The question is, rather, just what sort of formational economy does the creation have? If it is supposed to be able to generate all of its own complexity, or all of the entities it contains, and it does not, then it is "incomplete." But if it's not supposed to be able to do that, then it's not incomplete, just contrary to naturalist expectations.[30] Perhaps the creation is intended to be receptive to God's interaction, innovation and informational input. Perhaps it's supposed to "lack" the ability, to, say, generate new body plans by natural selection acting on random genetic mutations. If so, that's hardly a deficiency.

For instance, no one labels a Boeing 747 "incomplete" because it does not ski well or assemble itself without engineers or factories, or reproduce asexually.

The question-begging tendency of the Principle is evident in the other terms Van Till uses to describe its detractors. He speaks of "episodic creationism," "missing capacities," "an interventionist concept of divine creative action"[31] and "gaps." The concluding paragraph of his response portrays intelligent design theory as simply an argument based on ignorance of the creation's capacities:

> One of the theological reasons I am inclined toward [the Principle] ... is that form-imposing interventions appear to be instances in which God would overpower elements in the Creation, coercing them into configurations different from what they were equipped to actualize, thereby violating the being that was once given to them at the beginning.[32]

This argument assumes that the only alternative to the Principle is for God to first equip certain natural entities to actualize one thing, and then later to decide to reconfigure them—against their original purpose—to do something else. That's a false dilemma. In fact, no one holds the alternative Van Till proposes. The issue in question is whether God *intended* certain natural entities to have such capacities. Perhaps God did not intend for, say, inorganic chemicals to self-assemble into living things under certain law-like conditions. If that is the case, then it's not a violation of their initial being to be transformed into living things by divine activity, direct or indirect. (In any case, such a transformation is certainly not "overpowering" or "coercive." These terms are tendentious if not meaningless except when dealing with other intelligent agents.) In this way Van Till transforms his assumptions into conclusions, bypassing the actual arguments of his critics.

I agree with Van Till that we could reconcile much of Christian belief with the Principle, if we had very strong reasons for believing that the Principle obtains. We could even construct a *post hoc* theological argument to justify it. But, all things being equal, Christian theology alone gives it scant support. Seeing the world as a creation gives us no particular reason to expect it. What this means is that support for the Principle will have to come from elsewhere, however much one may insist on its theological pedigree.

The Principle and Salvation History

I argued previously that the Principle is especially difficult to square with God's acts in salvation history. For the sake of charity, in my original essay, I allowed Van Till the distinction between natural and salvation history, since, as a Christian, I assumed he would want to exempt certain biblical claims like Jesus' incarnation, resurrection and miracles from the constraints of his Principle. In his response, however, he indicated that he was in the process of reinterpreting these doctrines. "[I]n all candor, I do believe that traditional ways of depicting divine action in the Creation deserve a thoughtful re-examination in light of what we have come to know about the Creation since these traditional theological formulations were crafted."[33] Later, he seemed to abandon these "theological formulations" altogether.

This move unwittingly reveals that there is a tension between the Principle and traditional Christian theology, and that Van Till thinks the tension should be resolved in favor of the Principle. This strategy gave Van Till's view greater internal consistency but weakened its already tenuous claim to be grounded in the Christian tradition. As mentioned above, he wrote his response to my criticisms around the time of his retirement from Calvin College. He has since indicated that he had been bound by Calvin's "Form of Subscription," but no longer adheres to its tenets.[34] I mention this not to say "I told you so," but to illustrate that both Van Till's argument and his personal biography show how hard it is to introduce naturalism into theology without ultimately diluting the theology.

The Principle and Methodological Naturalism

Van Till objected to my claim that his Principle seems suspiciously well fitted to allow a theological rapprochement with "methodological naturalism." I do not intend to attribute bad motives to Professor Van Till or to tarnish him with guilt by association, but rather to describe what I think is the essence of his position. And I think his response bolsters my argument. For the reasons given above, I don't think that the Christian theist should expect the Principle to obtain. The near universal admonition to adopt methodological naturalism is not by itself a good reason to assume it is true, and is certainly not a good *theological* reason.

In his response, Van Till insisted that even if the Principle is true, it doesn't follow that naturalism can account for other features of the natural world. I agree. After all, there are other grounds for doubting the truth of naturalism. I would doubt naturalism even if I had never heard these arguments. But Van Till's argument for this point is perplexing. He claims that he holds that there are "empirical marks of [God's] creative activity" in creation, as well as evidence that nature is "*not* self-explanatory" and needs a "creative Mind to conceptualize it."[35] This suggests that he agrees with intelligent design theorists on the central point at issue, even though he officially disagrees with them.

This claim also contradicts what he has said elsewhere. For instance, in an appearance with physicist Alan Guth in April 2000 (The Nature of Nature Conference, Baylor University), Van Till admitted that he could not think of a way in which the world would look different if naturalism or theism was true. I don't know what Van Till's current view is, but he's clearly held contradictory views on the question of whether God's activity is detectable in nature.

In any case, my central claim still stands: the Principle is just what the naturalist would expect. For the theist, it's possible but hardly expected. To pretend otherwise is, as I previously argued, to make a virtue of a perceived necessity.

Unfortunately, it's difficult to make this point delicately, since someone might interpret it in an *ad hominem* fashion; but critical scholarship requires the ability to discern between a good reason and a rationalization. It seems to me that the empirical case for *making the Principle a first principle*[36] is a non-starter. At best, it could be a tentative conclusion based on detailed evidence, but not a starting assumption. And the theological case for it is extremely weak. In fact, Van Till's putatively theological arguments obscure rather than support the best (if not only) reason for treating the Principle as a given, namely, to accommodate methodological naturalism.

Since Van Till dislikes the malodorous qualities of "methodological naturalism" in this debate,[37] it's hardly surprising that he avoids defending it directly. Nevertheless, the Robust Formational Economy Principle and methodological naturalism fit like a hand in a glove. That's surely not a coincidence. Even Van Till's own explanation of the origin of his view hints at the role of methodological naturalism. He writes that he is inclined to adopt the Principle

> on the basis of the coherent convergence of my theological perspective, and my assessment of what the natural sciences are learning about the universe's formational history and the remarkable processes that have contributed to the form-actualizing drama that has played out in the last several billion years on planet Earth.

Since he includes "methodological naturalism" in his understanding of natural science, it is not an *ad hominem* argument to point out that he has devised a theological perspective compatible with it, and given theological arguments to justify the decision. Whatever else it is, the Principle should be seen as a "coherent convergence" of Christian theism and methodological naturalism, not as the development of themes intrinsic to Christian theology and the evidence of science. If one wants to assimilate methodological naturalism while remaining a nominal theist, the Robust Formational Economy principle is probably the best strategy available. Moreover, it has the added benefit—as Professor Van Till notes—of making the term "methodological naturalism" superfluous. ("Fortunately I need no such term as *methodological naturalism*.") It does this, of course, by annexing methodological naturalism to theology.

8. The Difference It Doesn't Make

Teleological Evolution? Evolutionary Creation?

Stephen C. Meyer

The term "theistic evolution" means different things to different people.[1]

Some theistic evolutionists affirm that God actively directs the evolutionary process by, for example, directing seemingly random mutations toward particular biological endpoints. On this view, God has actively created new organisms by directing mutational change to produce new forms of life. This view has the virtue for theists of being at least minimally compatible with orthodox Jewish and/or Christian doctrines of creation since it affirms that God is actively doing something to bring life into existence. On the other hand, this view contradicts the (scientifically) orthodox Neo-Darwinian view of the evolutionary process as a purely purposeless, unguided and undirected mechanism—a "blind watchmaker" as Richard Dawkins has called it. Indeed, as George Gaylord Simpson, one of the architects of Neo-Darwinism wrote in *The Meaning of Evolution,* the theory implies that "man is the result of a purposeless and natural process that did not have him in mind."[2]

Other theistic evolutionists see the evolutionary process—including both the origin and subsequent evolution of life—as a *purely* unguided and undirected process, just as orthodox Neo-Darwinists do. These theistic evolutionists conceive of God's role as much more passive. They conceive of God as merely sustaining the laws of nature which in turn allow life to emerge and develop as the result of otherwise undirected and unguided mechanisms such as mutation and natural selection. While this view comports nicely with scientific materialism and Neo-Darwinism, it seems to contradict religiously orthodox views of the creation of life by denying that God played any active role in creation or even that He knew what the evolutionary process would ultimately produce.

Some theistic evolutionists want to affirm both (scientifically) orthodox Neo-Darwinism and materialistic origin-of-life scenarios and (religiously) orthodox Jewish or Christian theology at the same time. But they face the logical difficulty of explaining how God can direct an undirected process. Typically, those attempting to reconcile their theological commitments with Neo-Darwinism will refer to God as using the evolutionary process to create without specifying whether "the evolutionary process" refers to the wholly undirected Darwinian process of random mutation or something else. Thus, attempts to reconcile standard forms of biological (and chemical) evolutionary theory and orthodox Jewish or Christian theology typically either lack specificity or logical coherence or both.

Perhaps in an attempt to split the horns of this dilemma, some theists who accept the adequacy of materialist explanations have proposed a new view, or at least a view with a new name. For example, Denis O. Lamoureux, an Associate Professor of Science and Religion at St. Joseph's College, University of Alberta, advocates a position he calls "teleological evolution" or "evolutionary creation."[3] He prefers the term "evolutionary creation" to "theistic evolution" because, as an evangelical Christian, he wants to emphasize his belief in creation by making the term "evolutionary" merely the modifier rather than the noun in the description of his position. For him, evolution refers merely to "the method through which the Lord made the cosmos and living organisms."[4] He prefers the term "teleological evolution" to "theistic evolution" because he wants to affirm, *contra* Simpson and other Neo-Darwinists, that evolution is "a planned and purpose-driven natural process."

But what exactly is meant by this idea? And does it provide an adequate scientific explanation for the origin and development of life? And if so, does it do so with enough specificity to be distinguished from standard materialistic theories of evolution and reconciled with a traditional Judeo-Christian understanding of God as the Creator of life?

According to Lamoureux, the theory of evolutionary creation affirms that "the Creator established and maintains the laws of nature, including the mechanisms of a purpose-driven teleological evolution."

He provides several illustrations to convey what he has in mind. For example, he suggests that "God organized the Big Bang, so that the deck was stacked"[5] to produce life. He also likens God to an expert billiards player who can sink all the balls on the billiard table in one shot. He likens the precise arrangement of the balls and billiard player clearing the table with one single shot to God's initial act of creativity in bringing the universe into being with a very precise arrangement of matter (or initial conditions). Just as the billiard player can clear the table with one shot, God can create everything (the universe as well as all forms of life) with an initial act of creativity in which he arranges matter just right at the beginning of the universe and then lets it unfold deterministically in accord with the laws that He also established in the beginning. Thus, God needs no additional shots (no further acts of creativity) to bring life into existence. Lamoureux also compares the process of biological evolution to embryological development in which an organism develops deterministically from a fertilized egg through time in accord with the laws of nature.[6]

Thus, Lamoureux seems to have in mind a kind of front-loaded view of intelligent design in which the initial conditions of the universe are arranged or designed in such a way that life will inevitably evolve without any additional input or activity of a designing intelligence. As he explains, "design is evident in the finely-tuned physical laws and initial conditions necessary for the evolution of the cosmos through the Big Bang, and design is also apparent in the biological processes necessary for life to evolve."[7] Thus, in some places he also refers to his view as "evolutionary intelligent design."[8]

Although he uses the term intelligent design to describe his own view, he objects to the contemporary theory of intelligent design, and indeed, to any argument that implies that a designing intelligence played a role in the origin or development of life after the universe itself first originated. According to Lamoureux, to invoke a specific instance of intelligent design or divine action after the initial creation of the universe would imply a violation of natural law by invoking the activity of "a God-of-the-gaps." As he explains:

> Intelligent Design Theory ... is a narrow view of design and claims that design is connected to miraculous interventions (i.e., God-of-the-gaps miracles that introduce creatures and/or missing parts) in the origin of living organisms. For example, parts of the cell like the flagellum are said to be "irreducibly complex," and as a result, they could not have evolved through natural processes. Since this is the case, ID Theory should be termed *Interventionistic* Design Theory.[9]

Lamoureux objects to intelligent design as an explanation for specific features of biological systems or events in the history of life because he wants to confine God's creative activity to the very beginning of the universe and biological theories of intelligent design, he thinks, imply that an intelligent agent (quite possibly God) may well have acted at discreet points in the history of life after the origin of the universe.

Three Problems with the Theory of Evolutionary Creation (or Teleological Evolution) and its Critique of Intelligent Design

Having developed an argument for intelligent design as an explanation for the origin of the information necessary to produce the first life (an event that happened well after the beginning of the universe), it should come as no surprise that I disagree with Lamoureux's critique of the theory of intelligent design and his theory of evolutionary creation. I do so for several reasons. Since several contributors to this volume ably explain why ID arguments are not God-of-the-gaps arguments, I will focus my response on three other problems with Lamoureux's position.

First, I see no reason to *assume* that the designing intelligence responsible for life and the universe (whom I personally believe was God), necessarily confined His activity to the very beginning of the universe. He may or may not have done so. I agree with Lamoureux that the laws of nature were established, and are maintained, by God. I also agree that the fine-tuning of these laws and the initial conditions of the universe provides evidence of intelligent design. Nevertheless, I see no reason to assume that this fine tuning is the only evidence of design in the natural world. Nor do I think that the cosmological fine-tuning accounts for everything we find in the biological world (see below).

Of course, Lamoureux finds appealing the concept of a God who had the wisdom to arrange matter so exquisitely at the beginning of the universe as to make any future actions on His part unnecessary. Others like to think of God as more actively involved in the process of creation. They find it appealing to think of God acting like a great composer who first establishes a theme at the beginning of his work and then adds new variations to that original theme at episodic intervals thereafter.

As a Christian, however, I affirm that God acted entirely freely, and was under no compulsion to act in a way that either appeals to, or affirms our, aesthetic sensibilities. So I think the question of when God acted should remain a matter for empirical investigation and not be determined by our aesthetic or theological preferences one way or another. As Robert Boyle often argued, the job of the scientist (or what he called the "natural philosopher") is not to assume beforehand what God must have done, but to study the world to find out what God actually has done.[10]

Second, unlike Lamoureux, I do not think that material processes and mechanisms of evolution are sufficient to account for the origin of living forms, whether the origin of the first life or the major innovations in body plan design that appear during the history of life thereafter. Yet, his attempted harmonization of theistic belief with evolutionary theory presupposes the adequacy and/or creative power of established evolutionary processes and mechanisms—mechanisms that supposedly demonstrate the "incredible self-assembling" capacity of "the natural world during the distant past."[11]

But what if the evolutionary mechanisms that Lamoureux extols do not have the creative power or self-assembly capacity long attributed to them, as many scientists now argue? Before we construct elaborate theological harmonizations of the doctrine of creation with Neo-Darwinism and other materialistic evolutionary theories, shouldn't we make sure that these theories are true? I think so. And for reasons that I have explained elsewhere, I (and even many leading evolutionary biologists) now doubt that known or even postulated mechanisms of undirected evolutionary change have the creative power long attributed to them.

In my book *Signature in the Cell: DNA and the Evidence for Intelligent Design*,[12] for example, I show that no undirected chemical evolutionary mechanism provides an adequate explanation for the origin of the information necessary to produce the first life. I have also shown elsewhere[13] that neither the mutation/selection mechanism nor any of the other leading proposed evolutionary mechanisms provide a sufficient explanation for the origin of the major innovations in biological form (in particular, the novel body plans) that arise during the history of life on earth. For this reason, I dispute Denis Lamoureux's view that known evolutionary mechanisms are the main "method through which the Lord made ... living organisms."

Third, I think Lamoureux's front-loaded view of design is scientifically problematic. I do so because of my own study of the problem of the origin of the first life and the critically related problem of the origin of biological information. In the remainder of this chapter, I would like to discuss this problem and show why Lamoureux's theory of evolutionary creation, with its front-loaded view of design, is insufficient to account for the problem of the origin of biological information—as well as the closely related problem of the origin of life.

The Origin of Biological Information: The Fundamental Mystery

SINCE THE ELUCIDATION OF the structure of the DNA in 1953 and the advent of the molecular biological revolution, biologists have recognized the fundamental importance of information to biological systems. When Watson and Crick discovered the structure of DNA, they also discovered that DNA stores information in the form of a four-character alphabetic code. Strings of precisely sequenced chemicals called nucleotide bases store and transmit the assembly instructions—the information—for building the crucial protein molecules and machines the cell needs to survive.

Crick later developed this idea with his famous "sequence hypothesis," according to which the chemical parts of DNA (the nucleotide bases) function like letters in a written language or symbols in a computer code. Just as letters in an English sentence or digital characters in a computer program may convey information depending on their arrangement, so too do certain sequences of chemical bases along the spine of the DNA molecule convey precise instructions for building proteins. Like the precisely arranged zeros and ones in a computer program, the chemical bases in DNA convey information in virtue of their "specificity." As Richard Dawkins has noted, "the machine code of the genes is uncannily computer-like."[14] Software developer Bill Gates goes further: "DNA is like a computer program but far, far more advanced than any software ever created."[15] Similarly, biotechnology specialist Leroy Hood describes the information stored in DNA as "digital code."[16]

But if this is true, how did the information in DNA arise? As it turns out, this question is related to a longstanding mystery in biology—the question of the origin of the first life. Indeed, since Watson and Crick's discovery, scientists have increasingly come to understand the centrality of information to even the simplest living systems. DNA stores the assembly instructions for building the many crucial proteins and protein machines that service and maintain even the most primitive one-celled organisms. It follows that building a living cell in the first place requires assembly instructions stored in DNA or some equivalent molecule. As origin-of-life researcher Bernd-Olaf Küppers has explained, "The problem of the origin-of-life is clearly basically equivalent to the problem of the origin of biological information."[17]

Understanding Lamoureux's View: Pick Your Poison

Denis Lamoureux does not directly address the problem of the origin of the first life or the origin of the information necessary to produce it. He doesn't say which specific naturalistic theory of the origin of life—if any—he favors. His metaphors for how God creates (deck stacked at the beginning; God as cosmic billiard player; evolution as embryological development, etc.) imply that deterministic laws caused life to self-organize or self-assemble from some highly-configured and therefore, information-rich, set of initial conditions at the beginning of the universe. Nevertheless, he does not say whether he thinks that (a) all the information necessary to produce the first and subsequent forms of life was present in the initial conditions of the universe, or whether (b) the laws of nature added new information during the subsequent "self-assembly" process. In any case, both proposals are scientifically problematic. So let's consider each in turn, starting with the second.

ARE LAWS CREATIVE?

LAMOUREUX, LIKE OTHER EVOLUTIONARY creationists, sometimes speaks as though he thinks the physical laws of nature might be *generating* the new information necessary to produce new forms of life. For example, he refers to evolution as "a planned and purpose-driven natural process" and affirms "that humans evolved from pre-human ancestors, and over a period of time the Image of God and human sin were gradually and mysteriously manifested."[18] Since Lamoureux disavows specific acts of divine creation as illicit appeals to a "God-of-the Gaps," and since he affirms that humans, at least, acquire new attributes and characteristics during the evolutionary process, it might be that he thinks that the laws of nature are *generating* the new information necessary to produce living systems and their unique attributes.

But do laws of nature generate information? There are good reasons to doubt this. To see why, imagine that a group of small radio-controlled helicopters hovers in tight formation over a football stadium, the Rose Bowl in Pasadena, California. From below, the helicopters appear to be spelling a message: "Go USC." At halftime with the field cleared, each helicopter releases either a red or gold paint ball, one of the two University of Southern California colors. The law of gravity takes over and the paint balls fall to the earth, splattering paint on the field after they hit the turf. Now on the field below, a somewhat messier but still legible message appears. It also spells "Go USC."

Did the law of gravity, or the force described by the law, produce this information? Clearly, it did not. The information that appeared on the field already existed in the arrangement of the helicopters above the stadium—in what physicists call "the initial conditions." Neither the force of gravity, nor the law that describes it, caused the information on the field to self-organize. Instead, gravitational forces merely transmitted preexisting information from the helicopter formation—the initial conditions—to the field below.

There is a deeper reason that laws can transmit, but not generate information. Scientific laws describe (by definition) highly regular phenomena or structures, ones that possess what information theorists refer to as redundant *order*. On the other hand, the arrangements of matter in an information-rich text, including the genetic instructions on DNA, possess a high degree of complexity or aperiodicity, not redundant order.

To illustrate the difference compare the sequence ABABABABABAB to the sequence "One small step for a man, one giant leap for mankind." The first sequence is repetitive and ordered, but not complex or informative. The second sequence is not ordered, in the sense of being repetitious, but it is complex and also informative. The second sequence is complex because its characters do not follow a rigidly repeating, law-bound pattern. (It is also informative because, unlike a merely complex sequence such as: "sretfdhu&*jsa&90te," the particular arrangement of characters is highly exact or specified[19] so as to perform a (communication) function. In any case, informative sequences have the qualitative feature of complexity (aperiodicity), and thus are qualitatively distinguishable from systems characterized by periodic order that natural laws describe or generate.

To say that the processes that natural laws describe can generate functionally specified informational sequences is, therefore, essentially a contradiction in terms. Laws are the wrong kind of entity to generate the phenomenon in question. The claim also betrays a categorical confusion. Physical laws do not generate complex sequences, whether functionally specified or otherwise; they describe highly regular, repetitive, and periodic patterns. This is not an insult to the laws of physics and chemistry. It's simply what they do.

And yet, some scientists claim that we must await the discovery of new natural laws to explain the origin of biological information. Manfred Eigen has argued that "our task is to find an algorithm, a natural law, that leads to the origin of information."[20] But there is another reason that we will not discover such a law. According to classical information theory, the amount of information present in a sequence is inversely proportional to the probability of the sequence occurring. Yet the regularities we refer to as laws describe highly deterministic or predictable relationships between conditions and events.

Laws describe patterns in which the probability of each successive event (given the previous event and the action of the law) approaches unity. Yet information content mounts as *im*probabilities multiply. Information is conveyed whenever one event among an ensemble of possibilities (as opposed to a single necessity) is specified. The greater the number of possibilities, the greater is the improbability of any one being specified, and the more information is transmitted when a particular possibility is specified. If someone tells you that it is raining, he will have conveyed some meaningful information to you since it does not rain (or have to rain) every day. If, however, he also tells you that today the raindrops are falling down, rather than up, he will not have told you anything informative since, presumably, you already know that rain always falls down (by natural law). As Fred Dretske has explained:

> As p(si) [the probability of a condition or state of affairs] approaches 1 the amount of information associated with the occurrence of si goes to 0. In the limiting case when the probability of a condition or state of affairs is unity [p(si)=1], no information is associated with, or generated by, the occurrence of si. This is merely another way to say that no information is generated by the occurrence of events for which there are no possible alternatives.[21]

Natural laws as a category describe situations in which specific outcomes follow specific conditions with high probability. Yet information is maximized when just the opposite situation obtains, namely, when antecedent conditions allow many possible and improbable outcomes. Thus, to the extent that a sequence of symbols or events results from a predictable law-bound process, to that extent the information content of the sequence is limited or effaced (by redundancy). Thus, natural laws do not generate or complex informational sequences and they cannot be invoked to explain the *origin* of information, whether biological or otherwise.

Does the Configuration of Matter at the Beginning of the Universe Explain the Origin of Biological Information?

To be fair, Lamoureux mainly seems to have a different scenario in mind—one in which the information necessary to produce living systems is entirely present at the beginning of the universe. Indeed, taken at face value, each of the metaphors he uses to describe his front-loaded view of evolutionary intelligent design emphasizes how the "deck was stacked" at the beginning. Recall, for example, that he likens the evolution of the universe and life from the initial arrangements of matter at the beginning of the universe to the process of embryological development. He affirms that everything necessary to produce an adult organism (including, presumably, the biological information) is presumably already present in the early embryo. It simply unfolds through time in accord with the laws of nature. In his billiard ball example, the precise arrangement of the billiard balls, the precise shot of the billiards player constitute information-rich initial conditions. The movement of the balls in response to the precise shot occurs deterministically in accord with the laws of momentum exchange.

In the same way, Lamoureux seems to think that the information necessary to produce the first (and subsequent) living forms is already present in arrangements of elementary particles just after the beginning of the universe. But is this view scientifically plausible? Was the information necessary to produce the first life present in the arrangement of elementary particles just after the beginning of the universe? For this to be true, there must be a law that can transmit whatever information was present in the configuration of elementary particles and use that information to produce the first living cell? Since the first living cell would, at the very least, require genetic information, that raises two more precise and analytically tractable questions by which Lamoureux's proposal can be evaluated:

- Was the information necessary to produce a functional gene (information-rich DNA molecule) present in the arrangement of elementary particles just after the beginning of the universe?
- Is there a physical law that could use the information present in the arrangement of elementary particles just after the beginning of the universe to produce a functional gene?

In both cases the answer is "No." Indeed, it turns out that even the biologically relevant chemical subunits of DNA themselves do not contain the information necessary for producing a functional gene or the information DNA contains. Nor is there a law that describes how these subunits self-organize into functional genes.

To see why, recall what we know about the structure of the DNA molecule. (See Figure 1 on page 164.)

What the Structure of DNA Reveals about the Inadequacy of Self-Organizational Models of the Origin of Life

DNA depends upon several chemical bonds, each of which is governed by laws of chemical attraction. There are chemical bonds, for example, between the sugar and the phosphate molecules that form the two twisting backbones of the DNA molecule. There are bonds fixing individual (nucleotide) bases to the sugar-phosphate backbones on each side of the molecule. There are also hydrogen bonds stretching horizontally across the molecule between nucleotide bases making so-called complementary pairs. These bonds, which hold two complementary copies of the DNA message text together, make replication of the genetic instructions possible—just as chemical laws allow ink to adhere to paper. However, there are no chemical bonds *between the bases along the vertical axis in the center of the helix*—any more than there are laws determining that an "A" must always follow a "B" in the written English language. Yet it is precisely along this axis of the DNA molecule that the genetic instructions in DNA are encoded.[22]

Further, just as magnetic letters can be combined and recombined in any way to form various sequences on a metal surface, so too can each of the four bases, A, T, G, and C, attach to any site on the DNA backbone with more-or-less equal facility, making all sequences equally probable (or improbable) given the laws of physics and chemistry. Indeed, there are no differential affinities between any of the four bases and the binding sites along the sugar-phosphate backbone. The same type of "n-glycosidic" bond occurs between the base and the backbone regardless of which base attaches. All four bases are acceptable; none is preferred. As Bernd-Olaf Küppers has noted, "a present day understanding of the properties of nucleic acids indicates that all the combinatorially possible nucleotide patterns of a DNA are, from a chemical point of view, equivalent."[23]

This fact about the structure of DNA was first noticed in 1967 by the physical chemist Michael Polanyi. In a seminal article titled *Life Transcending Physics and Chemistry*,[24] Polanyi showed first that the laws of physics and chemistry generally leave open (or indeterminate) a vast ensemble of possible configurations of matter, only very few of which could have any role in a functioning biological organism. Specifically, he noted that the chemical laws governing the assembly of the chemical subunits in the DNA molecule allow a vast array of possible arrangements of nucleotide bases, the chemical letters, in the genetic text. In other words, the chemical properties of the constituent parts of DNA (and the laws governing their arrangement) do not determine the specific sequencing of the bases in the genetic molecule. Yet, the specific sequencing of the nucleotide bases in DNA constitutes precisely the feature of the DNA molecule, namely its functionally specified information that origin of life biologists most need to explain.[25]

Self-Organization and Evolutionary Creation

In my book *Signature in the Cell*, I have shown why the chemical indeterminacy of the DNA molecule has devastating implications for self-organizational models of the origin of the first life and the origin of genetic information. Typically, these scenarios suggest that the forces of chemical necessity (as described by physical and chemical law) make the origin of life—and the origin of the genetic information that it requires— inevitable.[26] But if law-like processes of chemical attraction do not account for the specific sequencing of nucleotide bases that constitute the information in DNA (as shown above), then such processes cannot reasonably be invoked as the explanation for the origin of the information in DNA in the first place. (It turns out that the information stored in RNA also defies explanation by self-organizing forces of chemical attraction). Indeed, for those who want to explain the origin of life as the result of self-organizing properties or natural laws intrinsic to the material constituents of living systems, these elementary facts of molecular biology have devastating implications. The most logical place to look for self-organizing chemical laws and properties to explain the origin of genetic information is in the constituent parts of the molecules carrying that information. But biochemistry and molecular biology make clear that law-like forces of attraction between the constituents in DNA (as well as RNA and protein)[27] do not explain the sequence specificity of these large information-bearing bio-molecules. To say otherwise is like saying that the law-like forces of chemical attraction governing ink on this page are responsible for the sequential arrangement of the letters that give this chapter meaning.

What does this have to do with the adequacy of "evolutionary creationism"? Quite a lot. Evolutionary creationists such as Denis Lamoureux affirm that the laws of nature as "established and maintained" by God are sufficient to produce life from the initial configurations of matter at the beginning of the universe. Thus, the so-called evolutionary creationist position—with its emphasis on the deterministic unfolding of the history of life in accord with pre-established conditions and laws of nature—entails a commitment to some form of self-organizational origin-of-life scenario. In Lamoureux's billiard ball illustration, for example, a law of nature (the conservation of momentum) ensures a transition from an initial arrangement of matter (the balls arranged on the table) to a subsequent state (in which all the balls are resting in the pockets of the billiard table). Indeed, all his illustrations of evolutionary creation suggest that there must be some law governed process that ensures a *deterministic* transition from the initial arrangement of elementary particles present at the big bang to a subsequent arrangement of matter constituting a living cell. He does not specify whether the laws that affect this transition actually generate new information or just transmit information already present in the initial arrangement of matter in the universe. Either way, though, he seems to assume that some law-like deterministic process must generate a living organism starting ultimately from the elementary particles (or matter and energy) present at the beginning of the universe. As such, Denis Lamoureux's theory of evolutionary creation entails a kind of self-organizational theory of the origin of life. And, indeed, Lamoureux himself lauds the "self-assemblying character of the natural world."[28]

Yet, self-organizational scenarios fail to account for the origin of genetic information for the simple reason that it is now clear there are no self-organizing forces of attraction that can account for the sequence specificity of DNA and RNA bases—the carriers of genetic information in all known cells. Moreover, the irreducibility of genetic information to the chemistry of the DNA and RNA poses a particular difficulty for the front-end loaded evolutionary creationist view of Denis Lamoureux, Darrel Falk and others.

Evolutionary creationists insist that God's direct, discrete or special creative activity has played no role in the history of the universe since the initial moment of creation at the Big Bang. They imply, therefore, that the known laws of nature acting on (presumably information-rich) configurations of elementary particles were sufficient to organize matter into the information-rich structures we see today in living systems. Yet, if the chemical subunits of DNA lack the self-organizational properties necessary to produce the informational sequencing of DNA, it is difficult to see how far less-specifically configured and less-biologically specific elementary particles (present just after the Big Bang) possessed self-organizational capacity necessary to arrange themselves by natural law into fully functioning organisms. In other words, if the chemical subunits of DNA—nucleotide bases, sugars and phosphates, each constituting highly-specific and biologically-relevant arrangements of atoms—do not contain the information necessary to produce a functional gene, then the far less specifically-configured and biologically-relevant elementary particles present just after the beginning of the universe most certainly lacked the information to do so as well. But if this is so, then the deck was not stacked from the beginning as Lamoureux and other evolutionary creationists or teleological evolutionists have argued. Moreover, and if the laws of physics and chemistry merely transmit information but do not generate it, then some other sources of information must have arisen after the beginning of the universe in order to produce life.

THE DIFFERENCE IT DOESN'T MAKE

IN *SIGNATURE IN THE Cell*, I argue that intelligent design provides the best explanation for the information necessary to produce the first living cell. In the process of making that case, I critique the adequacy of self-organizational theories of the origin of biological information. Of course, neither self-organization, nor intelligent design, exhausts the logical possibilities for explaining the origin of information. One could also invoke contingency or chance, either of the directed or undirected variety. For example, many chemical evolutionary theorists have invoked chance (usually in conjunction with prebiological natural selection) in an attempt to explain the origin of information necessary to produce the first life. (In *Signature in the Cell*, I show why those theories fail.) Neo-Darwinists also invoke random and undirected mutations to explain the origin of the new biological information necessary to produce novel forms of life. (As mentioned above, elsewhere I have shown why those theories fail.[29]) On the other hand, some theistic evolutionists, such as Gordon Mills, have suggested that mutations generate anatomical novelty but that they are directed by a guiding intelligence.[30]

Unfortunately, none of these approaches involving contingency represent live options for the committed "teleological evolutionist" such as Lamoureux. For by invoking contingency or chance, whether directed or undirected, the evolutionary creationist or teleological evolutionist runs the risk of qualifying his position out of existence. If, for example, the teleological evolutionist seeks to avoid the information-theoretic difficulties discussed above by invoking *undirected* chance, rather than law, to explain the origin of genetic information, his position becomes indistinguishable from standard materialistic versions of evolutionary theory (whether biological or chemical). Indeed, if he invokes random and *undirected* mutational events to account for the origin of biological form and information, it is difficult to see how he can continue to affirm that evolution is "a purposeful process" (Lamoureux's words) in contrast to the orthodox Neo-Darwinian understanding of biological origins. If, on the other hand, the evolutionary creationists invoke directed contingency (the active and intelligent guidance of genetic mutations during the course of biological history, for example), then he violates his own self-imposed injunction against invoking intelligent or divine action as a cause during the history of life.

Thus, barring an empirically unsupportable and theoretically incoherent commitment to the view that natural laws *generate* specified biological information, it is difficult to see how Lamoureux's theory of "evolutionary creation" or "teleological evolution" differs in substance from conventional materialistic theories of evolution that rely on undirected contingency and deny any intelligent guidance or direction in the history of life. And, even if Lamoureux eschews any reliance on contingency in his theory, it is difficult to see how his theory would differ, in that case, from demonstrably inadequate theories of self-organization.

C. S. Peirce's maxim "for a difference to be a difference, it must make a difference" seems applicable here. Lamoureux purports to offer a novel theory of biological (not cosmological) origins. Yet, one must ask: does the word "teleological," in the phrase "teleological evolution," make any scientific difference within that context? Does the word creation in the label "evolutionary *creation*" designate any distinctive insight for our understanding of *biological* origins? Does Lamoureux's theory add anything to the scientific explanation of biological origins that a Neo-Darwinist or chemical evolutionary theorist or self-organizational theorist does not already have? Regrettably, the answer to these questions is "no." Indeed, since Lamoureux is unwilling to specify any role for his Creator (beyond the causally necessary, but insufficient, one of establishing and maintaining ordinary physical laws), "teleological evolution" and "evolutionary creation" seem to be little more than empty phrases describing a theory that, in the end, reduces to standard, and completely inadequate, materialistic modes of explanation.

Figure 1: The Chemical Structure of DNA showing its information-bearing properties and the bonding relationships of its chemical constituents. Note that no chemical bonds link the nucleotide bases (designated by the letters in boxes) in the longitudinal message-bearing axis of the molecule. Note also that the same chemical bonds link the different nucleotide bases to the sugar-phosphate backbone of the molecule (denoted by pentagons and circles). These two features of the molecule ensure that any nucleotide base can attach to the sugar phosphate backbone at any site with equal ease or probability, thus demonstrating that the properties of the chemical constituents of DNA do not determine its sequencing or account for its information content.

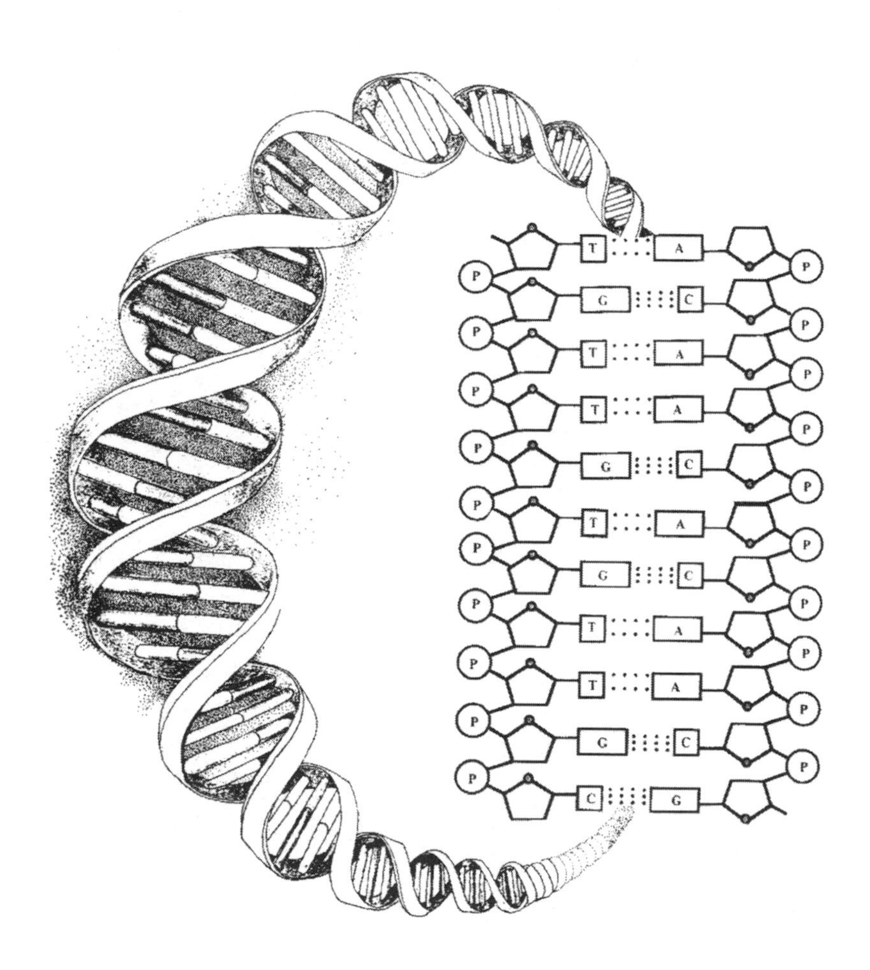

III.

Catholics and Evolution

9. Everything Old is New Again

The Older Catholic Apologists' Responses to Darwin

Denyse O'Leary

Introduction: How Should We Understand the Catholic Response to Darwinism?

For most Catholics, the Church's response to Darwin's theory of evolution is a puzzle. Though some official texts give important guidance, the Catholic Church is widely portrayed by the media—Catholic and secular—as deferring to the Pontifical Academy of Sciences.[1] And the Academy, for its part, assures us that Darwinism (survival of the fittest, without evidence of design) is the best way to interpret the creation[2] of the Lamb of God who was crucified to take away the sins of the world. At first glance, at least, this seems peculiar, since the Lamb never showed interest in determining who, exactly, the fittest were. When asked who should inherit the kingdom, "he called a little child and had him stand among them."[3] Was that child "the fittest"?

The Founder of Christianity went on to propound a value system that is entirely unlike Darwinism, and assumes that there is a design in life, that God is actively present in creation at all times, and can intervene when he thinks it right. Many millions of Catholics worldwide not only assume that that is true—and act on the assumption—but are quite ready to point to evidences of the fact. But among certain "Catholic" academics, things have currently moved so far in the Darwinist direction that Templeton Prize winner Francisco Ayala, formerly a Dominican, has informed us that the very idea of design is blasphemous.[4]

The puzzle is, why? Scripture, tradition, reason, and evidence[5] all endorse design. Besides, the idea of design, broadly construed, is hardly a novel doctrine. Few traditional religious views assume that blind chance or mechanical, unalterable laws are the origin of life, species, or humanity. Nor do they assume that humans cannot reliably detect design because "our brains are shaped for fitness, not for truth," in Steve Pinker's famous phrase. Humans have always thirsted for truth. If Pinker is right, we are not to have any further truth, because we have the Final Revelation already, in the form of Darwinism. Natural selection acts on random mutations to produce the living world in all its splendor, and no further revelation can ever follow.

Yet, despite the current somewhat ambivalent posture among Catholics, a century ago, Darwinism was vigorously denounced by key English language Catholic writers, such as literary lions G. K. Chesterton and Hilaire Belloc, as well as zoologist St. George Jackson Mivart. In general, their arguments were not based on sectarian religion or partisan politics, but on the use of reason in evaluating evidence, including the striking absence of key evidence for Darwin's beliefs. This absence continues to the present day, despite energetic statements to the contrary by lobby groups and partisan academics.

Catholic Apologists and Darwin: They Did Not Think Aquinas Would Approve

Contrary to a stereotype among some Catholics, the debate over Darwinism is not a new controversy stirred up by Protestant Creationists. Even when Darwin was writing his *Origin of Species* (1859), he blamed the poverty of the fossil record[6] for the absence of evidence for his belief that natural selection creates the massive information needed for new species. The Cambrian Explosion find in the Canadian Rockies[7] was a huge problem for Darwin's followers. After a century and a half of further excavation, the fossil record still fails to support Darwin's belief that natural selection (survival of the fittest/natural selection acting on random variations) creates intricate information resulting in new species. This is either bad news for the fossil record or bad news for Darwin's beliefs. One must choose.

The controversy has never been about the generic idea of change over time or common ancestry, but rather about how such things occur. As Catholic writer George Sim Johnston puts it:

> "Evolution" is the idea that all life-forms share common ancestors, and maybe even a single ancestor. This is a reasonable, if unproven, hypothesis. It is reasonable not because anyone has ever seen a species turn into a new one (one hasn't), but because all life-forms share certain genetic material and there are nested hierarchies of structure within the major animal groups. "Evolution" means common descent, but does not tell how it happened.[8]

Darwinism in its ideological form is essentially a struggle on the part of atheists and some theists to control the narrative of how "evolution" happened.

No later evidence has ever suggested that the Catholic writers of a century ago were wrong on this, their main point—that Darwinism generally lacks evidence. The puzzle is why Darwinism persists in school systems, including Catholic school systems, as well as in popular media, sometimes enforced by court decisions, and is often endorsed by well-meaning Catholic and other Christian voters.[9]

It is not hard to see why, in the early twentieth century, Darwinism was popular despite very minor evidence: it underpinned eugenics.[10] Many hoped that eugenics would solve social problems, by preventing the births of future "social problems." Indeed, there is a large—and largely untold—story about forced sterilizations in the cause of eugenics.[11] Darwin's own opinions were racist (although, in fairness, they were not radically so for a wealthy Englishman of his time). He thought, for example, that

> [t]he great break in the organic chain between man and his nearest allies, which cannot be bridged over by any extinct or living species, has often been advanced as a grave objection to the belief that man is descended from some lower form; but this objection will not appear of much weight to those who, from general reasons, believe in the general principle of evolution. Breaks often occur in all parts of the series, some being wide, sharp, and defined, others less so in various degrees.... At some future period, not very distant as measured by centuries, the civilised races of man will almost certainly exterminate, and replace, the savage races throughout the world.... At the same time, the anthropomorphous apes, as Professor Schaaffhausen has remarked, will no doubt be exterminated. The link between man and his nearest allies will then be wider for it will intervene between

> man in a more civilised state, as we may hope, even than the Caucasian and, and some ape as low as a baboon, instead of as now between the negro or Australian and the gorilla. [12]

That sort of racism—in which some people are really just animals or next door to being so—was one of the huge problems that traditional Catholics had with Darwin. The Catholic Church has an honorable history of opposing eugenics, forced sterilization, and eugenic racism. This is due to the assumption, emphasized by John Paul II, that God creates each individual human soul—that every individual has intrinsic value and should be accorded dignity. A human being could renounce his inheritance, but no one should deny it or take it away from him. In South Africa, for instance, Catholic schools came to reject apartheid long before such rejection was fashionable.[13]

The key question—and it still perplexes this author—is, why does Darwinism persist as a tax-funded, court-supported public orthodoxy, often backed up by ecclesiastical authorities, in the face of a racist pedigree and genuinely limited evidence, restricted to minor and easily reversible changes in finches and moths?[14] Can the older Catholic writers help us understand?

Those writers understood the situation very much better than many moderns. Then as now, the controversy over Darwinism is not the age of the Earth or whether some generic idea of "evolution" occurs, or whether natural selection plays a role in evolution. Darwinism in its modern, "Neo-Darwinian" form is a claim that natural selection, acting on random genetic mutations, can add vast amounts of information, thus creating new phyla and species, all by itself.

To say that Darwin's proposition has never really been demonstrated would be to shower it with undeserved praise. Yet, contrary to the older Catholic writers' expectation that Darwinism would die out in their lifetimes, Darwinism has triumphed not only in tax-funded school systems, but also the wider culture. To see how deeply it has penetrated culture, consider the London *Guardian*'s recent feature, the "evolutionary agony aunt,"[15] who treats humans, chimpanzees, and bonobos as equal in her attempts to offer advice—not that any chimpanzees or bonobos have ever written to her about their problems or that any humans in serious emotional pain would care much what chimps or bonobos do to resolve theirs.[16]

This approach to human problems is, for all practical purposes, insane. But how did things get to this point? First, we need to say a bit about Darwin, and what the older Catholic writers saw, that has been largely effaced.

The older authors sensed something that has only been firmly established in recent decades, *via* access to Darwin's private papers: Darwin was a materialist atheist, and his theory was developed precisely to support that view. Recently, science historian Michael Flannery republished *World of Life*, a book written by Darwin's co-theorist of natural selection Wallace.[17] Flannery also set out little publicized facts about Darwin. His key point is that Darwin had become a materialist atheist long before he wrote *On the Origin of Species* (1859). This established fact is entirely contrary to popular pundit proclamations that Darwin became a materialist atheist after he "discovered" natural selection or on account of the death of his daughter Annie, or else on account of animal suffering.[18]

Darwin, who had learned extreme philosophical materialism as a youth from Edinburgh radicals, was in fact looking for a theory of evolution that would support materialist atheism (though his argument in the *Origin of Species* is more modest on its face). The problem with evolution was not that it was disbelieved. It was widely believed among those educated enough to even know about it.[19] But no theory of evolution fully and obviously supported materialist atheism in the way that Darwin's did. Today, it is conventional to deny this fact, as in—for example—the popular staple of many liberal churches, Evolution Sunday.[20] But the older Catholic writers knew both the truth and the significance of Darwin's and his X Club's friends' intentions.[21] Thus, their responses are worth a look.

Essentially, there were two streams of evolutionism in the mid-Nineteenth century. One was materialist and the other was non-materialist.[22] The Catholic writers did not dispute the second, but they certainly disputed the first.

As noted earlier, Darwin argued that natural selection acting on random variations produces the intricate machinery of life. But, as theoretical physicist Leonard Susskind said recently, Darwin's masterstroke was to have "ejected God from the science of life."[23] If that, rather than following the evidence, is the goal, demonstrating the explanatory value of Darwin's theory is actually superfluous. The slender evidence base for Darwinism that Michael Behe outlines in *The Edge of Evolution* (2007) is irrelevant because Darwin's theory must by definition be true—unless one wishes to believe that space aliens did it, which surprising numbers of key scientists have considered.[24]

Michael Flannery has no time for the current Darwin hagiography. Assessing Darwin candidly in relation to Wallace, he considers the official history a calculated imposture. Historians have slowly begun to catch on, as more and more primary materials have surfaced. Admitting that Darwin often played dumb and was hardly the figure that Victorians made of him, sympathetic biographer Janet Browne correctly observes that his autobiography led the way in creating a smokescreen around key issues—a smokescreen that the older Catholic writers pierced.

For example, Flannery notes, late in life Darwin informed visiting atheists,

> "I never gave up Christianity until I was forty years of age." This was false, and Darwin knew it was false. He had been drawn to materialist atheism in university, as a member of the Plinian Society freethinkers' circle. At 29, he was making materialist statements in his notebooks, like "Why is thought, being a secretion of the brain, more wonderful than gravity, a property of matter? It is our arrogance, it is our admiration of ourselves...."

So, Flannery explains, by 1838, Darwin was a thoroughgoing materialist: "That was four years before his rough 30-page sketch on transmutation and six years before the first 230-page draft of his general theory—he had even sketched its main application to man 33 years before [publication of] his *Descent [of Man]*." Thus, far from embracing materialism on account of the implications of his theory, Darwin developed a theory to support the materialism he already believed. Darwin encouraged the legend that he lost his faith when he beheld animal suffering, but it wasn't true.[25]

But now, what of Wallace? His story helps us understand the older Catholic writers' qualms about Darwinism. He had twice Darwin's field experience, yet he was neglected and ridiculed, just as Darwin was lionized. There were two problems, Flannery writes. The first was that Wallace tended to defer to Darwin, and avoided taking credit for his original contributions. And the second was that Darwin's bulldog, Thomas Huxley, and his X Club managed an exceedingly successful public relations campaign against him, as subsequent generations of Darwinists have done to many other skeptics.[26] Well, there was a third problem too: Wallace accepted an "overruling intelligence" that fine-tuned the universe, so he argued for intelligent evolution (IE). That was enough to get him marginalized—though he was hardly a conventional believer in traditional Western religion.

In Flannery's work, Wallace emerges as a sympathetic figure. For example, compare Wallace's view of aboriginal peoples to Darwin's view[27]:

> When he observed and lived among natives, whether along the banks of the Amazon or on the islands of Malaysia, he saw human beings with the same innate capacities as himself. In contrast Darwin was horrified by the natives of Tierra del Fuego and thought them closer to animals than men. Wallace had a more intimate and therefore deeper appreciation of indigenous cultures. Among the Dyak head-hunters of Borneo Wallace noted that, aside from ritualistic violence, tribal crime was almost unheard of.... In the end, the very different reactions between Darwin and Wallace to the "uncivilized" translated into two radically different views of humanity. Darwin saw man a little above the beasts; Wallace saw man a little below the angels.[28]

Well, the Catholic writers knew all about this kind of thing, and were hardly passive subjects of public relations campaigns for Darwinism. So let's look at what they said.

G. K. Chesterton (1874–1936): He Tried Common Sense, but Common Sense Did Not Matter

Chesterton was a very well known journalist and author in Edwardian England.[29] He wrote about the things that mattered to ordinary people in his own day. He was also a serious Catholic, and he sought to explain a traditional view of many matters occluded by modernism. One of them was evolution. In his essay "Doubts about Darwin," Chesterton scoffed at the claim that unguided natural selection could gradually build fundamentally new organisms *via* thousands of transitional forms, pointing out that many of the supposed transitions likely would have been incompatible with survival. Take the evolution of a mouse-like creature into a bat:

> If you will call up the Darwinian vision, of thousands of intermediary creatures with webbed feet that are not yet wings, their survival will seem incredible. A mouse can run, and survive; and a flitter-mouse can fly, and survive. But a creature that cannot yet fly, and can no longer run, ought obviously to have perished, by the very Darwinian doctrine which has to assume that he survived.[30]

Writing elsewhere, Chesterton noted that "even if we suppose" each incremental change "serve[d] some other purpose" than building a functional bat wing, "it could only be by a coincidence; and this is to imagine a million coincidences accounting for every creature."[31]

However, Chesterton's special interest regarding Darwinism was the evolution of human beings, not bats.

For example, everyone knows the popular human evolution story: Early man bopped early woman with a club and dragged her off to his cave. The trouble is, the story probably isn't true. As Chesterton noted:

> [A]ll our novels and newspapers will be found swarming with numberless allusions to a popular character called Cave-Man. He seems to be quite familiar to us, not only as a public character but as a private character. His psychology is seriously taken into account in psychological fiction and psychological medicine. So far as I can understand, his chief occupation in life was knocking his wife about, or treating women in general with what is, I believe, known in the world of the film as 'rough stuff.' I have never happened to come upon the evidence for this idea and I do not know on what primitive diaries or prehistoric divorce reports it is founded.[32]

For one thing, our early human grandmothers had defenses as well, including attacking a sleeping man[33] and then just running back to the home tribe. Our most ancient literature—the best available insight into early minds—makes clear that most fathers were concerned that their daughters make advantageous marriages, and not get raped beforehand, limiting their social futures.[34]

In *Everlasting Man,* Chesterton made a number of cautionary points about early man that are worth considering. For example:

- One must be cautious about concluding that modern people groups who live in a technically primitive way replicate Stone Age ancestors exactly, rather than merely adopting obvious general patterns for survival outside organized civilizations. Some groups may simply shed technically advanced features, the way Europeans did for many centuries after the fall of Rome. Technically complex features can be too difficult to maintain in a climate of growing disorder, so sometimes simplicity is the price paid for survival. It does not mean that people do not guess that things could be different, in principle.

  Also, technically primitive lifestyles should not be confused with socially primitive ones. A person who tells time by the sun and moon and has never made use of a watch may recognize 14 degrees of kinship, whereas a person kitted out with an iPod and a charge card may be unsure who his father is, and has never remembered to visit his mother, nor does he know the names of his next-door neighbors. The technically primitive people might be just as appalled by him as he is by them.
- Traditional people groups tend to be democratic rather than autocratic, so forget the stout club. As Chesterton puts it, "democracy is the foe of civilization." He means that, in a hunter-gatherer group, anyone with something useful to contribute to the hunt or gather might speak up in front of the hunt chief or the chief's wife. But as societies become more civilized, power becomes more specialized, more distant, and more reserved. Whatever Cave Man was supposed to have invented, it was not "l'etat, c'est moi," the *generalissimo*, or *der Fuehrer*. Cave Tribe could not afford any of that.
- Eighty years ago, Chesterton also noted the contradictory stances of what is now called "evolutionary psychology": "In spite of all the pseudo-scientific gossip about marriage by capture and the caveman beating the cave-woman with a club, it may be noted that as soon as feminism became a fashionable cry, it was insisted that human civilization in its first stage had been a matriarchy. Apparently it was the cave-woman who carried the club."

Yet nothing much has changed in this area since Chesterton wrote *Everlasting Man*[35] over 85 years ago, except the huge growth in the "evolutionary psychology" literature and its ever-more-expansive attempts to explain everything away. The interesting point is not the relative antiquity of the objections to Darwinism but rather the confidence Chesterton, like other older Catholic writers, had that Darwinism would die out in his own lifetime. Indeed, he wrote,

> Darwin's theory ... has largely been abandoned by the latest scientific men; and indeed is only still accepted as a piece of Victorian respectability by old-fashioned people like Bishop Barnes. But in any case, it never went very far towards touching the primary problems; and Darwin himself hardly pretended that it did.[36]

But things did not work out at all as Chesterton hoped. Every download of science media releases offers more and yet more "evolutionary psychology" folly based on Darwinism, to say nothing about further paltry claims in biology. One sees endless variations on: "Yeast sample undergoes evolution to spur immunity to virus; dramatic demonstration of Darwin's thesis, researchers say." All it really shows is how yeast continues to be yeast by dumping a binding site used by the virus—a loss, not a gain, in information. Clearly, ideology rather than startling new evidence undergirds these narratives.

The popular Darwinism that the older Catholic authors opposed flies in the face of the increasing incredibility of Darwinism's central thesis—that natural selection and random variation is a source of massive gains in information and, in the case of humans, vast cultural changes. Yet the thesis lands in print again and again, regardless. Chesterton's French-born friend Hilaire Belloc tried, likewise, to stem the growing tide of popular nonsense. Let's look at how Belloc fared.

Hilaire Belloc (1870–1953): The Map was Right but Belloc was in the Wrong Territory

Of French descent and from a literary family, Hilaire Belloc spent his life in England, after the collapse of his late father's business. Belloc was one of four key figures on the early twentieth century literary scene in England, the others being H. G. Wells and George Bernard Shaw, as well as his close colleague G. K. Chesterton (Shaw called the two "Chesterbelloc").

Belloc was a capable wit. As a freelance writer most of his life, when asked why he wrote so much, he explained, "Because my children are howling for pearls and caviar."

Belloc was also a passionate Catholic. With respect to evolution, he was best known for a 1920s controversy with H. G. Wells. He criticized two key aspects of Wells's *Outline of History*: First, he objected to a secular bias which gave more attention to a forgotten Greek war than to Jesus Christ, and second, Wells's staunch but poorly evidenced support for natural selection as a means of creating new organisms. Belloc famously observed that Wells's book was powerful and well-written, "up until the appearance of Man, that is, somewhere around page seven." Wells responded with *Mr. Belloc Objects*.[37] Belloc promptly followed with *Mr. Belloc Still Objects*.[38]

Belloc, like key Catholic contemporaries such as Chesterton and Mivart, had no problem with the idea of generic common ancestry, and little use for interpreting the "days" of Genesis 1 as ordinary, 24-hour Earth days. In fact, as he pointed out in *Survivals and New Arrivals* (1929),[39] Catholic culture in general had entertained the idea of evolution as a purposeful and progressive development over time for centuries.

> When the immensely ancient doctrine of growth (or evolution) and the connection of living organisms with past forms was newly emphasized by Buffon and Lamarck, opinion in France was not disturbed; and it was hopelessly puzzling to men of Catholic tradition to find a Catholic priest's original discovery of man's antiquity (at Torquay, in the cave called "Kent's Hole") severely censured by the Protestant world. Still more were they puzzled by the fierce battle which raged against the further development of Buffon and Lamarck's main thesis under the hands of careful and patient observers such as Darwin and Wallace.

So Belloc dismisses most familiar arguments we would now call "creationist." But he insists on a critical point, familiar to anyone who must deal with popular Darwinism:

> What are you to do with a man who uses the same word in different senses during the same discussion? As, for instance, who says he "be-

> lieves in Evolution," meaning growth (which all men believe in), and in the same sentence makes it mean:
>
> (a) The bestial origin of man's body—which is probable enough,
>
> (b) Darwin's theory of Mechanical Natural Selection, which is as dead as a door-nail.
>
> What are you to do with a man who puts it forth as a foundation for debate that the human reason is no guide, and who then proceeds to reason through hundreds of pages on that basis?
>
> Yet all that, and hundreds of derivatives therefrom, make up the horrible welter of the "Modern Mind."

Reading Belloc's penetrating insights on many subjects in *Survivals and New Arrivals* is, it must be said, a depressing experience. Belloc got so much right—and so much wrong. He hoped for a revival of the Catholic Church in Europe, against the current of the times. He, like Chesterton, thought Darwinism would die—"dead as a doornail"—for lack of confirming evidence.

It didn't die, and it couldn't, because Darwinism does not actually require evidence, any more than astrology does. Consider New York mayor Michael Bloomberg recently lauding the Ida fossil (supposedly the famed "missing link"), only months before it was retracted by science journals.[40] The principal question is why Mr. Bloomberg—whose office presumably suggests radically different responsibilities—thought he should even address the Ida fossil at all. He did so because Darwinism has become integral to our popular and elite culture. It shapes priorities, concerns, and values for growing numbers of people.

Like Chesterton, Belloc grossly underestimated the power of publicly funded beliefs hatched by tenured professors at universities, fronted in Sunday paper features and documentaries, and litigated by pressure groups. The thesis can tell us how anything from ample bosoms to gay lifestyles is somehow selected by Darwinian means.[41] Lack of evidence is no barrier to belief.

Against Belloc's hopes, here is a very short list of the behavior supposedly encoded by Darwinian evolution:

It can be shopping, voting, or tipping at restaurants. It can also be: Why children don't like vegetables (nothing to do with young 'uns preference for sweet things); why hungry men prefer plump women (not just because those women probably know where the kitchen is); why we have color vision (mainly to detect blushing); why we are sexually jealous (not fear of abandonment, but "sperm competition"); why toddlers are Neanderthals (not just immature); why we don't stick to our goals (evolution gave us a kludge brain); why women prefer men with stubble (except for those who don't); why gossip is good for you (despite wrecked relationships due to slander); why moral behavior is based on primitive disgust (not rational evaluation); why music exists (to "spot the savannah with little Pavarottis"); why art exists (to recapture that lost savannah); why art exists (to spread selfish genes); why altruism is really a form of sexual display; why altruism is really just selfishness; why a child must have a selfish motive for saving her sister's life; why right and wrong don't really make sense; why we don't eat Grandma (because she might baby sit the kids); why we don't (usually) hurt ourselves to hurt others; why we can't help behaving badly (it is programmed into our genes); oh, and religion is a sort of replicator or "meme" in our brains; and why we believe in God (because he is a supernatural cheat detector); or else why we believe in God (because belief is and is not adaptive at the same time; or, finally, why we believe in God (because we have a genetic predisposition to communicate unverifiable information).

There is a dark side too. Our Stone Age ancestors are deputized to explain, as Sharon Begley observes in *Newsweek*, why we rape, kill and sleep around.[42]

As both Chesterton and Belloc noted, there is little evidence for most such Darwinian propositions. They advance on naked speculation. But publicly funded, court-ordered Darwinism was the future that the carefully reasoning Belloc and Chesterton dreaded yet, alas, were unable to prevent.

St. George Jackson Mivart: Playing the Middle Against Both Ends

St. George Jackson Mivart (1827–1900) is a tragic figure, compared with the "Chesterbelloc" Catholic luminaries of the Edwardian literary era. He was born a Protestant, but converted to Catholicism in 1844, perhaps due to the influence of the Oxford Movement.

A physician, an outstanding zoologist—and a defender of Darwin—he got to know Thomas Huxley in 1859. His authorship of "On the Appendicular Skeleton of Primates" resulted in his election to the Royal Society in 1867,[43] but he wrote many other zoological works as well. In 1876, Pius IX made him a doctor of philosophy. Yet Mivart was later blacklisted by both the Darwinists and the Catholic Church. What went wrong?

Mivart initially supported natural selection enthusiastically as a factor in evolution. But he thought that evolution involved many other factors as well, and that species vary only within a certain space, not randomly or indefinitely, as Darwin proposed. Put another way: nothing will cause your house cat's progeny to become, within the lifetime of Earth, a catfish, nor the catfish's progeny a cat. At a certain point, the door usually just closes on radical further variations. If we consider life since the Cambrian Explosion, Mivart's view is actually closer to the observed history than Darwin's. But Mivart's view was much less popular, because it placed limits on natural selection as a design-free explanatory principle.

Those limits had major social implications, of course. They meant that eugenics might not produce a better human, survival of the fittest was not a suitable social principle,[44] and an image of merely glorified apes is not the best way to understand human beings.

Moreover, as the *Catholic Encyclopedia* puts it,

> [H]e unhesitatingly and consistently asserted the irreconcilable difference between the inanimate and the animate, as well as between the purely animal and the rational. By maintaining the creationist theory of the origin of the human soul he attempted to reconcile his evolutionism with the Catholic faith.[45]

The Darwinists strongly objected to Mivart's Catholic interpretation of evolution. As a result, he was, as Dom Paschal Schotti notes, "a naturalist to whom Darwin paid a watchful and nervous attention, whom Huxley brought out his biggest bludgeon to belabour in the arena of Evolution."[46]

In January 1871, Mivart published *On the Genesis of Species*,[47] a searing critique of the ability of Darwin's natural selection to build substantially new biological forms and features. Acknowledging that natural selection could account for the maintenance and extension of *already useful* biological features, Mivart declared that it "utterly fails to account for the conservation and development of the minute and rudimentary beginnings, the slight and infinitesimal commencements of structures, however useful those structures may afterward become."[48] Mivart's objection to the efficacy of natural selection became known as "The Incompetency of 'Natural Selection' to Account for the Incipient Stages of Useful Structures." Stephen Jay Gould explained Mivart's objection thus:

> We can readily understand how complex and fully developed structures work and how their maintenance and preservation may rely upon natural selection—a wing, an eye, the resemblance of a bittern to a branch or of an insect to a stick or dead leaf. But how do you get from nothing to such an elaborate something if evolution must proceed through a long sequence of intermediate stages, each favored by natural selection? You can't fly with 2 percent of a wing or gain much protection from an iota's similarity with a potentially concealing piece of vegetation. How, in other words, can natural selection explain the incipient stages of structures that can only be used in much more elaborated form?[49]

Although he disagreed with Mivart, Gould admitted the force of Mivart's objection, saying that "[t]his argument continues to rank as the primary stumbling block among thoughtful and friendly scrutinizers of Darwinism today."[50] Mivart's discussion of the kinds of coordinated changes that would have been required during embryological development to build new features like the eye and ear has, if anything, grown in relevance with recent discoveries in molecular biology, biochemistry, and genetics.

Apart from his scientific criticisms of natural selection, Mivart was also becoming increasingly concerned by the moral implications of Darwinism, especially the growth of "shallow thinkers"[51] promoting political agendas. Presumably, he had eugenics and social Darwinism in mind, among other things. It was his critique of Darwin's *Descent of Man*,[52] which is a long and detailed argument for racism and materialism, which sealed his breach with Darwin's circle. In later editions of *On the Origin of Species*, Darwin addressed Mivart's objections at some length. But, in truth, Mivart was already banished from the elite circle around the aristocratic Darwin, so Mivart's objections were irrelevant, whether well-founded or not.

Mivart was not excommunicated from the Catholic Church for his relations with Darwin and Huxley, which had soured anyway, but rather for his effort to apply the idea of evolution to the Catholic doctrine of Hell.[53]

Mivart's trouble with the Church started about 1885, with his increasingly free-floating speculations supporting the idea that science should determine matters of faith. In the Catholic Christian tradition, many key matters of faith are considered the outcome of either revelation or reason. In either case, the contributions of science must be demonstrated rather than assumed. For example, if we believe that God can act in nature, miracles may happen at times. Of course, we do not attribute to miracles matters that can be easily explained by nature, but we remain open to the possibility of God's intervention. That idea has been central to the Judeo-Christian tradition for thousands of years.[54]

The 1892 articles that landed Mivart in serious trouble argued in the highbrow Victorian periodical, *Nineteenth Century*, for "Happiness in Hell." He asserted that the traditional doctrine of hell caused loss of faith, and argued instead for

> evolution in hell and that the existence of the damned is one of progress and gradual amelioration—though never, of course, to the extent of raising the lost to supernatural beatitude, for the tenants of hell are its tenants eternally.[55]

Well, that is hardly Dante's ("abandon hope, all ye who enter here") *Inferno*.

A little background might help put Mivart's problem in context. Only in the nineteenth century did denial of Hell become common. Moreover, the doctrine is slightly different for Catholics than for Protestants, since Catholics accept the doctrine of Purgatory where the deceased but saved may be detained for purification.

So, as Scotti observes, Mivart "entered the thicket of theology with more goodwill than good sense." In July 1893 his articles were put on the Index of forbidden materials.[56] Mivart accepted the decree because, in a classical fine distinction, only the articles—not his actual views—had been condemned. That amounted to saying that he should not have made his views public, lest less educated people be misled by them.

The matter might have ended there except that in 1899, Mivart's articles again appeared on the Index. He then recanted his acceptance. He began to attack the Church, saying things like:

> I have no more leaning to atheism or agnosticism now than I ever had; but the inscrutable, incomprehensible energy pervading the universe and (as it seems to me) disclosed by science, differs profoundly, as I read nature, from the God worshipped by Christians.[57]

For this sort of thing, Mivart was excommunicated in January 1900. He died on April 1, 1900, the night before he was to give a lecture, and was buried six days later in unconsecrated ground. Mivart's friends bailed him out (literally) post-mortem. They provided evidence that he was of unsound mind due to his final illness (diabetes)[58] at the time he wrote the articles considered most objectionable. This evidence was accepted, and he was formally re-buried as a Catholic in 1904.[59]

Darwinism Should Have Died A Quiet Death, So Why Didn't It?

Despite its stunning history of survival, George Sim Johnston optimistically writes that Darwin's theory is due for retirement shortly, not because creationists oppose it but because many scientists themselves now have serious reservations. But some scientists have always had serious reservations, as Johnston clearly sets out in *Did Darwin Get It Right?*[60]

That was true in Darwin's day and has been true ever since. Yet it does not seem to affect the theory's staying power. By far the most common form of abusive mail that this writer gets is from Darwinists announcing that Darwinism is conclusively proven, based on paltry evidence such as changes in Galapagos finch beaks' size, depending on seed availability. It is hard to think of a change that could more easily reverse itself, and probably will, once seed availability changes again.

So the question returns yet again and again, unanswered: Why doesn't lack of serious evidence make any difference? In such cases, it is best to ask: What nonscientific function is Darwinism serving in society today?

One explanation for the survival of Darwinism today is sociological: It has enormous status as an elite and increasingly pop culture icon. Darwinism is easy. It requires little knowledge of anything beyond cultural ideas illustrated in tax-funded museum displays, iterated in science media releases grabbed for mainstream media (and easily read because they never tell you anything that pop culture could not), and taught in publicly funded schools. A Darwin supporter can look clever while knowing very little, and maybe even get an income from it.

The complexity of a true history of life would not be nearly as interesting: For example, there are many possible causes of evolution.[61] Scientists do not usually know how to assign a weight in particular cases. It may never be possible, due to lost information.[62] So Darwinism promises certainty in matters that should engender both reasonable doubt and discretion as to what to believe—and extra discretion about what to teach.

But there is more: Darwinism is, as the early Catholic writers recognized, the creation story of atheism. It enables atheism to function as the normal stance of science. Today, 78 percent of evolutionary biologists are pure materialists: No God and no free will.[63] Darwinism also birthed the pseudo-discipline of "evolutionary psychology,"[64] noted above, which has been spectacularly successful in popular culture, while illuminating very little. So, clearly, the older Catholic writers failed in their quest to bring reason and evidence to the table—almost.

A third cause of Darwinism's success is deliberately sponsored confusion. As Johnston usefully writes:

> There is one point in the debate over evolution that should be clarified from the start: The layman has to make a distinction between "evolution" and "Darwinism," something that is seldom done in the English-speaking world.[65]

The usual way that the Darwinist protects his atheist materialist or fellow-traveler beliefs from falsification is by confusing "Darwinism" with "evolution," and pretending that everyone who doubts the claim that natural selection and random variation can create significant amounts of functional information doubts either that evolution occurs or that the Earth is more than six thousand years old. But even if the Earth were six billion years old,[66] Darwinism would still face insuperable difficulties.

Could the Older Catholic Writers Have Succeeded in Some Way?

Whether they could have succeeded is a difficult question. Taxpayers funded Darwinism through most of the twentieth century, and they still do. And, as commentator Mark Steyn has pointed out in another context, government can indeed change hearts and minds. For one thing, tax money is money that can only be spent on government-approved projects. When a great deal of such money is raised and directed at a project, a great many resources can be bought and then directed at changing people's hearts and minds. For example:

> Once the state swells to a certain size, the people available to fill the ever-expanding number of government jobs will be statists—sometimes hard-core Marxist statists, sometimes social-engineering multiculti statists, sometimes fluffily "compassionate" statists, but always statists. The short history of the post-war welfare state is that you don't need a president-for-life if you've got a bureaucracy-for-life: The people can elect "conservatives," as the Germans have done and the British are about to do, and the Left is mostly relaxed about it. . . .[67]

Steyn is here talking about health care issues, but his point is generally applicable: Once a bureaucracy gets established and persists, its views tend to prevail no matter who is elected or what the people think. The parallel with Darwinism in the academy should be obvious. The remarkable thing is that—for once—it did not work nearly as well as expected in North America. Most North Americans still do not believe in either atheist materialism or its creation story, Darwinism.[68] So maybe the failure of the early resistance was not complete.

Having followed the intelligent design controversy for a decade, this writer has noticed a recent key change. Last year, despite being the 150th anniversary of the publication of *On the Origin of Species*—which should have continued Charles Darwin's century and a half of triumph—revealed that his followers' accolades are greeted with increasing incredulity, among both serious scientists and the general public, despite all the taxpayer funded support. For example, a group of serious scientists and thinkers convened last year at Altenberg, Austria,[69] to consider alternatives to Darwin's theory of evolution. And a recent Zogby poll[70] showed that most Americans really don't believe that natural selection and random change creates information, after countless years and dollars spent to convince them.

In any event, Darwinism was probably dealt a heavy blow by the growth of "evolutionary psychology," as noted above. Eventually, some intelligent people who pass through the checkout counters at supermarkets had to notice the obvious close relationship between that supposed discipline and the tabloid press. And one question leads to others....

Could the old Catholic writers have been on to something that the Pontifical Academy cannot consider just now? If so, why? Why does the imposture continue, with the appearance of sanction by the Church?

10. Can a Thomist Be a Darwinist?

Logan Paul Gage

In a little-recounted episode of the 2005 *Kitzmiller v. Dover* intelligent design trial, the plaintiffs (objecting to a four-paragraph statement read in biology class) summoned a curious expert witness: John F. Haught, former chairman of Georgetown University's theology department. Asked to identify the antecedents of intelligent design Professor Haught pointed to Thomas Aquinas' five arguments for the existence of God, "one of which was to argue from the design and complexity and order and pattern in the universe to the existence of an ultimate intelligent designer."[1]

Intelligent design (ID) was on trial because (in biology at least) it conflicts with Darwin's theory as taught in the classroom: modern Darwinian evolution claims that random mutation and natural selection, defined as purposeless and unguided processes, are sufficient to explain the stunning features of living things, while intelligent design claims there is evidence that at least some things are better explained by an intelligent cause.[2]

Haught slightly mischaracterized Thomas' argument, which says nothing about complexity *per se*. But Thomas certainly made a design argument by appealing to features of the natural world, as do contemporary ID theorists. Despite these similarities, however, some modern Thomists claim that Thomism is compatible with Darwinian evolution and *in*compatible with intelligent design. So which is it? Are Thomas' writings a precursor to intelligent design as even such design-critics as Haught claim? Or is Darwin compatible with—and ID irreconcilable with—Thomas' philosophical and theological framework? Or is there some third possibility? For the Catholic, these are important questions, because St. Thomas is the gold standard of Catholic thought—not infallible, to be sure, but trustworthy.

Thomism and Darwinism

In a typical discussion of Darwinian evolution, Christian philosophy, and intelligent design one is likely to hear that St. Thomas had no problem with secondary causes operating in nature and that St. Augustine knew that the Bible is "not a science textbook." Both of these are true, as far as they go. Unfortunately, such platitudes only obscure deeper sources of tension between Darwinism and Thomistic thought. Here I would like to explore three intimately related sources of tension: the problem of essences, the problem of transformism, and the problem of formal causation.

The Problem of Essences

First, the problem of essences. G. K. Chesterton once quipped: "Evolution ... does not especially deny the existence of God; what it does deny is the existence of man."[3] It might appear shocking, but in this one remark the ever-perspicacious Chesterton summarized a serious conflict between classical Christian philosophy and Darwinism.

In Aristotelian and Thomistic thought, each particular organism belongs to a certain universal class of things. Each individual shares a particular nature—or essence—and acts according to its nature. Squirrels act squirrelly and cats catty. We know with certainty that a squirrel is a squirrel because a crucial feature of human reason is its ability to abstract the universal nature from our sense experience of particular organisms.

Think about it: How is it that we are able to recognize different organisms as belonging to the same group? The Aristotelian provides a good answer: It is because species really exist—not as an abstraction in the sky, but they exist nonetheless. We recognize the squirrel's form, which it shares with other members of its species, even though the particular matter of each squirrel differs. So each organism, each unified whole, consists of a material and immaterial part (form). ("Species" here is a more encompassing concept than in modern biological definitions. For example, wolves and domesticated dogs might share a common essence.)

One way to see this form-matter distinction is as Aristotle's solution to the ancient tension between change and permanence debated so vigorously in the pre-Socratic era. Heraclitus argued that change is the underlying reality. Everything constantly changes, like fire which never stays the same from moment to moment. Philosophers like Parmenides (and Zeno of "Zeno's paradoxes" fame) argued exactly the opposite; there is no change. Despite appearances, reality is permanent. How else could we have knowledge? If reality constantly changes, how can we know it? What is to be known?

Aristotle solved this dilemma by postulating that while matter is constantly in flux—even now some somatic cells are leaving my body while others arrive—an organism's form is stable. It is a fixed reality, and for this reason it is a steady object of our knowledge. Organisms have an essence which can be grasped intellectually.

Enter Darwinism. Recall that Darwin sought to explain the origin of "species." Yet as Darwin pondered his theory he realized that it destroyed species as a reality altogether. For Darwinism suggests that any matter can potentially morph into any other arrangement of matter without the aid of an organizing principle. He thought cells were like simple blobs of Jell-O, easily rearrangeable. For Darwin, there is no immaterial, immutable form. In *The Origin of Species* he writes:

> I look at the term species as one arbitrarily given, for the sake of convenience, to a set of individuals closely resembling each other, and that it does not essentially differ from the term variety, which is given to less distinct and more fluctuating forms. The term variety, again, in comparison with mere individual differences, is also applied arbitrarily, for convenience's sake.[4]

Statements like this should make card-carrying Thomists shudder. This is an extreme expression of the anti-Aristotelian (and anti-Thomist) philosophy of nominalism. Nominalism (stemming from the Latin "nomen," or "name") suggests that the individual is the only reality—not the universal, form, or essence. The mind invents universals in order to group together similar objects. But the universal is not a reality in which the individual in some way participates.

But Thomas embraced form and, following Augustine, even maintained that a creature's form reflects the second member of the Trinity. For, "as it [the creature] has a form and species, it represents the Word as the form of the thing made by art is from the conception of the craftsman."[5]

The first conflict between Darwinism and Thomism, then, is the denial of true species or essences. For the Thomist, this denial is a grave error, because the essence of the individual (the species in the Aristotelian sense) is the true object of our knowledge. As Benjamin Wiker observes in *Moral Darwinism,* Darwin reduced species to "mere epiphenomena of matter in motion."[6] What we call a "dog," in other words, is really just an arbitrary snapshot of the way things look at present. If we take the Darwinian view, Wiker suggests, there is no species "dog" but only a collection of individuals, connected in a long chain of changing shapes, which happen to resemble each other today but will not tomorrow.

Now we see Chesterton's point. Man, the universal, does not really exist. According to the late Stanley Jaki, Chesterton detested Darwinism because "it abolishes forms and all that goes with them, including that deepest kind of ontological form which is the immortal human soul."[7] And if one does not believe in universals, there can be, by extension, no human nature—only a collection of somewhat similar individuals.

Classical notions of ethics were radically dependent upon this notion of a real, knowable human nature. Aristotle and others often argued for what is ethical in terms of what leads to human flourishing and fulfillment. Yet if there is no human nature, how can we know what human fulfillment looks like in general? Tim and Tom might, then, flourish under different moral codes. Lack of a human nature may leave us with "different strokes for different folks."

As philosopher Alasdair MacIntyre showed in *After Virtue,* the way out of this modern dilemma is to recognize that if something's nature includes purposes or proper functions, then "oughts" follow from that which "is."[8] For if man is a certain sort of being, if he has a certain formal nature, then there are facts about how man ought to behave. There are objective criteria by which we can judge a human being good or bad. This kind of *telos*-infused nature cannot be sustained by Darwinism, however, for Darwinism denies that organisms have formal natures or are purposefully made.

But the Darwinian will say, "We believe in function, too!" True, the Darwinian knows of function—that ears hear, for example. But to say in the

Darwinian sense that the function of ears is to hear, notes philosopher Lydia McGrew, is only to say that the information encoding ears was passed to progeny because ears happened to hear—and that hearing, presumably, gave these organisms some survival advantage. If in 10,000 years humans walk on their hands because this somehow aids survival, the Darwinian cannot claim that hands are meant for walking, only that hands in fact *do* walk at this time.[9] That is, the Darwinian cannot support the notion of *proper* function.

This is not a mere abstract point. The dilemma is playing itself out in contemporary debates in bioethics. Who are bioethicists like Leon Kass (neo-Aristotelian and former chairman of the President's Council on Bioethics) sparring with today if not thoroughgoing Darwinians like Princeton's Peter Singer who deny that humans, qua humans, have intrinsic dignity? Singer even calls those who prefer humans to other animals "speciesist," which in his warped vocabulary is akin to racism.[10]

So what justifies the excessive expense and effort required to keep a Down's syndrome baby alive? For the traditionalist, it is the baby's membership in the human species. This gives the baby intrinsic value. For the utilitarian like Singer such expense is not justified; one would do better to contribute to the World Wildlife Fund, for species' differences are not essential but accidental. As Singer notes:

> All we are doing is catching up with Darwin.... He showed in the 19th century that we are simply animals. Humans had imagined we were a separate part of Creation, that there was some magical line between Us and Them. Darwin's theory undermined the foundations of that entire Western way of thinking about the place of our species in the universe.[11]

If one must choose between saving an intelligent, fully developed pig or the Downs Syndrome baby, Singer thinks we should opt for the pig. Perhaps this is why natural law theorist J. Budziszewski writes, "If any contemporary scientific movement holds promise for the furtherance of the natural law tradition, it is not the stale dogma of natural selection, but frank recognition of natural design."[12]

The Problem of Transformism

The second conflict is very similar to the first. The Thomist, as we have seen, is committed to the reality of universals, for universals are the objects of higher knowledge. But it is not only the existence of species which Darwinism destroys; it is also their stability.

Darwinian Theory posits that all living things are related through one or very few ancestors (referred to as "Universal Common Ancestry") *via* solely material processes. But if living things have unchangeable essences, how can these living things change (or "transform") into other living things through mere material causes?

One Thomist recently put it this way to a gathering of the American Maritain Association: "For those defending at least some aspects of the classical idea of essences, the problem can be stated as follows: how can one kind of living substance with its own unique essence change into another kind? And beyond the how, why would this happen in the natural world? What intrinsic end or ends would it serve?"[13]

For Darwin, there was no problem to solve, for there are no *essential* differences between living things. We see this assumption at work in every new primatological study finding that apes have an inner mental life, use sign language, or form hierarchical social structures "just like we do!" The Thomist should see this as hyperbolic, for his starting point is our everyday experience of the world. And as David Berlinski sardonically observes, the first and most obvious fact about apes is that they are "behind the bars of their cages and we are not." Put plainly, "Beyond what we have in common with the apes, we have nothing in common, and while the similarities are interesting, the differences are profound."[14]

We should not be too flippant about this, however. Supporters of Darwin's theory are no doubt right that apes' capacities are more similar to ours than are, say, alpacas'. But sometimes these similarities serve to hide real transitional difficulties. British literary critic A. N. Wilson gives a fine example from his atheist days:

> A materialist Darwinian was having dinner with me a few years ago and we laughingly alluded to how, as years go by, one forgets names.

> Eager, as committed Darwinians often are, to testify on any occasion, my friend asserted: "It is because when we were simply anthropoid apes, there was no need to distinguish between one another by giving names."
>
> This creedal confession struck me as just as superstitious as believing in the historicity of Noah's Ark. More so, really.
>
> Do materialists really think that language just "evolved," like finches' beaks, or have they simply never thought about the matter rationally? Where's the evidence? How could it come about that human beings all agreed that particular grunts carried particular connotations? How could it have come about that groups of anthropoid apes developed the amazing morphological complexity of a single sentence, let alone the whole grammatical mystery which has engaged Chomsky and others in our lifetime and linguists for time out of mind? No, the existence of language is one of the many phenomena—of which love and music are the two strongest—which suggest that human beings are very much more than collections of meat.[15]

For the Darwinian, complex biological realities exist for the sake of their smaller units of composition. Richard Dawkins has gone so far as to suggest that we are the pawns of our selfish genes. In this view, biological reality is a continuum, and the smallest units of composition run the show. Species' differences—indeed, even individual organisms—are mere accidents of environment and mutation.

But for Thomas, "the elements are for the sake of the compounds, the compounds for the sake of living things."[16] That is, reality is decidedly discontinuous, hierarchical, top-down. The entire point of essences is that they are stable realities; they cannot change and thus can provide real knowledge. The differences between species (intelligible essences) are differences of *kind*. Thus those defending the tradition of natural philosophy found in Aristotle and St. Thomas simply cannot accept transformism—at least not without introducing teleological conceptions of change, which would transform Darwinian Theory itself.

The Problem of Formal Causation

Finally, before moving to consider intelligent design, there is the problem of formal causation. It is here that we find St. Thomas' unique contribution, illuminating the insights of Aristotle with the light of Christian knowledge.

St. Thomas argued against the Islamic scholars of his day who held that God is the direct cause of everything in nature, a view known as occasionalism. Put negatively, occasionalism denies that creatures exercise their own causal powers. It is God who always acts as the *only* cause; creatures only appear to cause effects. "On the contrary," as Thomas is fond of saying, God created creatures with real natures that have real powers. Thus, ants act in an ant-like fashion. Ants themselves cause effects.

God is, of course, also a true cause of ant behavior: He created ants, he sustains ants in being, and he concurs (co-operates) with every ant action. According to Notre Dame philosopher Alfred Freddoso, this last aspect was extremely important to medieval Aristotelians: "It cannot be emphasized enough that the position being rejected here (viz., that God's action in the world is exhausted by creation and conservation) is regarded as too weak by almost all medieval Aristotelians."[17] These medieval thinkers would be scandalized by the claims of those modern Christian thinkers who exclude God from nature except as the First Cause and a merely bureaucratic role as sustainer of the universe.

So Thomas believed in true secondary causes. In a certain sense it is true that God causes everything. But in the act of creation God *also* delegates to creatures the power to act as true causes of their creaturely behavior, according to their natures. Because Aristotle is so well known for recognizing teleology *intrinsic* to living things, and because Thomas is so well known for this view of secondary causation, some of today's Thomists think that their tradition can whole-heartedly embrace Darwinian evolution. After all, Darwin just claimed nature is due to secondary causes, right? Nature just "does its own thing." It is this drastic over-simplification which lies at the heart of the casual acceptance of Darwinism among some classically thinking people today. We must dig deeper.

Recall that for Thomas, creatures are a combination of form and matter. The question that must be answered, then, in any version of Thomistic evolution, is where form comes from. Darwin, denying Aristotelian essentialism, saw organisms' traits as accidental properties of living things that change with the winds of time. Not so St. Thomas.

In his recent book *Aquinas on the Divine Ideas as Exemplar Causes*, Catholic University philosophy professor Gregory T. Doolan gives the most extensive treatment to date of Thomas' notion of "exemplar causation."

Exemplar causes are an integral part of Thomas' metaphysics. An exemplar cause is a type of formal cause—a sort of blueprint; the idea according to which something is organized. For Thomas, these ideas exist separately from the things they cause. For instance, if a boy is going to build a soap-box derby car, the idea in his mind is separate from the form of the car; yet the car's form expresses the idea, or exemplar cause, in the boy's mind. Exemplar causes actually do something. They are "practical ideas," writes Doolan.[18]

For Thomas (and here is the important point) a creature's form comes from a similar form *in the divine intellect*. In other words, the cause of each species's form is extrinsic. In fact, writes Thomas, "God is the first exemplar cause of all things."[19] Creatures do possess the causal powers proper to the nature God granted them, but creatures most certainly do not possess the power to create the form of their or any other species.

For instance, frog parents have the proper ability to generate tadpoles. They are able to bring out the natural form that is present in the potentiality of matter. However, the frog parents cannot create the form "frog." After all, Thomas reasons, if frog parents could create the form "frog" they would be the creators of their own form, and this is clearly a contradiction. Natural things can *generate* forms of the same species, but they cannot *create* the form of a species in general.

Thus natural agency is not eliminated, yet God is still actively involved in nature. Specific forms originate and reside in his mind, though God allows creatures the dignity of acting in this creative drama. Still, Thomas is careful to note that while secondary causes are real, "God ... can cause an effect to result in anything whatsoever independently of middle causes."[20]

By now it should be clear how different Thomas' philosophy of nature is from Darwinism. For Thomas, form is not merely an apparent reality that can be molded into any other form. Rather, a natural form originates in God's mind. He directly create it. It is a forethought, so to speak, not an afterthought. Species, then, come to be because of his will and power (either successively or all at once). They are neither the product of a trial and error process of natural selection nor the mere intrinsic unfolding of secondary causes. Secondary causes have their place, but they are inherently impotent to create novel form.

Let's face it: Thomas Aquinas was not evolutionist, let alone a Darwinist, in any sense.

Thomism and Intelligent Design

Given the active role of God in nature in Thomas' system, one might think today's Thomists would encourage the pursuit of signs of intelligent design in nature. Yet in recent years, some Thomists have shied away from ID. They do so not only because of lax scrutiny of the tensions just discussed but also because of three common misperceptions of ID: First, that ID is "mechanistic" and even embraces a "modernist" view of science; second, that ID is a "God of the gaps" theory; and third, that ID is inherently "interventionist." While many Thomists harbor doubts about the more extravagant claims of Darwinian science, taken together these three factors make it difficult for some Thomists to embrace intelligent design. That is as unnecessary as it is unfortunate.

Is ID Mechanistic and Modernist?

One of the defining hallmarks of modern Thomism is its strong rejection of early modern philosophy as seen in René Descartes and Francis Bacon. To simplify, modernists reduce Aristotle's four causes down to only two causes, and, as a result, reduce all knowledge to empirical knowledge. Both moves strike directly at Thomistic philosophy, so it is no surprise that they have aroused Thomists' ire.

"Causes" in Aristotle's sense explain why something is the way it is, and as Thomas explains, "there are four kinds of cause, namely, the material, efficient, formal and final."[21] Aristotle and Thomas would explain a marble statue by reference to its material cause (the marble), its efficient cause (the sculptor), its formal cause (the shape of the statue), and its final cause (the purpose of honoring Athena). A modernist, in contrast, sees only material man and marble at work. Ultimately, all is explained by atoms in motion—not by immaterial ideas, forms, or purposes. Thus for the modernist, knowledge is necessarily and exclusively knowledge of the empirical.

Some Thomists insist that ID is methodologically flawed because they think that ID, like modernism, rejects formal and final causation. This is incorrect. Far from rejecting final causation, ID theorists argue that there is empirical evidence of purpose or teleology in nature. They argue that at least some features of nature are best explained by intelligent activity, since such features exhibit evidence of foresight and planning.

By reintroducing intelligent causes as a legitimate scientific pursuit, and by rejecting the Darwinian notion that material and efficient causes suffice to explain nature, ID theorists may well open the door for renewed attention to formal and final causes. Thomists should welcome ID as a partner.

Still, some Thomists insist that ID inherently views nature mechanistically. Those who say this consistently have in mind Michael Behe's argument for the "irreducible complexity" of what are referred to in the scientific literature as "molecular machines." They seem to forget that Thomas repeatedly used analogies between living objects and man-made artifacts. So they should hardly be offended that Behe would compare some aspects of microbiological structures to machines.

Besides, ID arguments propose the very opposite of mechanism—agency. Consider Stephen Meyer's argument concerning the informational content of DNA. In *Signature in the Cell*, Meyer argues that blind material causes are insufficient to produce the *immaterial* information content of DNA. An *immaterial* mind, Meyer claims, is a better explanation than any mindless, material cause.[22]

Some Thomist critics go one step further and claim that ID concedes a modernist, Enlightenment view of science. Perhaps this is because ID proponents insist that ID arguments fall within the domain of natural science.

But this criticism has things precisely backward: ID theorists challenge the Enlightenment notion that only matter matters, that science cannot take immaterial concepts like mental causation seriously. ID challenges this directly, noting that while materialist science may have seemed plausible in the age of steam, it is hardly plausible in today's world of the information super-highway—run on the power of the invisible and the immaterial. According to ID theorists, accounting for nature in all its richness requires that we appeal not just to material but to personal causes as well.

Moreover, the claim that design is empirically detectable concedes nothing to the modernist idea that reason is limited to the empirical realm. Nor does anything in ID imply that only science can provide real knowledge. One can argue for empirical evidence of design and also defend, say, knowledge of divine revelation, moral knowledge, knowledge of abstract essences and knowledge derived from philosophical arguments for the existence of God.

Does ID Promote the "God of the Gaps"?

The second confusion regards the claim that intelligent design is a "God of the gaps" argument. As Thomist Edward Feser writes, "Aquinas does not argue in this lame 'God of the gaps' manner.... Paley did, and 'Intelligent Design' theorists influenced by him do as well."[23] Expressed more formally, a "gaps" argument is known as an argument from ignorance. These arguments base claims upon what one *does not* know rather than upon what one *does* know. Critics misconstrue contemporary ID arguments (and perhaps Paley's as well) as, "I do *not* know how this feature of the natural world arose via material causes; therefore God did it!"

Yet this too is simply a misunderstanding. ID is not an argument for God's existence. Rather, it is an inference to an intelligent cause. Some people think ID theorists are being coy, but they just want to avoid overstating their argument. Thomas drew the same distinction in *Summa contra Gentiles*:

> For seeing that natural things run their course according to a fixed order, and since there cannot be order without a cause of order, men, for the most part, perceive that there is one who orders the things that we see. But who or of what kind this cause of order may be, or whether there be but one, cannot be gathered from this general consideration.[24]

So there's certainly nothing anti-Thomistic in distinguishing between a generic argument for design and an argument for God's existence—even if the former might provide evidence for the latter.

Furthermore, ID—whether true or false—is not an argument from ignorance. ID proponents argue from the known features of natural objects, the *known* causal capacities of agents in our everyday experience and the known limits of certain material causes. In fact, this is the same method that makes Thomism so appealing. Experience teaches us that some effects in our everyday observation of the cause-and-effect structure of the world always come from intelligent agents. Material causes simply do not suffice to explain some things.

If, for instance, I come home and find that the magnetic letters on the refrigerator say "I love Daddy," I know that a mind rather than material causes alone (e.g., strong winds blowing through the kitchen window) produced the message. I already have numerous experiences with written language; I know the limits of material causes in this arena. ID merely formalizes this common experience with analytic rigor.

Take Stephen Meyer's argument mentioned previously. Meyer argues that DNA, which contains the same semantic quality as human language, also comes from a mind. He surveys today's most prominent materialistic theories for the origin of DNA's specified complexity and concludes they lack the causal resources to explain this salient property of DNA. But intelligent agency does not. Thus, judged by standard modes of reasoning in the historical sciences, intelligent agency is a better explanation. The form of Meyer's argument is precisely the same as Darwin's. (Darwin learned it from Sir Charles Lyell, the founder of modern geology.) The method involves looking to presently operating causes to explain past events in natural history. DNA is often called a "code," and if Meyer is correct the metaphor runs deeper than materialist philosophy ever dreamt.

Is ID Interventionist?

Finally, as we have already seen, in arguing against the occasionalists St. Thomas affirmed that God has given nature causal capacities of its own. They are bounded, of course, by certain actions of which only God is capable, but nature has its role nonetheless. And this fact has led certain Thomists to an aesthetic preference for scientific theories that do not involve God's "interference" in nature. They are wary of ID's seeming "interventionism."

Whereas materialists must be non-interventionists, theists have more explanatory resources at their disposal. Thus it seems that the evidence should decide the matter for theists. Perhaps it is logically possible that God limited himself to secondary causes in natural history, but we cannot deduce that beforehand. If the fossil record remains discontinuous despite the occasional media hype of a new "missing link,"[25] and if field studies of natural selection continue to show that natural selection merely keeps populations healthy, then so be it.[26] Maybe God acted as a primary cause at different periods in life's history.

Christians *already* believe this. They recite it in every creed. As Avery Cardinal Dulles—an advocate for teleological evolution—wrote:

> Christian Darwinists run the risk of conceding too much to their atheistic colleagues. They may be over-inclined to grant that the whole process of emergence takes place without the involvement of any higher agency. Theologians must ask whether it is acceptable to banish God from his creation in this fashion.... [God] raised Jesus from the dead. If God is so active in the supernatural order, producing effects that are publicly observable, it is difficult to rule out on principle all interventions in the process of evolution. Why should God be capable of creating the world from nothing but incapable of acting within the world he has made?[27]

For Christians this is surely a needed warning against swallowing popular prejudices. But even so, is the intelligent design proponent necessarily committed to God's repeated intervention in the natural world? Absolutely not. Postulating intelligent agency as a necessary causal ingredient for certain features of nature does not commit one to exactly when or how this feature arrived on the scene. Just recall the letters on my refrigerator: I cannot be certain who put them there, or how, or when, but I surely know the arrangement is intelligently designed.

Catholic biochemist and ID proponent Michael Behe, for one, thinks it unlikely that God intervened directly in the development of the biological realm. Rather, he speculates that God may have front-loaded the information and laws necessary to humanity's development into the beginning of the universe. Behe thinks

> ... the assumption that design unavoidably requires "interference" rests mostly on a lack of imagination. There's no reason that the extended fine-tuning view ... necessarily requires active meddling with nature.... One simply has to envision that the agent who caused the universe was able to specify from the start not only laws, but much more.[28]

Intelligent design by natural laws and initial specifications is still intelligent design, and it may be detectable in the same way as the "fine-tuning" of the laws of physics. The detectable effects of intelligent design could be the same, however that design was implemented.

Thomas himself, far from being worried about intervention, thought there was good reason to think that God purposefully "intervenes" in nature, writing that

> the divine power [can] at times work apart from the order assigned by God to nature, without prejudice to His providence. In fact He does this sometimes to manifest His power. For by no other means can it better be made manifest that all nature is subject to the divine will, than by the fact that sometimes He works independently of the natural order: since this shows that the order of things proceeded from Him, not of natural necessity, but of His free will.[29]

Thomas' way of speaking here is more helpful than speaking of "intervention," which is often used pejoratively. In Thomas' view, when God acts directly in nature he is not invading foreign territory, tampering with something he should have fixed earlier, or violating natural laws established in opposition to his will. He is acting within the world that he created and that he sustains from moment to moment. If he sometimes chooses to act independently of the natural order, to bring about results that would not have happened if nature were left to its own devices, that is his prerogative. Thus those Thomists who decry "interventionism" may not be as Thomistic as they think.

Conclusion

Still, St. Thomas' argumentation differs at times from modern design arguments. For one thing, Thomas is more concerned with ontology than biology or other natural sciences. His chief concern is why something should exist at all, not, say, the intricate features of particular biological organisms or the fine-tuning of physical constants. For another, Thomas preferred deductive arguments. ID proponents prefer newer forms of argumentation, especially "inference to the best explanation"—the method common in the historical sciences whereby one must not only weigh the strengths and weaknesses of a given hypothesis but also compare hypotheses with each other. In this fashion a scientist can decide which theory currently explains the data better than all rivals and yet remain open to new data or hypotheses which might change the equation.

While they don't provide the certainty of deductive conclusions, one advantage to these arguments is that they recognize that this finite world often requires tradeoffs: One cannot sit satisfied having raised questions about an ID argument; rather, he must show that his own hypothesis is *better* at explaining relevant data.

As Alexander Pruss, an analytical Thomist and former Georgetown colleague of John Haught, writes:

> On the compatibility between Thomism and ID, the answer is surely positive. Thus, one might think that the irreducible complexity types of arguments provide a strong probabilistic case for design *and* that the existence of teleology provides a sound deductive argument for a first cause."[30]

Nevertheless, despite the different subject matter and styles of argumentation, Thomists and ID theorists have, as we have seen, *much* in common. The dismissal of intelligent design by some contemporary Thomists is unfortunate. For if reality is a unified whole, that is, if it stems from the divine mind as Thomas believed, would it not be odd if good philosophy concluded that life is designed but good science concluded that life was not?

11. Straining Gnats, Swallowing Camels

Catholics, Evolution and Intelligent Design, Part I

Jay W. Richards

The Catholic Church teaches that all human beings can know that God exists by "natural reason," and in particular, by observing the design manifest in nature. Though we need special revelation to know, for instance, that God is triune, the Church concurs with St. Paul, who wrote that "from the foundation of the world, God's invisible qualities—his eternal power and divine nature—have been clearly seen from created things" (Romans 1:20).

The early Church Fathers generally shared this conviction. For instance, in the fourth century, St. Athanasius (c. 296–373) said: "The rationality and order of the universe proves that it is the work of reason or Word of God."[1] St. Gregory Nazianzen (c. 329–389), one of the so-called Cappadocian Fathers, said much the same thing: "Now our very eyes and the Law of Nature teach us that God exists and that He is the Efficient and Maintaining Cause of all things."[2]

As these quotes suggest, the Church Fathers tended to think that God's existence should be obvious to everyone; nevertheless, they often offered so-called teleological or design arguments, since not everyone could see the obvious. Such arguments begin with some evidence of "order" from nature that is best explained by a Great Intelligence, Designer, or Creator.

Gregory Nazianzen, for instance, compares the creation to a lute:

> For every one who sees a beautifully made lute, and considers the skill with which it has been fitted together and arranged, or who hears its melody, would think of none but the lutemaker, or the luteplayer, and would recur to him in mind, though he might not know him by sight. And thus to us also is manifested That which made and moves and preserves all created things, even though He be not comprehended

> by the mind. And very wanting in sense is he who will not willingly go thus far in following natural proofs.... For what is it which ordered things in heaven and things in earth?... Is it not the Artificer of them Who implanted reason in them all, in accordance with which the Universe is moved and controlled?[3]

Various forms of design arguments have appeared throughout Church history. In the Middle Ages, for instance, St. Thomas Aquinas provided five famous arguments for the existence of God (the "five ways"), the fifth of which is a design argument. Thomas argues that the purposive orderliness of nature is an effect of a prior cause, which is God.

In the nineteenth century, William Paley proposed a related but distinct design argument, in which he compared the ordered complexity found in the biological world to such human technology as a watch. Paley's argument was deeply influential in the English-speaking world. In the middle of that century, however, Charles Darwin proposed an alternative to Paley—an impersonal process of natural selection and random variation—that could mimic the work of an actual designer. It is widely reported that Darwin's theory, particularly in its contemporary form (Neo-Darwinism), has now fully supplanted the earlier certainty that God's design could be seen clearly in nature.

Standing athwart this storyline, however, in recent years, an ecumenical group of scholars—mostly scientists and philosophers—have been developing new design arguments, using new intellectual tools and new evidence from natural science. By common appellation, they are called "intelligent design" or "ID" theorists. These include scientists such as Michael Behe, Jonathan Wells, and Guillermo Gonzalez, philosophers such as William Dembski, Stephen Meyer, this author, and many others, such as legal scholar Phillip Johnson. ID theorists argue that nature, or certain aspects of nature, are best explained by intelligent design. They tend to focus on the detectable and isolatable effects of intelligence. As William Dembski, one prominent ID proponent, puts it: "Intelligent design (ID) is the study of patterns in nature that are best explained as the result of intelligence."[4]

ID proponents often distinguish this "detectable" design from both chance and natural law, as we all do in ordinary experience. When we look at Mt. Rushmore, for instance, we distinguish the sculpted faces of the four Presidents from the work of gravity, wind, and erosion. Similarly, some modern ID arguments distinguish repetitive, law-like explanations such as the gravitational force and natural selection, which explain some aspects of nature, from objects such as the coding regions of DNA or a bacterial flagellum, which arguably go beyond the ordinary capacities of natural, repetitive forces. Michael Behe points to "irreducibly complex" molecular machines like the bacterial flagellum, which seem to require foresight and so are probably inaccessible to the Darwinian mechanism. Stephen Meyer points to the "specified complexity" in the coding regions of DNA, which he argues (with William Dembski) is a reliable indicator of intelligent design. These ID arguments imply that while natural law explains some things, a complete explanation of the natural world will include intelligent agency as well.

At the same time, many ID theorists have argued that the physical laws, constants, and initial conditions that obtained at the beginning of our temporally finite universe are *themselves* the evidence of intelligent design, even if they are not, arguably, adequate to explain everything in nature. Guillermo Gonzalez and I have argued, for instance, that the eerie coincidence between fine-tuning for life in the universe, and the requirements for scientific discovery, suggest that the universe is designed not just for life but also for discovery.[5]

Since ID theorists focus on the effects of intelligence rather than on the specific locations or modes of design within nature, most ID arguments are consistent with a variety of views about how and when God acts in nature. In fact, ID arguments are not explicitly theological. One could pursue research within an ID framework without ever asking follow-up questions; nevertheless, these arguments have positive theological implications. As a result, they can be used to construct arguments for God's existence.

Given such affinities, orthodox Catholics should recognize ID theorists as important allies. Some Catholic intellectuals, however, have accused ID of a multitude of sins. Many of these same intellectuals seem relatively less troubled by Darwinian Theory, which attempts to explain away the teleology in nature. This follows decades of confusion among ordinary Catholics about what the Church teaches about evolution.

In this and the following two chapters, I hope to clear up some of this confusion and to show why orthodox Catholics should embrace and, indeed, contribute to work on intelligent design.

Is There a Catholic Position on Evolution?

In his inaugural papal Mass on April 24, 2005, Pope Benedict XVI observed: "We are not some casual and meaningless product of evolution. Each of us is the result of a thought of God. Each of us is willed, each of us is loved, each of us is necessary."[6] As far back as the 1960s, the new Pope—Joseph Ratzinger, a German theologian and former head of the Church's Congregation for the Doctrine of the Faith—had shown an interest in the subjects of creation and evolution. In the early 1980s, he preached a series of Lenten sermons, later published as a book, intended to make the doctrine of creation a priority for theology.[7] In the preface to his sermons, he calls out for special criticism one Catholic book of theology, in which the authors go "beyond one's worst fears. The reader discovers here phrases such as 'Concepts like selection and mutation are intellectually much more honest than that of creation.'" In response, he asks: "Would it be too harsh to say that the continued use of the term 'creation' against the background of these presuppositions represents a semantic betrayal?"[8]

In 1985, he hosted a symposium in Rome called "Evolutionism and Christianity." So his statement in 2005 was not off-the-cuff. For decades prior to becoming Pope, Benedict had been a critic of the Darwinian materialist view that life, and in particular human life, are the result of purposeless chance and necessity.

Despite this record, his 2005 comment gained attention, especially by reporters who thought the new Pope was contradicting his predecessor Pope John Paul II. (The previous pontiff had seemed more conciliatory on the subject of evolution. In a widely reported statement in 1996 to the Pontifical Academy of Sciences, John Paul had said that "new knowledge has led to the recognition of the theory of evolution as more than a hypothesis."[9])

In July 2005, just a few months after the inauguration of Pope Benedict, Christoph Cardinal Schönborn published an op-ed in the *New York Times,* arguing that Pope John Paul II's statement had been misinterpreted. While some forms of "evolution" might be true, Cardinal Schönborn insisted that strict Neo-Darwinism was both false and contrary to Catholic faith. While Schönborn was giving his own opinions, as Archbishop of Vienna and the chief editor of the *Catechism of the Catholic Church,* his comments carried weight.

Such frank comments were bound to lead to trouble, and they did. The public controversy that followed is well known. The controversy behind-the-scenes, in both Austria and Rome, is mostly the subject of speculation. What we know is that Schönborn, at the Pope's instruction, soon gave a series of lectures on creation and evolution, which were then published as *Chance or Purpose?* Then, in 2006, Pope Benedict held his annual conference at the papal summer residence, Castel Gandolfo, on the same subject. The conference included a number of Benedict's former students, along with the cock-sure, Darwinian origin-of-life theorist Peter Schuster, president of the Austrian Academy of Sciences. The proceedings of the conference were recently published as *Creation and Evolution,*[10] with a foreword by Cardinal Schönborn.

The arguments by Benedict and Cardinal Schönborn in these books are consistent with the Church's reasonably clear teachings on evolution. Unfortunately, only the public statements by Benedict, John Paul II and Schönborn are widely known. And out of context, they have created cognitive dissonance for reporters unaccustomed to interpreting Vatican statements on hard subjects. Unfortunately, the cognitive dissonance is not confined to religiously illiterate reporters. It persists in the minds of millions of Catholics.

On the one hand, many Catholic school students are told that the Church doesn't have a problem with "evolution." On the other hand, much of what they hear about evolution—even from their Catholic teachers—doesn't seem to jive with their Catholic beliefs. As a result, there is now wide spread confusion and ignorance among Catholics about what the Church does and does not teach on the subject. As it happens, while there is need for more frankness and clarity, the Church's general teaching is actually a lot less ambiguous than many popular, and even academic statements, suggest.

CREATION

TO UNDERSTAND THE CHURCH'S teaching on the subject of evolution, and the current controversy over intelligent design, you first have to understand what she affirms concerning creation, especially in light of the influence of the great medieval theologian Thomas Aquinas (c. 1225–1274).

St. Thomas on Creation

ST. THOMAS WAS THE first Christian theologian to make full systematic use of the Greek and pre-Christian philosopher Aristotle, most of whose works weren't directly available for centuries to Europeans after the fall of Rome. But contrary to stereotype, Thomas did not strictly follow Aristotle. His views are actually a sophisticated synthesis of Neo-Platonism and Aristotelianism in light of the biblical witness and sacred tradition.

"Cause"

Thomas drew heavily on the distinction between potency and act, which was foundational for Aristotle's metaphysics.[11] He also used the philosopher's famous four causes: material, efficient, formal, and final, though he modified them for his own purposes. Unfortunately, the way Aristotle and Thomas used the word "cause" can be confusing to modern readers. Simply put, by "cause," they were referring to positive factors that explain or are responsible for something else. To fully grasp how Aristotle and Thomas understood the four causes, you have to get inside their metaphysical systems. I'm not sure how to do that for readers, at least without a lot of verbiage, so we'll have to settle for a common translation of these concepts into the modern vernacular: the material cause explains what something is made of; the efficient cause explains where something came from—who or what produced or moved it; the formal cause explains what something is; and the final cause explains the ultimate purpose toward which something tends.

For Aristotle, not just organisms, animals, and people, but all physical objects tend toward the perfection of their nature or essence. So Aristotle had a teleological view of nature, though his view, counterintuitively, did not require that natural "teleology" reflect the intentional purposes of a Creator. Though some interpreters see Aristotle as approaching theism, Cambridge philosopher David Sedley contrasts Aristotle's view with the view of his teacher, Plato: Aristotle's "project was to retain all the explanatory benefits of [Plato's] creationism without the need to postulate any controlling intelligence...."[12]

In explaining the doctrine of creation, Thomas appeals to all four causes, and uses "formal" cause in several different ways. God is the ultimate cause of everything other than himself, and did not create indirectly through the mediation of angels or other beings: "Since God is the efficient, the exemplar [a type of formal cause] and the final cause of all things, and since primary matter is from Him, it follows that the first principle of all things is one in reality" (*Summa Theologica* I:44:4).

In his *Summa Theologica*, Thomas argues that God directly creates every human soul and he directly made the original body of Adam. He insists: "The rational soul can be made only by creation" (I:90:2). To be clear, he means that each human soul, which is the formal principle of the body, is created directly from nothing: "[I]t cannot be produced, save immediately by God" (I.90.3).

Thomas considers creation *ex nihilo* to be the pre-eminent meaning of the word "create." As he puts it: "To create [in the unique sense attributable to God] is, properly speaking, to cause or produce the being of things" (I:45:6). In other words, God doesn't just take pre-existing stuff and fashion it, as does the Demiurge in Plato's *Timaeus*.[13] Nor does he use something called "nothing" and create the universe out of that. Rather, God calls the universe into existence without using pre-existing space, matter, or anything else. So when he creates the universe from nothing, God's creative act is an act in eternity that does not involve changing one thing into another.[14]

Beyond Creation Ex Nihilo

At the same time, Thomas knows that words can be used in more than one sense. So when talking about the nature of the "light" God creates on day one, he says: "Any word may be used in two ways—that is to say, either in its original application or in its more extended meaning" (I:67:1). The same is true with the word "creation." In the proper sense, it refers to God's creation of everything from nothing. But in the extended sense, most of us apply it to other types of acts by God and other agents. Thomas tended to use the word more strictly, but he certainly didn't limit God's activity to this one type of act. For instance, Thomas referred to God "making" or "producing" different things on different days—light, the firmament, plants, animals, etc. He also referred to God changing something he has already created into something else—"fashioning" it as it were.

For instance, Thomas argues that God did not create Adam's body from nothing, but produced it directly using, as he puts it, the "slime of the Earth": "The first formation of the human body could not be by the instrumentality of any created power, but was immediately from God" (I:91:2). And God made Eve immediately, but not *ex nihilo,* using Adam's rib (I:92:2–4). Similarly, God produces the fish and birds from water. So while clearly using Aristotelian categories, Thomas otherwise adheres to a quite straightforward, literal reading of the second chapter of Genesis. And he freely talks about God's direct causal activity in nature in several different ways.

In treating creation in general, Thomas holds that everything other than God gets its being from God. Article 1 of Question 44 (in part one of the *Summa Theologica*) even answers affirmatively the following: "Is God the Efficient Cause of All Things?" (I:44:1). Thomas goes on to argue that, through his ideas or "exemplars," which are themselves formal causes, God "is the first exemplar cause of all things" (I:44:3) and the final cause of everything. (In this way, incidentally, he departs from Aristotle, who did not have a view of a transcendent God with pre-existing ideas distinct from particular natural entities, and toward whom everything tends.)[15]

Despite this philosophical language, where the views of philosophers appeared to conflict with the plain reading of Scripture, Thomas was perfectly happy to say: "On the contrary, Suffices the authority of Scripture."

It's vital to remember that for Thomas, God both can and does act in the created order in a variety of ways. Certainly he creates *ex nihilo*—indeed, that's the purest, most proper sense of the word "create" since it sets God apart from all his creatures—but that's only part of the story of God's diverse causal activity.

Though Catholics are not obligated to share all of St. Thomas' views on science and his interpretation of the early chapters of Genesis, the Magisterium still concurs with him that God is the ultimate Creator of everything and is free to act in various ways within his creation. In keeping with the historical Christian teaching on the subject,[16] for instance, the First Vatican Council (1869–1870) affirmed the doctrine that God is the Creator of everything, material and immaterial, other than himself. God's creation is *ex nihilo*, from nothing. That means that God didn't use any pre-existing thing to create the universe.

In its article "On God the Creator of All Things," the Council condemned certain ideas contrary to this doctrine, especially materialism, which reduces everything to matter, and pantheism, which identifies God and nature.

Moreover, the Council insisted, in keeping with the tradition, that we can know that God exists by observing the creation and using our natural reason:

> 1. The same Holy mother Church holds and teaches that God, the source and end of all things, can be known with certainty from the consideration of created things, by the natural power of human reason: ever since the creation of the world, his invisible nature has been clearly perceived in the things that have been made.[17]

The recent *Catholic Catechism* affirms all these points.[18]

Historical Catholic Response to Darwin's Theory of Evolution

The Church's teaching on the doctrine of creation is quite clear, at least in general terms. The Church's response to "evolution" is a more complicated story.[19] Charles Darwin first proposed what he called his theory of "descent with modification" in 1859 in *On the Origin of Species*. His central idea was that natural selection and random variation, and some other secondary processes, can produce the diversity and adaptive complexity of life. Darwin intended for his theory to displace the then-current idea that the species had been specially created by God.

Despite its obvious theological relevance, there has never been an official Church pronouncement on the book. *The Origin* was never placed on the Index of Prohibited books, although various speculative books on evolution, such as Henri Bergson's *Creative Evolution*, spent decades on the list. And Darwin's contemporary, John Henry Cardinal Newman, seemed unperturbed by the naturalist's theory. He wrote to a friend, "Mr. Darwin's theory need not then be atheistical, be it true or not."

If you knew nothing else, you might think that Catholicism has no problem with Darwinian evolution. But that would be a serious misunderstanding.

It's certainly true that the Catholic Church, as a whole, has not had the same struggles with the issue that some Protestants have had. Though the Church affirms that the universe has a finite age, the length of the "days" of Genesis 1 and the precise age of the universe, for instance, have not been burning issues for most Catholics.[20]

Still, biological evolution in the Darwinian sense has always been a subject of concern, even if that concern is not so obvious to the outside observer. Unlike Cardinal Newman, late nineteenth century Catholic intellectuals such as G. K. Chesterton and Hilaire Belloc both saw Darwin's theory as fundamentally at odds with Catholic belief (see Denyse O'Leary's chapter in this volume). The Church's official response has not been so different; but it's not easy to discover that fact. Father Martin Hilbert has summarized the situation quite well:

> The church does not pretend to give scientific answers to biological questions. But it does point out that some Darwinist claims are mere materialist metaphysics pretending to be science, because it knows that were it to remain silent on a truth—the nature of man—that has been entrusted to it by God, that truth would soon disappear, only to be replaced by the ever-changing dogmas of a materialist science.
>
> Even so, the Catholic Church has been surprisingly sparing in its pronouncements on the subject, given that Darwin's theory has been used to underpin some fairly disastrous worldviews, such as Nazism and communism. The church has never been very comfortable with the theory, but, perhaps fearing bad press of the kind that arose from its condemnation of Galileo, has usually preferred to deal with theo-

logians who were enthusiastic about evolution in more discreet ways than by magisterial interventions.[21]

Unfortunately, such discretion does little to resolve the key question that a careful observer wants to ask: Exactly which "Darwinist claims" are "mere materialist metaphysics pretending to be science," and which are solidly based on the evidence of nature? Ask the question that precisely, and the answers get fuzzier.

Part of the trouble is that the word "evolution" is a wax nose—it means different things to different people. Catholics have wide latitude to entertain various "evolutionary" hypotheses, including the cosmic evolution of stars and planets after the "Big Bang," and the common ancestry of various organisms. And in the encyclical *Humani Generis* (1950), Pope Pius XII granted Catholics permission to consider whether the human body came from "pre-existent and living matter."

At the same time, Pius XII dogmatically established that the human soul did not evolve from pre-existing life but is specially created by God. In the same encyclical, moreover, he insisted that Adam and Eve were historical figures, that the fall into sin was an historical event, and that all human beings descended from the original pair of human beings (rather than from different ancestors). The *Catholic Catechism* reaffirms this point: "The account of the fall in Genesis 3 uses figurative language, but affirms a primeval event, a deed that took place at the beginning of the history of man. Revelation gives us the certainty of faith that the whole of human history is marked by the original fault freely committed by our first parents."[22]

The new *Catechism* says little about evolution explicitly. And as the details above suggest, the Church has been more concerned with the origin and nature of human life, than with the origin and evolution of non-human life. Still, the statements from Pope Benedict above make clear that a materialistic version of evolution, which proceeds without plan or purpose, is contrary to Catholic faith. One might assume that that implicates *Neo-Darwinian* evolution (as opposed to mere change over time or common ancestry). Neo-Darwinism, as defined by its proponents, replaces real design and purpose with a designer substitute: the "mechanism" of natural selection and random genetic mutations. Schönborn emphasized this point in his piece in the *New York Times*. And in his third Lenten homily on creation, Benedict responds directly to the idea that life can be reduced to blind chance and necessity, to a series of "haphazard mistakes":

> What response shall we make to this view? It is the affair of the natural sciences to explain how the tree of life in particular continues to grow and how new branches shoot out from it. This is not a matter for faith. But we must have the audacity to say that the great projects of the living creation are not the blind products of chance and error. Nor are they the products of a selective process to which predicates can be attributed in illogical, unscientific and even mythic fashion. The great projects of the living creation point to a creating Reason and show us a creating Intelligence, and they do so more luminously and radiantly today than ever before. Thus we can say today with a new certitude and joyousness that the human being is indeed a divine project, which only the creating Intelligence was strong and great enough to conceive of. Human beings are not a mistake but something willed; they are the fruit of love.[23]

If that's not a rebuke of Neo-Darwinism's "blind watchmaker thesis," it's hard to see what else it could be.

But, alas, in official statements (as opposed to sermons and op-eds), the Church has been less explicit in distinguishing different senses of "evolution" and in connecting the dots between Neo-Darwinism and materialism. The distinctions that do get made are sometimes obscure. Rather than distinguishing between changes in gene frequency, or common descent, and Neo-Darwinism, Catholic writers often will distinguish between the legitimate "science of evolution" and the broader, questionable philosophy of "evolutionism." For instance, in his introduction to *Creation and Evolution* (the collection of papers presented at Pope Benedict's 2006 conference at Castel Gandolfo) Cardinal Schönborn says:

> Whereas faith today no longer has any difficulty in allowing the scientific hypothesis of evolution to develop in peace according to its own methods, the absolute claim of the philosophical explanatory model "evolution" is an all the more radical challenge to faith and theology.[24]

Schönborn seems to imply that the problem is merely with a philosophical gloss, called "evolution," that is parasitic on the otherwise perfectly legitimate scientific hypothesis that goes by the same name. But the problem, which the Cardinal himself recognized in his *New York Times* piece, is more specific and more acute, namely, that the Neo-Darwinian theory of evolution *is materialistic in its essence*. Between 2005 and 2006, the Cardinal's distinctions seem to have gotten rather less precise.

A related problem is that the word "evolution" itself is pre-Darwinian and, in a sense, anti-Darwinian. This semantic slippage is a source of all manner of confusion. The word, both by etymology and by initial use in the 19th century, implied progress—an unfolding of something already present from the beginning—like an acorn growing, over time, into an oak tree. That's why Darwin didn't use the word "evolution" in early editions of the *Origin of Species*, and only later adopted it.

Pope Benedict's language in the quote above sounds purposive rather than Darwinian. Earlier in the same homily, he speaks of creation and evolution as "complementary—rather than mutually exclusive—realities." How can he say that? Because he defines evolution as purposeful. In the nineteenth century, he says, "It became evident that the universe was not something like a huge box in to which everything was put in a finished state, but that it was comparable instead to a living, growing tree that gradually lifts its branches higher and higher to the sky."[25] And elsewhere he has said: "The Christian idea of the world is that it originated in a very complicated process of evolution but that it nevertheless still comes in its depths from the Logos. It thus bears reason in itself."[26] Understood in this way, of course, "evolution" means practically the opposite of what the typical Darwinist means by the word.[27]

Many European Catholics were already accustomed to thinking in terms of a purposeful unfolding of nature when Darwin proposed his theory of "descent with modification" in the mid-nineteenth century. ("Evolution," incidentally, is a Latinate word.) In the English-speaking world, however, the word "evolution" has long since become a simple synonym for Neo-Darwinism. So if a German Catholic theologian speaks positively about "evolution" in an English-speaking context, he may have in mind a non-Darwinian, purposeful unfolding of a pre-existing reality, but he probably won't be understood that way.

Teleology

It is easy for Catholic thinkers to interpret evolution as purposeful and progressive because of the centrality of the concept of teleology in the Catholic tradition, drawn largely from the work of Thomas Aquinas and other medieval scholastics. Deriving from the Greek word *telos* meaning "end" or "purpose," teleology, according to the *Catholic Encyclopedia*, is "the doctrine that there is design, purpose, or finality in the world, that effects are in some manner intentional, and that no complete account of the universe is possible without reference to final causes."[28] And insofar as Neo-Darwinism is an attempt to explain finality and purpose as merely apparent or illusory, it is contrary to the longstanding Catholic commitment to teleology in nature. So using private meanings of the word "evolution" will not, in the long run, solve the problem.

To sum up the essentials, then, Catholics affirm that God is responsible for the creation in its entirety, that everything exists for God's purposes, that God directly creates human souls, that all human beings are descended from a human pair that fell into sin, that we can discern, indeed, know the reality of God's existence and design from nature and by reason apart from special revelation. Catholics are free to hold a diversity of opinions on many of the details at the boundary between creation and evolution. They may entertain various evolutionary hypotheses in cosmology and biology and weigh the empirical evidence for and against them without fear of censure by the Magisterium.[29]

The Church has consistently taught that natural science has its own, legitimate autonomy as a discipline. At the same time, there is only one reality. Theology and science are not, as the late Stephen J. Gould claimed, "non-overlapping magisteria." A proposition cannot be true in biology but false in theology. So it's possible for an *assertion* drawn from science to conflict with a Church teaching. And some meanings of the word "evolution" do conflict with certain Catholic truth claims. For instance, no faithful Catholic can affirm, with George Gaylord Simpson, that "[m]an is the result of a purposeless and natural process that did not have him in mind. He was not planned."[30]

If Simpson's starkly anti-Catholic and anti-Christian position were merely his quirky opinion, we wouldn't have a problem. Unfortunately, Simpson's understanding of biological evolution is precisely the mainstream view and is almost universally taught in textbooks.[31] So with respect both to what is true and what we can know, the orthodox Catholic and orthodox Neo-Darwinian positions are irreconcilable. At most one of them can be true.

Catholic Critics of Intelligent Design

Still, there's no detailed and unique Catholic position on the generic question of "evolution." Nor is there one Catholic response to modern arguments for intelligent design. Some Catholics, like biochemist Michael Behe, neuroscientist Michael Egnor, and yours truly, affirm intelligent design. Others are critical. Many aren't sure what to think, since they see apparently orthodox Catholics arguing about it.

Catholic critics of ID, such as there are, fall roughly into three groups: the dissenters, the accommodationists, and the uncharitable. Not everyone fits neatly into one of these three categories, but it's important to realize that not every "Catholic" critic of ID has the same perspective. In fact, some of them are not orthodox Catholics at all.

The Dissenters

Dissenters plainly depart from core teachings of the Magisterium. Some are "cafeteria Catholics." They accept part of the package, but disagree with Church teaching on what they think are isolated issues like divorce, priestly ordination or contraception. More thoroughgoing dissenters have made their peace with naturalism and modernism, and interpret Catholic doctrine along those lines. Those who fully surrender to materialistic thinking don't believe that God raised Jesus from the dead, that the consecrated bread and wine really becomes the body, blood, soul, and divinity of Jesus under the appearance of bread and wine, or that Jesus was born of the Virgin Mary. In many cases, they are basically deists with vestments.

Some dissenters, like biologist Francisco Ayala, cease to be visibly Catholic—though, like Ayala, a fervent ID critic, they may reap benefits from their past Catholic associations. For instance, a 2007 cover story in *U.S. Catholic* magazine featured an interview with Ayala criticizing ID as "bad science" and "bad theology."[32] Other dissenters self-identify as Catholic, but what they really believe is another matter. In any case, just because a Catholic dissenter embraces Neo-Darwinism or dislikes ID, that doesn't mean that a believing Catholic should do the same.

The Accommodationists

Besides the outright dissenters, who abandon key Catholic doctrines rather than seek to preserve them, there are accommodationists who want to make peace with Darwinism and a materialist view of science but remain more or less faithful Catholics. They either live with the tension unresolved or try to integrate Darwinism in creative if inconsistent ways into their theology. Biologist Ken Miller is an example. He's an ardent defender of Neo-Darwinism and the principle of "methodological naturalism" that bars design arguments entry into scientific jurisdictions and allows Darwinism to win by default. But he also proclaims himself a believing Catholic.[33]

Unfortunately, few accommodationists identify themselves as accommodationists. On the contrary, some do their best to make their arguments sound traditional. In the current debate, an especially popular strategy is to try to marshal St. Thomas for the cause, since he is held in such high regard. This makes it much harder for the casual observer to notice the accommodation.

One such critic is Gonzaga University philosopher Michael Tkacz. Tkacz argues that God's creative activity never involves a change from one thing to another. So he objects to ID arguments that he thinks (incorrectly) imply that God "intervenes" in nature:

> This is the view that nature, as God originally created it, contains gaps or omissions that require God to later fill or repair. Given the Thomistic understanding of divine agency, such a "god of the gaps" view is clearly inconsistent with a proper conception of the nature of creation and, therefore, is cosmogonically fallacious.[34]

This is a familiar caricature of ID, which ID proponents have corrected many times. In truth, ID *per se* is neutral with respect to how and when intelligent design is implemented (though God can do what he wants to do). Logically, detecting design within a framework in some particular spot within nature (the subject of ID arguments) is a different issue from determining how that design came about.

Tkacz attributes this "god of the gaps" view to Michael Behe (who is a conservative Catholic):

> Now, a Thomist might agree with Behe's knowledge claim that no current or foreseeable future attempt at explanation for certain biological complexities is satisfactory. Yet, a Thomist will reject Behe's ontological claim that no such explanation can ever be given in terms of the operations of nature.

But Behe has never made any such claim and Tkacz provides no reference to substantiate his characterization of Behe's argument. On the contrary, in his book *The Edge of Evolution*,[35] Behe suggests just the opposite—that all the design in biology might trace back to the fine-tuning of physical constants and cosmic initial conditions.[36] This creates a problem for Tkacz's conclusion: "Insofar as ID theory represents a 'god of the gaps' view, then it is inconsistent with the Catholic intellectual tradition." Since the "insofar" clause isn't satisfied, the central thesis for the article dissolves.

Rather than belabor his portrayal of intelligent design, however, let's focus on Tkacz's representation of St. Thomas and the Catholic tradition, which I think falls short of the mark.

Creation Ex Nihilo Isn't the Whole Story

As we saw earlier, it's true that Thomas considers creation *ex nihilo* to be the pre-eminent meaning of the word "create." It distinguishes God's creative power from the kind of "creation" of which human beings are capable. When God creates the universe from nothing, he's not changing one thing into another,[37] as Tkacz notes.

But as we've already seen, for Thomas (and Christianity for that matter), this isn't God's only mode of action. As part of his initial causal activity, God also fashions certain things he has created (though God doesn't need hands and fingers to do this of course). Or, to put it differently, he uses his transcendent power to change creatures, even to change them into something else entirely. This may be embarrassing to some; but Thomas couldn't be clearer on this point. He believed that God made Adam, not *ex nihilo*, but from "the slime of the Earth" (Thomas' words). And God made Eve, according to St. Thomas, not *ex nihilo*, but from Adam's rib, which obviously pre-existed Eve. These actions may not be "creation, properly speaking," but they involve God exercising his creative power in a different but still direct way within the created order.

According to Thomas, God takes a material object and does something with it that it wouldn't do on its own. That's the key point. Call it "making," "crafting," "producing," "quasi-creating," "fashioning," "fiddling," "tinkering," "breaking the rules," or whatever you like. But contrary to Tkacz's assertion, the "Thomistic understanding of divine agency" (assuming that locution refers to Thomas' view of the matter, in conformity with the settled teaching of the Church) includes God creating *ex nihilo* both initially and subsequent to his initial creation of the world, his acting directly in nature—sometimes using pre-existing material—and his acting through secondary causes. And by implication, this would include every permutation of these options.

Must All Organisms Have "Natural" Explanations?

Tkacz also implies that Thomas' view of creation entails that everything in nature must have a "natural" cause:

> Unlike the causes at work within nature, God's act of Creation is a completely non-temporal and non-progressive reality. God does not intervene into nature nor does he adjust or "fix up" natural things. God is the divine reality without which no other reality could exist. Thus, the evidence of nature's ultimate dependency on God as Creator cannot be the absence of a natural causal explanation for some particular natural structure. Our current science may or may not be able to explain any given feature of living organisms, yet there must exist some explanatory cause in nature. The most complex of organisms have a natural explanation, even if it is one that we do not now, or perhaps never will, know.[38]

But is it a dogmatic truth for Catholics and other Christians that for "any given feature of living organisms ... there must exist some explanatory cause in nature"? No, it's not. Does the *Catechism* say that the "absence of a natural causal explanation for some particular natural structure" can't be even one piece of evidence that nature depends on God? No. Did Thomas make this claim? Hardly. Do we know empirically that every aspect of every organism in nature has an explanation within nature? No.

While Tkacz's claim may be a logical possibility (though I'm not sure about that, since the world isn't eternal), it's not a logical truth. And more to the point, there is nothing in St. Thomas' own words that commends it. It contradicts Thomas' views about the origin of the world and it blatantly contradicts settled Catholic doctrine, which holds, for instance, that God directly creates each human soul, which obviously is a feature of those living organisms known as human beings. Besides, there's nothing in Catholic doctrine that prevents God from, say, turning a banana into a bonobo—though we have no reason to think he has done so. In fact, that ability follows straightforwardly from the fact that God is transcendent and omnipotent. Tkacz's sweeping assertion is plainly unwarranted.

Tkacz's assertions look like deductions from naturalism, rigid Aristotelianism, or a hybrid of the two, not like implications of Thomism. Naturalism and orthodox Aristotelianism require that everything in nature have a cause within nature, because there aren't any other possibilities. There isn't anything transcending nature and there is no "beginning" at which a transcendent God created the world—in whatever fashion he chose to do so. Of course Aristotle had a concept of an unmoved mover eternally wedded to the cosmos (itself the subject of much scholarly dispute), which goes farther than modern naturalism; but that entity alone fails to capture the concept of a transcendent God free to create or not create the universe, or to act freely within it. Thomas incorporated Aristotle's idea, but *his understanding of God vastly outstripped it.*

A God who acts within nature has always been an embarrassment to large strands of Greek philosophy. Today, many theologians or Scripture scholars, not to mention average Christians, have sought to avoid the embarrassment of the physicality of Christianity by providing natural explanations for the miracles in the Bible. Whether it was a "miracle of generosity" that allowed five thousand people to appear to be fed by a few loaves and fishes, or merely a psychosomatic effect that allows someone crippled to walk, a "natural" explanation for something apparently miraculous might appear to reconcile our ordinary experience of the regularity of nature with the "spiritual" comfort from faith.

In the cases above, we're dealing with modernist and naturalist philosophy that overrides the claims of the faith. Orthodox Catholics are usually quick to sniff out that sort of thing. In Tkacz's article, however, we seem to be dealing with an overbearing Aristotelianism refracted through modern naturalistic science and then identified with the views of Thomas Aquinas. An orthodox Catholic is less likely to see the problem here, unfortunately, since it appears to be traditional.[39]

"Intervention"

The real question, in any case, is not whether God has to "fix up" or "adjust" natural things (as Tkacz says). Nor is it whether God "intervenes,"[40] which is a pejorative way of putting the point, since it implies that God, if he acts within nature, is breaking in from the outside, and upsetting some natural balance. The real question is whether, according to Thomas and the Catholic tradition, God acts only in the way Tkacz dictates.[41] As we've seen, the answer to that question is a resounding no. God creates the world from nothing and continually upholds it, to be sure, but he also acts directly within it for his purposes. To affirm the latter is not to deny the former.[42]

Catholic theology maintains that God can and does act within nature in a variety of ways, including ways that almost seem designed to scandalize certain philosophical rules of propriety. To imply that God acts only in one way—the way that just happens to cause naturalists the least consternation—misrepresents the Catholic tradition and erects, subtly, stumbling blocks to faith.

The Uncharitable and Misunderstanding

Besides dissenters and accommodationists is another group of critics that are uncharitable. By "uncharitable," I mean that they mistakenly attribute implausible views to ID proponents. The views of these critics derive largely either from an uncharitable reading of the ID literature or from a failure to study it at all. Sometimes, their response is snobbish, treating the debate over Darwinism and intelligent design as a silly little disagreement in a backwater Protestant subculture of no particular interest to Catholics, whose understanding of teleology is so rich and profound that the outcome of the debate over Darwinism isn't especially interesting. These critics may even associate ID with young earth creationism.

Trailing off from the uncharitable critics, however, are many orthodox Catholics not wedded to the spirit of the age, who have no truck with naturalism or Darwinism, and who are not dogmatically invested in a rigid interpretation of one Greek philosopher. These include some of those who are uncharitable but not dogmatically so. They also include many Catholics who feel uneasy about ID. Some imagine that ID is contrary to the Catholic spirit, that ID resuscitates the supposed "deism" of William Paley, or that it's basically a Protestant phenomenon. In the past, ID theorists have often greeted these criticisms with incomprehension. They know the critiques are unfair, if not completely off base. But so far, ID proponents have failed to provide a full response.

In what follows I hope to do so. I believe that, once we introduce the proper corrections and clarifications, we can show why orthodox Catholics should be enthusiastic supporters of ID.

12. Separating the Chaff from the Wheat

Catholics, Evolution, and Intelligent Design, Part 2

Jay W. Richards

The Whipping Boy: Mechanism

Thomist Ed Feser has said: "From an Aristotelian-Thomistic point of view, one of the main problems with 'Intelligent Design' theory is that it presupposes the same mechanistic conception of nature that underlies naturalism."[1] Other critics have said similar things. What's going on here? To answer that question fully, we'll need to do a lot of unpacking.

Orthodox Catholics have long opposed the overreaching of the so-called "mechanical philosophy" that came to prominence in the seventeenth century with René Descartes (1596–1650) and Francis Bacon (1561–1626). In the foreword to an important text by French Catholic philosopher Etienne Gilson (*From Aristotle to Darwin and Back Again: A Journey in Final Causality, Species, and Evolution*), Christoph Cardinal Schönborn calls mechanism "the dominant form of reductionism in science."[2] As critics of the Aristotelian philosophy that had come to dominate the thinking in medieval Europe, Descartes and Bacon banished formal and final causation (which explain, respectively, what something is and its purpose, or the end toward which it tends) from science for leading to dead ends and sterile explanations. Bacon continued to affirm that formal and final causes existed, while Descartes seemed to deny them altogether. In fact, Descartes departed so far from Aristotle's "qualitative" way of describing the natural world that he reduced matter to mere extension. This foreshadowed a tendency in modern science to reduce every material object to mere quantity.

Unfortunately, the word "mechanism," like "evolution" and "creation," has always meant different things to different people.[3] Defined etymologically, a "mechanical" natural philosophy would be one that treats certain natural objects as machines—as various parts or systems arranged to perform a certain function. So, according to Merriam-Webster, a "machine" can be "a constructed thing whether material or immaterial," "an assemblage of parts that transmit forces, motion, and energy one to another in a predetermined manner." Secondarily, it can refer to "a living organism or one of its functional systems."[4]

But mechanism is often summarized more narrowly. Here's how Cardinal Schönborn describes the problem of "mechanism" in his foreword to Gilson's book:

> While no one can question the methodological value of treating natural things as if they were nothing but an agglomeration of simpler parts "all the way down" ... the ontological question remains ... : Is a stable natural whole—whether atom, or molecule, or bio-chemical, or cell, or plant, or animal—truly nothing but an arbitrary combination of "indifferent" parts? In other words, is it not really a *whole* at all, but only a label we give to a relatively stable interaction of parts?[5]

When push comes to shove, the *Catholic Encyclopedia* defines mechanism in the same way. The author, like Schönborn, admits that mechanistic explanations are useful for understanding natural objects, and even allows that emphasizing such explanations was a legitimate reaction to a "decadent scholasticism" that used fruitless appeals to formal and final causation. It's widely agreed that this tendency in Aristotelian philosophy became an obstacle to exploring and understanding the material aspects of nature and needed to be challenged. Mechanism was a chief form of the challenge. Unfortunately, the challenge often came in the form of an over-reaction, and in Catholic intellectual surveys, the legitimate challenge is often defined, confusingly, in terms of the overreaction. So despite affirming the legitimacy of the mechanistic challenge, the *Catholic Encyclopedia* ultimately defines mechanism as the project of reducing wholes to parts, physical objects to mere quantity, and, in the end, objects to mere mathematical abstractions.[6] In other words, it identifies "mechanism" with its most extreme manifestation.

Mechanism and Reductionism

But why is this procedure called "mechanism" rather than simply reductionism, which is surely the more apt term? Either some natural wholes are greater than the sum of their parts, or they're not. If you affirm the first proposition, then you're not a complete reductionist. If you affirm the second, then you are.

Mechanism, on the other hand, involves a cluster of ideas much broader than whole-to-part reductionism. As a result, identifying these two ideas as synonyms is bound to be misleading, and seriously so, for several reasons.

First, machines *are* more than the sum of their parts. *To claim otherwise, ironically, is to be highly reductionist about the reality of a machine.* Some critics of "mechanistic" philosophy deny this obvious point. Ed Feser, for instance, says:

> Take a few bits of metal, work them into various shapes, and attach them to a piece of wood. Voila! A mousetrap. Or so we call it. But objectively, apart from human interests, the object is "nothing but" a collection of wood and metal parts. Its "mousetrappish" character is observer-relative; it is in the minds of the designer and users of the object, and not strictly in the object itself. "Reductionism" with respect to such human artifacts is just common sense. We know that cars, computers, and cakes are objectively "nothing but" the parts that make them up—that their "carlike," "computerlike," or "cakelike" qualities are not really there inherently in the parts, but are observer-relative—precisely because we took the parts and rearranged them to perform a function we want them to perform but which they have no tendency to perform on their own.

Notice where the reductionism lies here. Feser assumes that unless a whole is already present inherently in its parts, then there really is no whole. "[T]he object," we're told, "is 'nothing but' a collection of wood and metal parts."[7] Such reasoning is a universal acid. One could just as well argue that Handel's *Messiah* is nothing but a compilation of musical notes and words to accompany them!

The reasoning is equally absurd when applied to machines. In truth, even the simplest human machines, like a mousetrap or a cotton gin, are greater than the sums of their parts. You can lay out the parts of a mousetrap on a table and they won't do anything useful. They certainly won't reliably trap mice. Indeed, in even the simplest human machine, the parts are taken up, as it were, in service of a *function* imposed on them by an agent. Practically everything interesting about the machine is its arrangement for a function. That function is distinct from the parts, *it is real*, even if when separated from them, it doesn't exist except in the mind of the builder. That's why we issue patents and have intellectual property laws. The function defines the purpose of a machine—its end. Since such machines aren't even reducible to their parts, they certainly aren't reducible to particles, laws, extension, or matter.

This purposive nature of the machine is especially obvious in high technology—such as computing and communications technology—in which the role of intelligence and information predominates over the material substrate in which it is embedded. So the concept of a machine does not imply, let alone entail, reductionism.

Second, whether organisms *simply* are machines, or are *nothing but* machines, is a separate question.

Third, even if parts of organisms are literally machines, it doesn't follow that complete organisms are so, any more than it follows that because Van Gogh's *Starry Night* is made of paint, it's just paint.

Fourth, while organisms are far more than mere machines built by humans, they are surely not less.

Fifth, if mechanism is the belief that natural wholes are reducible to their parts, then not all thinkers frequently identified as "mechanists" deserve the label. Conversely, if mechanism is defined broadly enough to encompass figures as diverse as Descartes, Bacon, Robert Boyle, Galileo Galilei, Johannes Kepler, Gottfried Leibniz, Isaac Newton, and William Paley—all of whom are often called "mechanists"—then the word can't plausibly be identified with pure, whole-to-part reductionism.

Sixth, "mechanism" is often wrongly *contrasted* with teleological explanations, such as final causation, and different thinkers are then lumped together under the label. Edward Feser, for instance, says that "the founders of modern science—Galileo, Descartes, Boyle, Newton, et al...." held "that final causes and the like ought to be eschewed in favor of 'mechanical' (i.e. non-teleological) explanations."[8] But, this is simply not true for Newton, Boyle (as we'll see below) and several other founders of modern science. Besides,

Galileo, Kepler, and others were Renaissance Neo-Platonists, not mere reductionists.[9] They sought formal patterns, and especially mathematical patterns in the physical world, but this has little to do with the idea that wholes are nothing more than the sum of their parts.

What is happening, I think, is that one historical figure—Descartes—is being used to represent the different and even contradictory views of some who came after him, such as Isaac Newton (1643–1727). This has led to a simplistic stereotype of "mechanists" in some Catholic literature that is hard to correct because it is so widespread. This wouldn't matter, except that the stereotype creates blind spots that become obvious in the ID debate. Thinking clearly on these subjects requires more refined categories.

Teleo-Mechanism

Etienne Gilson can help in this regard. What distinguishes his book, *From Aristotle to Darwin and Back Again*, from the pack is that it treats the so-called "mechanists" individually, rather than lumping them together.

Ironically, the more stereotypical view of intellectual history appears in Cardinal Schönborn's *Foreword* to Gilson's book. There, Cardinal Schönborn claims that Newton held a "particles and laws" view of nature: "Others provided much of the tinder for revolution," he says, "but it was Isaac Newton who set the fire blazing." By bringing "into stable alliance the opposing intellectual poles of empiricism and rationalism under a 'particles and laws' model of all physical reality, these two poles were yoked together in a new synthesis expected to reach and explain all of physical reality!"[10]

To establish this point, the Cardinal quotes Newton's stated wish at the beginning of the first edition of his *Principia* (1687): "I wish we could derive the rest of the phenomena of Nature by the same kind of reasoning from mechanical principles." And, from this, Schönborn can then say that Darwinism followed from the ideas Newton set in place:

> Darwinism represents the final triumph of the Newtonian revolution, a "particles and laws" mode of explanation with an added element of chance to provide an entirely mechanistic solution to the problem of biological origins.[11]

Now, Newton certainly wanted to explain as much as possible in mathematical terms, and he studied an aspect of nature—planetary orbits—highly submissive to such explanations. This can lead to a type of reductionism in which everything from physics to psychology gets shoehorned into a mathematical box. But Newton never proposed, as did Descartes, that animals were mere automata. Nor did he ever imply that all wholes are reducible to their parts. Besides, Darwin didn't even propose a mathematical idea: he proposed a design-substitute. Clearly, there are several distinct concepts getting confused here.

Schönborn's explanation, though quite representative of many Catholic summaries of intellectual history, not only fails to capture Newton's views, but also elides the most glaring difference between Newton and Darwin: Newton defended the reality, indeed, the indispensability of real design in explaining the natural world.

Gilson, though, is a careful reader of Newton. He recognizes that while Newton (successfully) extended mathematical explanation of nature farther than anyone who preceded him, nevertheless, in his overall approach, Newton was much closer to Thomas Aquinas than he was to either Bacon or Descartes. "The first great and indisputable triumph of mechanism was the astronomy of Newton," Gilson writes. "However, Newton gave proof of more prudence than Bacon or Descartes in his philosophy of nature."[12] This is an important observation, by an eminent Catholic philosopher. Would that it would trickle down to the textbook summaries! For contrary to stereotype, Newton didn't observe the boundaries supposedly laid down for science by Descartes and Bacon.

In fact, Newton's appeal to a gravitational force—which implied that objects influence other objects instantaneously from a distance—seemed like occultism to those Cartesians whose intuitions were more decidedly materialistic. If you view reality as a bunch of little sticky balls bouncing around in the void, then Newton's idea of gravity—instantaneous action at a distance—seems downright spooky. In his *Opticks* (1704) Newton tried to persuade the Cartesians as best he could by appealing to the idea of ether through which rays of light could propagate. But in doing so, Gilson notes, he

> gave as proof of it something which appears today to be a curious process of scientific reasoning. Speaking of those who denied his theory of a gravitational force, he reproached recent philosophers for banishing "the consideration of such a cause out of natural philosophy, feigning hypotheses for explaining all things mechanically, and referring other causes to metaphysics; whereas the main business of natural philosophy is to argue from phenomena without feigning hypotheses, and to deduce causes from effects, till we come to the very first cause, which certainly is not mechanical."
>
> There follows then in Newton's text a long series of questions that mechanist science leaves without answer, or in view of which, in order to find answers to them, the mechanists invent gratuitous explanations.[13]

According to Gilson, Newton asks questions that would have delighted Aristotle. Indeed, Newton was optimistic about reasoning from effects within the physical world back to a first cause, as Thomas did in his famous "fifth way," the last of his five arguments for the existence of God.

To speak of Newton as a mechanist in the mold of Descartes or, worse, Darwin, then, is surely to obscure his actual view, which is better described as "teleo-mechanist."[14] This is the view also held by William Paley, the English author of *Natural Theology* and of watch-resting-on-a-heath fame. Charles Darwin read Paley as an undergraduate. And it was Paley's view that was to serve as a foil in Darwin's *Origin of Species*.

In truth, the distance between Aristotelian and teleo-mechanist reasoning is not nearly as great as is often maintained. Again, Gilson recognizes this:

> [C]ontrary to what we most often imagine, the substance of finalist reasoning is exactly the same as that of mechanist reasoning. The most attentive mechanists recognize the fact after their fashion, which is, not to deny teleology; but to try to give it mechanist explanations, taking the risk of falling back in the last resort on chance as an explanation of the living organism....[15]

Even among the "most attentive mechanists," however, there are profound differences. Gilson recognizes this reality, though not quite clearly enough. Darwin, in effect, denied *real* teleology. He accepted the thing needing to be explained in biology—a certain type of adaptive, integrated complexity that functions for the benefit of an organism. This, he agreed, looks designed. And he could not talk about such things without talking of functions, which seem to imply purpose (modern Darwinists have the same problem). But he tried to explain these realities, or explain them away, with an impersonal process that lacks foresight—natural selection plus chance variation. Teleo-mechanists like Newton and Paley, whatever the current value of their arguments, defended *real* teleology, as did Thomas Aquinas and to a certain extent Aristotle (though Aristotle's concept of teleology didn't imply ultimate consciousness or intention).

Now the common response to this line of argument goes something like this: "Yes, yes. Newton thought God created and designed the world. He even thought that God tinkered with the orbits of the planets. But then Laplace came along and showed that God wasn't needed to keep the planets in orbit. And then Darwin came along and showed that God wasn't needed to explain the adaptations in living organisms. And all this made it look like God was out of a job."

As the famous story goes, when Napoleon asked Laplace of God's role in his nebular theory of the planets, Laplace replied: "I have no need of that hypothesis." This seems to crystallize the complaint against "mechanism": we have gone from a poorly conceived interventionism to a functional deism or even atheism, all as a result of a trajectory set in motion by mechanists like Newton and perpetuated by Paley. Thus Schönborn refers to "the deistic conception of God and the static conception of living things implicit in Paley's mechanistic 'watchmaker' arguments."[16] Newton is often charged with such deism as well.

Now let's set aside the merits of Newton and Paley's arguments, and consider this charge that their conception of God is deistic. The charge is as common as it is mistaken. Deism is the belief that God sets up everything at the beginning and then lets nature run itself. This is obviously not Newton or Paley's view. Newton at least thought God did all sorts of things in the created order. He once wrote: "[W]here natural causes are at hand God uses them as instruments in his works, but I do not think them sufficient alone for ye Creation."[17]

For Newton, God maintained and interacted with everything in the physical universe, which Newton called the "divine sensorium." Since God is infinite, Newton supposed the universe, too, was infinite. He argued that God's ordering activity was apparent in the orderly orbits of the planets. The "mechanism" of Newton and others like William Paley was not a particles and forces view; it was at the very least a particles, forces, and design view, in which material objects somehow have purposive form imposed on them. That form determines *what they are and what they do*. Natural wholes, on this view, are still greater than the sum of their parts.

Schönborn (along with many others) is apparently using "deism" to refer to the idea that God lets nature run its course except when he occasionally intervenes. That's a problematic view, to be sure. But it's not Newton's view, nor is it an implication of his view. And it's obviously not even deism. It's akin, perhaps, to Olympian polytheism, in which gods act within nature to change the weather and the outcomes of battles, but don't create, transcend or uphold nature itself.

Moreover, to claim—along with Newton, Thomas, Augustine and every orthodox Christian from time immemorial—that God sometimes acts directly in nature to effect certain outcomes, implies nothing about what God is doing at every other time. It does not imply, as is sometimes claimed, that God is merely another member of the universe.[18] God can act directly in nature, while still upholding and transcending it.

Schönborn knows this, surely, since he writes elsewhere about Newton's "vehement critique of deism" in the *General Scholium* in his *Principia*.[19] And the Cardinal also knows, despite his slip-ups in the introduction to Gilson's book, that Newton was no mere mechanist, and that he steadfastly opposed the creative capacities of chance, contrary to Darwin. In *Creation and Evolution: A Conference with Pope Benedict XVI in Castel Gandolfo*, Schönborn writes:

> For [Newton] the symmetry and regularity of the planets' orbits is a phenomenon that cannot be explained by "mere mechanical causes." This "most elegant" system can have arisen only through the counsel and dominion of a supreme intelligence. From natural phenomena we arrive at certainty about the Creator.[20]

This inconsistent treatment of Newton (and those like Paley who shared his view), I would suggest, is the result of using words like "mechanism" and "deism" ambiguously, and acting as if the only options are Aristotle and Descartes (or worse, Aristotle and Darwin). Again, grappling with the current debate over intelligent design requires that we be much more precise.

Of course, when Newton (purportedly) proposed that God tweaked the orbits of the planets from time to time to prevent them from spinning out of control, he was a bit hasty. It's certainly not the best design argument ever devised. In his desire to avoid what he saw as the complete "mechanistic reductionism" of Descartes, he appealed to God's direct activity; but he did so in an area of ignorance. Once the planetary orbits were better understood, and other planets and bodies discovered, the problem Newton tried to solve, disappeared. This was truly an example of an argument from ignorance and God of the gaps reasoning. There's nothing logically wrong with pointing out gaps in certain forms of explanation. If materialist explanations cannot explain, say, altruism, there's nothing whatsoever wrong in pointing that out. Of course it's imprudent to invoke God's action directly if the "gap" is simply a gap of ignorance; but truly bad God of the gaps arguments actually are extremely rare in Judeo-Christian Western history.[21] The God of the gaps accusation is mostly a myth used by atheists to exclude teleological explanations—no matter how reasonable or solidly based on knowledge rather than ignorance. It's troubling to see certain Catholics use the same rhetorical ploy.

Besides, just because Newton mistakenly invoked divine action is no reason to reject the possibility that purpose could be the best explanation for some things in nature, including nature herself. On that general point, the Christian tradition, including especially Thomas Aquinas, agrees with Newton. Newton should not be blamed for the view of Laplace and others, who added another idea, which philosophers refer to as the "causal closure" of the universe. According to this idea, the workings of the physical universe are purely deterministic and not open to "outside" influence. Everything that happens is the necessary outcome of the initial conditions set up at the beginning.[22]

Now, the teleo-mechanistic philosophy of Newton and Paley is not without its dangers. In fact, it could lend itself to deism, but at the level of metaphor rather than logic. If nature is conceived of as a watch, for example, as William Paley suggested in his famous story of a watch resting on a heath, then it might eventually occur to someone that the best kind of watch would be self-winding. And then people would get the impression that God, if he's really powerful, would set the whole thing up at the beginning. And then someone would argue that it would impugn his dignity and power to be tinkering with his creation after he's got it running. And suddenly everyone would have the impression that it would be "best"[23] if God just got everything started correctly and then retired. So you'd go from teleo-mechanism to a highly deterministic deism.[24]

That is an unsavory suggestion; but not one step in this chain of reasoning is entailed by the previous step. It all proceeds by impression and supposition, as a result of the promiscuous use of a very limited metaphor, and a single bit of carelessness by Newton. It's hardly inevitable. So the first take-home lesson is that we should avoid arguments from ignorance. The second is that we should always mind our metaphors.

The third take-home lesson is this: there's no need to wrestle on the horns of a false dilemma: Aristotle and Descartes do not represent the only logically possible views of nature's teleology.

Handmaidens or Slavemasters? The Role of Philosophy in Theology

Traditionally, theology was seen as the "Queen of the Sciences," since it involves the totality and greatest of truths. Philosophy and science, while valuable in their own right, have been understood as "handmaidens" to theology.[25] For theology, philosophy at its best can help the Christian explain and articulate the core truths of the "deposit of faith." This deposit includes the truths described in the Apostles' Creed: that there is a Father, Son, and Holy Spirit, all of whom are one God; that the Son was born of the Virgin Mary, suffered under Pontius Pilate, was crucified, died, was raised from the dead, and so forth. This content of the faith precedes our philosophizing about it.

As important as philosophy can be to theology, however, no single, non-Christian philosophical system is either identical with the Christian faith or can completely capture Christian theology. Indeed, though some philosophies are more congenial to the faith than others, in his encyclical *Fides et Ratio*, Pope John Paul II said: "The Church has no philosophy of her own nor does she canonize any one particular philosophy in preference to others."[26]

Still, it's easy to underestimate Aristotle's influence in Roman Catholicism, due to his influence on the "Angelic Doctor" Thomas Aquinas. The Greek philosopher's contributions are invaluable.[27] Partially for this reason, however, we've sometimes failed to keep critical distance between the pagan philosopher and the faith itself. Traditional Catholics are much more likely to have an Aristotelian blind spot than, say, an Epicurean blind spot.[28]

Probably the most unaccommodating element of Aristotle's thought to Christian theology is the idea that the universe is eternal rather than created as the free act of a transcendent God. *The implications of this belief in an eternal universe permeate not just Aristotle's physics but his metaphysics as well.*[29]

This is one of the reasons that when Aristotle's major writings were first introduced to the Christian West in the thirteenth century, they were not received with universal warmth.[30] St. Bonaventure, Thomas' contemporary in Paris, identified a key danger with Aristotle. A Franciscan friar, Bonaventure preferred the Neo-Platonism of Augustine and the other Church Fathers to Aristotle, though he drew on Aristotle's work and was far more Aristotelian than the Church Fathers themselves.[31] Of course he rejected Aristotle's doctrine of the eternity of the world. But he was also deeply troubled that some Catholic scholars followed Aristotle in rejecting the doctrine of divine ideas or exemplars, which function as a sort of blueprint by which God creates the world.[32] He saw serious consequences in this trajectory, since it implied that God did not know individuals.[33]

These matters might now seem like obscure scholastic disputes of the angels-resting-on-the-head-of-a-pin variety, but they relate directly to the difference between Aristotle's philosophy and Christian theology in general, and also *between the thought of Aristotle and St. Thomas*. And they bear directly on the problems certain contemporary Thomists have with ID.

Nature and Art: Thomas Or Aristotle?

Regrettably, the error Bonaventure identified in the thirteenth century has seen a resurgence (no doubt unwittingly) among some contemporary Thomists. This becomes clear in their treatment of Aristotle's view of so-called "immanent teleology"—yet another term that means different things to different people.[34] In its pure Aristotelian form, for teleology to be immanent means that the purposes or ends of objects and organisms are within the world, and are not the result of an intentional, transcendent agent. They do not reflect a pre-existing blueprint in the mind of a transcendent God.

When speaking of organisms, the basic idea of immanent or intrinsic teleology is that natural objects, unlike human artifacts, are directed toward the perfection of their nature rather than toward ends imposed on them externally. Though both Aristotle and Thomas freely compared organisms with art and human artifacts, neither thought that the ordinary organisms we encounter were literally the same as human artifacts. (The *origin* of organisms, however, is a different matter and one on which Aristotle and Thomas part company—we'll discuss that below.)

First, consider Aristotle. In Book VI of *Metaphysics*, he says: "Art is a principle of movement in something other than the thing moved; nature is a principle in the thing itself." So for Aristotle, there was a qualitative difference between nature and art, especially with respect to the teleology they exhibit.

We can get some sense of Aristotle's point by thinking of organisms, for there are obvious differences between the organisms we encounter and human artifacts.

First of all, organisms, unlike windmills and statues, can reproduce. Setting aside the question of ultimate origin, organisms aren't built part-by-part as human artifacts are. New organisms are born rather than produced or constructed.

Second, unlike a windmill, an organism seems to be internally directed. It is following an end, its own good, according to its own nature rather than being directed externally as is, for example, a tractor, a violin, or an arrow. This is true even of organisms that, so far as we know, are not literally acting for conscious purposes. The self-directedness is present at the earliest stages of development. To put it in contemporary terms, the information needed to build an adult human being is present, in a nascent form, in a tiny human embryo. The process of development, though dependent on its environment, directs itself toward an end determined by its nature. It is not pushed externally like a wagon.

Third, organisms are far more functionally integrated than are human machines. The "parts" of a living organism seem directed toward the end of the entire organism far more completely than the metal and wood parts of a windmill, for instance, are directed toward being aspects of a windmill. Though they don't bring themselves into existence, in a sense, organisms self-assemble during development.

Fourth, organisms (arguably) have natures. We can speak of natural kinds, such as plants and animals, flowers and dogs; but windmills and violins, we suppose, are *artificial*. They exist only because of human intentions.

This all seems correct so far as it goes, and is hardly problematic. But the problems surface quickly when you begin to ask follow up questions. Have organisms always been going on as they are now? Can we extrapolate what we observe in the ordinary course of things, *all the way back*? Although they are, in a sense, internally directed, do organisms, other natural objects, and the universe itself also have an extrinsic purpose? Ultimately, whence comes not just the *purpose* of organisms but also their *form*? Do these reflect the purposes and intentions of an agent? Where did reproducing organisms come from in the first place? Is every aspect of organisms derived from physics or chemistry? Can biological organisms be explained purely in terms of physics and chemistry? Does chemistry have within it the active potential, on its own, to give rise to organisms? In other words, are organisms immanent within lower orders like physics and chemistry? And are fundamentally different organisms immanent within the organisms that now exist?

In short, a generic notion of immanent teleology doesn't resolve much interesting in our discussion, since the Christian, unlike Aristotle, distinguishes the ordinary *generation* of offspring from parent from the original creation and production of life, organisms, and the universe itself. As Catholic philosopher Vincent Torley puts it: "The intrinsic finality we find in all living things cannot account for the *coming-to-be* of the *first* living things."[35]

How is all this relevant to the ID debate? Well, to state the obvious, ID arguments focus on the origin of the natural world, and of various features of nature or organisms, and don't imply (and certainly don't require) that organisms lack natures or that every individual organism is literally an artifact built part by part; only that each is similar in important ways to artifacts as, ultimately, the products of intelligent agency. So on that point, they agree with Thomas, though not with Aristotle.

Since Aristotle viewed the world as eternal, he could hold teleology to be immanent, that is, within the world and individual forms, "all the way back." (He did believe in the spontaneous generation of life, but how that fits with his general view is unclear.) Since he supposed the world had gone along as it does in the present from all eternity he didn't feel the need to resolve the problem of where that teleology came from. He could just extrapolate from ordinary observation of frogs giving rise to frogs.

Of course, he did have a Prime Mover (or a series of movers) at the top story of the universe, but he did not think of the Prime Mover as a fully personal, transcendent, purposeful God (though at times he moved in that direction, and Thomas charitably interpreted Aristotle as a theist). What is clear, despite tough questions over how to interpret Aristotle, is that in his view, the world being moved is every bit as eternal as the Prime Mover. The Prime Mover is a final cause but not the efficient cause of the world.[36] That the world might not have existed, and might have come into existence in the finite past, seemed like an impossibility to him.[37] And the business of a God acting in nature, calling a people to himself, leading them through the desert, revealing his law, calling us by name, becoming incarnate and dying on a cross—all this would be quite bizarre to attribute to his Prime Mover.

For Thomas, the creation and its constituents ultimately fulfill not merely their own purposes, but God's. They have both natural and supernatural ends. This goes beyond anything Aristotle believed, and contradicts Aristotle at key points. Indeed, it strikes at the very foundation of Aristotle's view of natural teleology.

To understand what is at stake, consider the pagan philosophical account of creation known to the Church Fathers. In the *Timaeus*, Plato describes, in anthropomorphic language, a Craftsman or Demiurge arranging the entire cosmic order. Plato thus suggested (he calls it a "likely story") that the universe is the product of intention and intellect (*nous*). In this way, the account has close affinities to the biblical picture.[38]

Of course, from a Christian perspective, Plato's Demiurge is highly deficient. It works with a pre-existent substrate (either space or chaotic matter) and so lacks the sovereignty and transcendence of God. It does not create the universe from nothing, without resistance, as God does. Moreover, the Demiurge is subordinate to the Forms, by which it models the universe. The Demiurge is, ultimately, more a mere fashioner than a creator. Finally, the idea of the Demiurge later came to describe a malevolent being that was subordinate to a higher God (see John West's chapter in this volume for more on this). Still, *Timaeus* shows that, for Plato, the ultimate purpose of the world comes from outside it rather than within it.

Donald Zehl explains the difference between Aristotle and Plato succinctly:

> What is immediately striking ... is the absence from Aristotle's natural philosophy of a purposive, designing causal agent that transcends nature. Aristotelian final causes in the formation of organisms and the structures of the natural world are said to be immanent in nature (i.e., the nature or "form" of the organism or structure) itself: it is not a divine Craftsman but nature itself that is said to act purposively. Such an immanent teleology will not be an option for Plato. Aristotle's teleology is local, not global: while it makes sense to ask Aristotle for a teleological explanation of this or that feature of the natural world, it makes little sense to ask him for a teleological explanation of the world as a whole. Moreover, for Aristotle the development of an individual member of a species is determined by the form it has

> inherited from its (male) parent: the goal of the developing individual is to fully actualize that form. For Plato the primeval chaotic stuff of the universe has no inherent preexisting form that governs some course of natural development toward the achievement of some goal, and so the explanatory cause of its orderliness must be external to any features that stuff may possess.[39]

Aristotle rejected the picture described in *Timaeus,* since he thought it implied both a beginning of the universe and a beginning of time.[40] And he proposed a form of teleology that did not point to a purposive agent. So, despite its obvious deficiencies, the *teleology* (though not the details) that Plato describes in *Timaeus* is closer to the biblical narrative of creation in Genesis 1 than is anything in Aristotle. For Aristotle's "immanent teleology" to be useful for Christian theology, then, it must be significantly modified.

And Thomas did modify it. Like Bonaventure, Augustine, and many other Christian theologians, Thomas held to a doctrine of divine ideas or exemplars.[41] According to Thomas, the form and *telos* of natural kinds, contrary to Aristotle, must ultimately be *extrinsic,* "impressed" on them (his word) by God and reflecting the blueprints, ideas, or "exemplar causes" in the eternal mind of God.[42] In his definitive study, *Aquinas on the Divine Ideas as Exemplar Causes,* Catholic University philosopher Thomas Doolan explains Thomas' view this way:

> [T]he [divine] idea is formal cause of the thing that it exemplifies even though it is not intrinsic to that thing.... Thomas considers the fundamental characteristic of form to be that it is a pattern by which something is the kind of thing that it is. For this reason, he concludes that form is not limited to an intrinsic mode of causality: it can be extrinsic as well. And this is what he considers an exemplar to be, namely, an extrinsic formal cause.[43]

Now this doesn't mean that Thomas thought that God constructed every individual organism separately (though he did think that God directly created each human soul). It is the nature of organisms to reproduce after their kind. But the same reasoning doesn't apply to their *origin,* either ultimately or in time (say, on the fifth "day" of creation).

So Thomas *reworks* Aristotle along Christian (and Neo-Platonic) lines. In a sense he splits the difference between Platonic ideas or forms, which existed independently of individuals, and Aristotelian forms, which did not.[44] As a result, he has to rework Aristotle's category of formal causation.[45]

For Thomas, the forms as exemplar causes exist independently of *our* minds, but not of God's. While natural objects are composites of matter and form, the individual forms in the world reflect the exemplars in the divine mind, which are the pre-eminent formal causes of natural forms. According to the doctrine of divine ideas, the "idea of 'man' in the divine mind is prior to its instantiation in an individual and the common nature accounts for why all things that participate in that nature have the properties that they do."[46]

As Andrew Haines explains, "Thomas believes that individual objects are given their form *both* by an internal and external principle of intelligibility.... [His] doctrine of realism is perhaps one of the most intriguing, since it falls in the middle of two extremes."[47]

It's hard to overestimate the importance of this point in the present discussion. *A misinterpretation of Thomas here is one source of the opposition of some Thomists to ID.* Thomas, simply put, was not strictly an Aristotelian. In fact, in his writings, he cites Augustine and Pseudo-Dionysius (both Neo-Platonic in their orientation) more in total than he cites Aristotle.[48] And his fondness for Boethius—also a Neo-Platonist—is beyond dispute. Even his Aristotelianism came in part from Neo-Platonism, which integrated Plato, Aristotle, and Stoicism.[49] To explain away or dismiss this, that is, to force Aristotelianism on him, is quite simply to do violence to Thomas' views.

Because Thomas' views were a creative synthesis of different philosophical views in light of revelation, he could, despite all the qualifications and the obvious influence of Aristotle, still speak of the natural world as an artifact:

> [A]ll creatures are related to God as art products are to an artist.... Consequently, the whole of nature is like an artifact of the divine artistic mind. But it is not contrary to the essential character of an artist if he should work in a different way on his product, even after he has given it its first form. Neither, then, is it against nature if God does something to natural things in a different way from that to which the course of nature is accustomed.[50]

In other words, Thomas moves beyond Aristotle's categories of "nature" and "art." Pope Benedict recognizes this. God's creation, he argues, falls between nature and artifact in the Aristotelian senses: "If creation cannot be recognized as the metaphysical middle term between nature and artificiality, then the plunge into nothingness is unavoidable."[51] So here, Thomas parts from Aristotle. While Thomas agreed with Aristotle on many things, to speak of the "Aristotelian-Thomistic" view of immanent teleology—as does Ed Feser, for instance—is as confusing as speaking of the "Jeffersonian-Hamiltonian" view of federal power.[52]

Recall that Thomas affirms that God even sometimes uses pre-existing objects to produce living things, for instance, in producing Adam (from the slime of the Earth) and Eve (from Adam's rib). So natural objects such as bats and bees and beluga whales have artifact-like features. In fact, we might speak (though carefully) of the original creatures that God creates or produces as *pre-eminent* artifacts since, unlike mere human artifacts, not only their form but also their matter depends on the power and purposes of God alone. Moreover, unlike crude human artifacts, the teleology of organisms is both intrinsic, in one sense, but ultimately extrinsic, in another sense. Finally, their purposes are anchored in the eternal mind of God, not merely in the minds of finite artificers.

Complementary Handmaidens

I mean none of this to disparage the use of Aristotle for theology. Aristotle is brimming with insights. If I seem to be pushing especially hard against him here, it is because he is a key source of the blind spot that prevents some Thomists from seeing the promise in intelligent design, and in perpetuating the myth that ID is contrary to Catholic belief.[53]

My view is that Neo-Platonism, Aristotelianism, teleo-mechanism, and other philosophical frameworks are useful when held together and used to cast light on the faith, but that all should be held critically. All have advantages and disadvantages. We should never too easily identify a philosophical system with Christianity *per se*. At the same time, we must guard against developing an eclectic and inconsistent hodge podge of a philosophy.

For contrast with Aristotle, let's consider the view of a prominent teleo-mechanist, Robert Boyle.

Robert Boyle (1627–1691), a teleo-mechanist like Newton, objected to Aristotelian philosophy because he thought it tended to make nature autonomous and self-sufficient. Boyle, in contrast, used the atomistic (or corpuscularian[54]) philosophy that was as fashionable in his day as Aristotelianism had been in previous centuries. (Descartes had revived this view from the ancient pagan philosopher Epicurus, who viewed the basic constituents of material reality as tiny, indestructible bits of matter that could be combined into myriad forms. "Atom" literally means "not cuttable.") Epicurus was a materialist who reduced everything to atoms moving blindly in an infinite and eternal void; so his philosophy, taken as a whole, was hardly a Christian-friendly resource. Unlike Epicurus, however, Boyle used atomism to resuscitate teleological explanation—final causation if you will—even though Descartes had sought to banish it from science. "The Epicurean atomist hypothesis—once reviled as the most materialistic and atheistic philosophy of antiquity—was turned into a theory of which God was an essential component."[55]

For Boyle, a virtue of this teleo-mechanist view was precisely that it showed that the physical world depended on an intelligent Creator. Since matter could hardly do anything on its own, Boyle could argue that nature's manifest design could only be the product of a transcendent Creator. The apparent teleology of nature pointed clearly *beyond* nature for its source, as it did in Plato, and was not "immanent" within nature itself, as it was for Aristotle.

Boyle emphasized that physical laws and properties of matter (he preferred to speak of "rules" rather than "laws") are the result of God's will and not his nature, and so had to be discovered rather than deduced from reason.[56] They quite explicitly reflect the purposes of God, and must constantly be upheld by God. "Laws," for Boyle, implied an active and independent Lawgiver.[57]

He could speak of God as constantly involved in the outcome of physical events and of fashioning human beings from a "Lump of Stupid matter."[58] Though not the most felicitous way of speaking, the "stupidity" of matter was not, for him, an insult to God or to the matter he created. God simply chose to create and use a quite limited material to create much more admirable beings.

In hindsight, with our knowledge of chemistry and quantum physics, Boyle's theistic atomism looks like an over-reaction to the Aristotelian tradition. Still, Boyle was right to suspect that Aristotelian philosophy lends itself to a type of naturalism, albeit one with an Unmoved Mover at the top story. Whereas Boyle was inclined to give matter few innate propensities, Aristotle's inclinations were just the opposite. As Cambridge philosopher David Sedley puts it, Aristotle's "project was to retain all the explanatory benefits of [Plato's] creationism without the need to postulate any controlling intelligence."[59] Aristotle is hard to nail down on the question of theism; but Sedley has clearly captured a tendency in Aristotle's thought.

Held together, the contrasting views of nature in Boyle and Aristotle provide a valuable lesson: if we want to know the innate (and God-given) capacities and limits of nature, we need to look at nature itself. That is, we need to discover them. For instance, whether chemistry can give rise to life on its own is at least partly an empirical question, rather than a philosophical or theological one. Unfortunately, neither Boyle nor Aristotle was in a position to give the question much empirical traction. But we know a great deal more about the natural world—about both its capacities and limits. Rather than merely analyzing various ancient and early modern philosophies of nature to answer our questions, then, we should look at nature. We should not hold a philosophy of nature that effectively dictates what God must have done, but one open to the evidence of what God *has* done.

This is a robustly Thomistic attitude. In using Aristotle, Thomas was using what he took to be the most up-to-date and rigorous "science" then available, though he was careful to separate the wheat from the chaff. Certain modern Thomists, however, seem to become seized with the internal logic and systematic grandeur of Aristotelian philosophy, and unwittingly opt for that logic over the complex but intentionally Christian synthesis of St. Thomas. Rather than engaging ID and the empirical evidence on which it is based, they prefer instead to dismiss it for departing from Aristotle.[60]

At the same time, Catholics following closely in the actual theology and *spirit* of Thomas Aquinas—for whom the deposit of faith took precedent over even the greatest pagan philosophers—will be able to consider with an open mind the question: What are nature's capacities and limits? They will also have far fewer problems with ID than do those whose philosophical intuitions are more rigidly Aristotelian.

13. Understanding Intelligent Design

Catholics, Evolution, and Intelligent Design, Part 3

Jay W. Richards

Is ID Teleo-Mechanistic?

The discussion of the previous chapter might suggest that, while contemporary ID arguments may not be "mechanistic" in the stereotypical sense, perhaps they are teleo-mechanistic. Is ID inherently tied to the teleo-mechanist view of Newton, Boyle, and Paley?

There are certainly similarities between ID and teleo-mechanist arguments from figures like Newton and Paley; but as we've seen, there are also similarities between Newton and St. Thomas. In truth, it's impossible to pigeonhole ID theorists so neatly. ID, after all, is a big tent intellectual program. Not all ID proponents share an identical philosophy of nature or the same theological views.

Moreover, much of the technical work by ID theorists purposefully avoids proposing a full-blown philosophy of nature. William Dembski's analysis of the "design inference," for instance, is an attempt to rationally reconstruct the elements by which we infer design in everyday experience.[1] His reconstruction is intended to provide statistical and scientific precision in analyzing a certain group of systems in microbiology. As he has stated many times, it is a complete misunderstanding to see this statistical analysis as supplying or implying a philosophy of nature. And yet certain critics continue to perpetuate just this misunderstanding.

In *The Design Revolution*, written in 2004, Dembski notes that

> in focusing on the machinelike features of organisms, intelligent design is not advocating a mechanistic conception of life. To attribute such a conception of life to intelligent design is to commit the fallacy of composition. Just because a house is made of bricks doesn't mean that the house itself is a brick. Likewise, just because certain biologi-

> cal structures can properly be described as machines doesn't mean that an organism that includes those structures is a machine. Intelligent design focuses on the machinelike aspects of life because those aspects are scientifically tractable and are precisely the ones that the opponents of design purport to explain by physical mechanisms.[2]

Moreover, in his other theological and philosophical writings, Dembski has complained about the limits of mechanistic philosophy. (This attitude is shared by other philosophically inclined ID proponents, most of whom think that modern philosophies of nature need to be radically rethought.) For instance, Dembski has referred to Paley's teleo-mechanical philosophy as a "kludge of divine interventionism and material philosophy." He goes on to say:

> I am a much bigger fan of the Church Fathers than I am of William Paley. I like Paley and think he has a lot of good insights. But I think the watch metaphor was in many ways unfortunate. It is faulty, because the world is not like a watch.
>
> The Church Fathers did not use the watch. Instead, they spoke about a musical instrument. Gregory of Nazianzus ... makes a design argument which is virtually parallel to William Paley's, except in place of a watch he has a lute. ... I think this is a much better metaphor than a watch. As Christians, we believe that God is not an absentee landlord. God creates the world but then he also interacts with it.
>
> The watch metaphor is the type of metaphor that we get from a mechanical philosophy.... But with a lute ... you need a lute player; otherwise it is just sitting there. In fact, it is incomplete without the lute player.[3]

Dembski is not a Catholic but it is very hard to see how his point contradicts Catholic theology.

ID as a Tertium Quid

Of course, like Aristotle and St. Thomas, design theorists often compare biological structures to human technology. Biochemist Michael Behe, for example, has compared the "irreducibly complex" properties of the bacterial flagellum to a mousetrap. But only the most unfair critic would take him to mean that God literally puts every bacterial flagellum together piece-by-piece, or that a bacterial flagellum is nothing more sophisticated than an isolated mousetrap. And even if Behe did claim that, it still wouldn't be a reductionist argument—that the whole isn't greater than the sum of its parts. Quite the opposite. All the parts could exist simultaneously without producing the "machine." So the arrangement of the parts constitutes something in addition to the parts themselves. (Note also that Behe is focusing on sub-systems of organisms, not whole organisms.)

Similarly, ID theorists enthusiastically compare biological systems with information and communications technology. In his important book *Signature in the Cell*, Stephen Meyer speaks of the coding regions of DNA as "software" and "blueprints," and refers to cells as "information-processing systems." Such software contains information, not just information in the sense of complexity or information capacity, but specified, sequenced, coded information, like human language.

Computer technology differs from earlier, "mechanical" technology. Our very way of speaking expresses the difference. In describing computer technology, we combine both mechanical and linguistic analogies, as the word "information" suggests. Meyer and other design theorists like William Dembski often appeal to language, written and spoken, in describing the design of the natural world. In his theological reflections, Dembski draws heavily on the Bible's way of describing God's interaction with the world—through the spoken and written word. God *speaks* the world into existence in Genesis 1. And the Apostle John, echoing the opening chapter of Genesis, refers to the Son as the Logos, the "Word," of God. It is through this Word that "all things are made." Rather than describing God as an artificer who "moves particles around," Dembski describes God as speaking or "imparting information" to the world. (Of course, he still recognizes that if God wants to move particles around, he can do that too.) This way of speaking has connections not only in biology, but in quantum physics as well, where information increasingly is seen as fundamental to understanding the nature of the physical world.[4]

I have spoken of nature as a book (echoing Galileo, Kepler and others), with a text that can be "read through" to the intentions of its Author. And in *The Privileged Planet*, Guillermo Gonzalez and I argue that the eerie correlation between the requirements for complex life and for scientific discovery suggests that nature is meant to be "read"—that it is designed not just for life, but also for discovery.

Earlier design arguments tended to work with only two categories: order and chance/chaos. As a result, Aristotle and St. Thomas Aquinas could speak of a rock falling to Earth and an insect moving toward its food as examples of the same phenomenon—end-directed order. But the simple, repetitive order evident in, say, physics, is quite different from the integrated complexity of, say, written text and the coding regions of DNA. Such complexity is similar to randomness since it is not repetitive. (See Stephen Meyer's chapter in this volume for more details.) But we still recognize that it is qualitatively different from mere randomness or chaos, and cries out for a different type of explanation.

Moreover, a great deal of the evidence to which ID theorists appeal is of very recent vintage—revealed with the sensual prostheses of telescope and microscope. No one knew about the cosmic microwave background radiation or the bacterial flagellum, or the fine-tuning of the fundamental forces until well into the twentieth century. So ID arguments appear quite modern compared to the broad, metaphysical arguments contained in St. Thomas' *Summa contra Gentiles*.

At the same time, contemporary ID arguments represent a return to a very traditional and biblical way of speaking. Linguistic metaphors have *closer* affinities to the Bible's way of describing God's relation to the world than to either the Aristotelian or teleo-mechanist ways of speaking. Rather than high tech teleo-mechanistic philosophy, ID draws on elements from both the Aristotelian and teleo-mechanist traditions, but doesn't fit wholly in either camp. It is in many ways a *tertium quid*, a third possibility.

Once we realize this, it becomes obvious that, contrary to the suppositions of some Catholic critics, ID is highly anti-reductionist. Unlike materialists, who seek to ground everything in matter or physical laws, ID theorists argue that such laws point beyond themselves to a purposive agent. And unlike Neo-Darwinians, who tend to have a simplistic "beads-on-a-string" view of DNA, and to reduce life to the coding regions of DNA, Stephen Meyer and other design theorists argue emphatically for the priority of the whole to the parts: it is only in the context of the entire cellular system, for instance, that the coding regions of DNA code for proteins. To speak in medieval scholastic terms, the nucleotide pairs that make up the "letters" in DNA have a passive potentiality to be arranged specifically, so that they can code for proteins. But they don't have the power by themselves to perform that function or to specify the order in which they are arranged. In fact, if they did, they would be poor carriers of information.

The cell, like the organism, contains a nested hierarchy of orders, in which the lower orders are taken up by the higher orders. We know empirically that organisms are far more than mere matter in motion. We have every reason to doubt that they can be reduced to simple particles and repetitive laws. They are constrained by rich and purposive "in-formation" that includes not just the one-dimensional instructions in DNA (in both coding and non-coding regions) but also the three dimensional architecture of the cell as a whole. And in multicellular organisms, the cell itself is but one part of a larger, exquisitely integrated, three-dimensional architecture that is irreducible to the lower levels. Most design theorists are convinced that we are just beginning to scratch the surface of understanding the irreducibly informational nature of organisms, and that we can only begin to grasp that nature when we think in a top-down rather than bottom-up fashion.[5]

While the refined concept of information is not drawn from Aristotelian philosophy, it's no coincidence that the word "information" contains the word "form." In fact, I suspect that the rich concept of information common in the ID literature captures some of the intuition that inspired Aristotle's concept of "formal" cause. (It also distinguishes ID from teleo-mechanists like Paley, who had a sense of a final causation, but not formal causation.) Information may be "in" matter, but it is decidedly not matter, energy, particles, or laws. Information orients and transcends all these things, as written text transcends ink and paper, and as form transcends matter. The concept of information is essential for understanding the biological world. And its presence in nature bespeaks purpose, that is, a final cause. That means that modern biological discoveries have already pushed science beyond the default position of particles-and-laws reductionism (as the eerie discoveries of quantum physics pushed physicists beyond the vision of atoms as little balls in the void). Contemporary ID arguments follow that trend and extend it.

Interestingly, in its discussion of evolution, the International Theological Commission's Vatican Statement on Creation and Evolution, when headed by Joseph Ratzinger, who would become Pope Benedict XVI, drew just this parallel:

> Modern physics has demonstrated that matter in its most elementary particles is purely potential and possesses no tendency toward organization. But the level of organization in the universe, which contains highly organized forms of living and non-living entities, implies the presence of some "information." This line of reasoning suggests a partial analogy between the Aristotelian concept of substantial form and the modern scientific notion of "information." Thus, for example, the DNA of the chromosomes contains the information necessary for matter to be organized according to what is typical of a certain species or individual. Analogically, the substantial form provides to prime matter the information it needs to be organized in a particular way. This analogy should be taken with due caution because metaphysical and spiritual concepts cannot be simply compared with material, biological data.[6]

Read one sentence again: "This line of reasoning suggests a partial analogy between the Aristotelian concept of substantial form and the modern scientific notion of 'information.'" In my view, that is exactly right. The authors of the document clearly have in mind the integrated, specified information found in DNA and the three-dimensional architecture of organisms, and not so-called "Shannon" information, a concept that does not discriminate between random noise and specified information. This is a point at which Catholicism could benefit from the work of ID theorists, and vice versa.

Unfortunately, on these matters, some Thomists, even those who ought to sympathize with ID, frequently misconstrue ID arguments. Edward Feser, for instance, in his excellent introduction to St. Thomas, shows skepticism of Darwinist "attempts to discard final causality and explain biological phenomena entirely in terms of efficient causality." And he reflects on the inadequacy of materialist explanations of DNA and the genetic code, that

> seem teleological through and through. Descriptions of this famous molecule make constant reference to the "information," "data," "instructions," "blueprint," "software," "programming," and so on contained within it."

He then speaks of DNA as "directed toward" a sort of "goal" or "end."[7] The ID theorist might sense an ally, but Feser quickly clarifies:

> It is important to note that this has nothing whatsoever to do with the "irreducible complexity" that "Intelligent Design" theorists claim certain biological phenomena exhibit; the Aristotelian need not take sides in the debate between Darwinian biologists and "Intelligent Design" theorists (who generally accept the mechanistic view of nature endorsed by their materialist opponents). Final causality is evident in DNA not because of how complex it is, but because of what it does, and would be equally evident however simple in physical structure DNA might have been.[8]

Really? These ideas have "nothing whatsoever to do" with one other? And Aristotelianism is neutral in the debate between anti-teleological Darwinism and teleological ID? If so, then something is seriously wrong with Feser's brand of Aristotelianism (note that though Feser is a Thomist, he speaks of the "Aristotelian," which is not the same thing).

From our discussion above, it should be clear why this critique misfires. Behe's argument is not based on the idea of "how" complex a bacterial flagellum is. In his arguments concerning systems like the bacterial flagellum, Behe points both to the *type* of complexity exhibited—which is out of reach to Darwin's "numerous, successive slight modifications"—and to the fact that the system has a function, a discernible purpose. It is the combination of these observations—inaccessibility to the Darwinian mechanism and the presence of a purposive function—that justifies the design inference.

Function, in Behe's argument, is a type of what William Dembski calls a "specification." Dembski's insight, widely used by ID theorists, is that it is *specified* complexity rather than mere complexity that reliably indicates an intelligent cause (it doesn't follow that specified complexity is the only such indicator). While in some cases complexity can be analyzed and measured mathematically, specifications—meaningful, independent patterns—are qualitative rather than quantitative.

Steve Meyer's argument, which focuses mainly on the coding regions of DNA and the information processing at the cellular level, proceeds along similar lines. We observe in organisms—and in their organs—purposiveness, function, adaptation, and the like. Everyone, including the most dedicated Darwinists, recognizes this fact. Now, though we don't nearly understand all the details, we know that organisms use proteins to build organs. And we know that proteins are constructed from sequential strings of amino acid chains built and folded inside cells, and the information for their sequencing (and much more besides) is coded linearly in DNA. This process is extraordinarily difficult to explain briefly, but when we get to the "coding regions" of DNA, we can then isolate information into the one dimensional-vertical axis of the DNA molecule, just as we can isolate the information in written English by following the text from the left to right across a page. So far, it is only at this *very* low level, focused very narrowly, that we can quantify the "complexity." But it is the fact that that high complexity contributes to an end-directed *function* at higher levels that allows us to discern a specification. Otherwise, for all we would know, the sequence of base pairs, like the random letters from a Scrabble game, might just be a jumble.

So again, for the most common ID arguments in biology, the quantitative and qualitative—complexity and a special kind of end-directed pattern—*together* allow us to discern purpose and intelligence, if not in the organism itself, then in its ultimate origin. It's not mere "complexity" alone.

This ID insight has obvious affinities with the teleological view of St. Thomas Aquinas, despite the fact that he lacked the category of "specified complexity." In any case, it should be clear that it is erroneous to treat ID arguments as if they were merely a rerun of the teleo-mechanist, let alone the reductionist arguments of previous centuries.

Translation Error: Chance, Law, and Design

Is ID a Zero-Sum Game?

Catholic physicist Stephen Barr is a frequent critic of ID. In an article in *First Things*, for instance, he says:

> The ID movement's version [of the design argument] is hostage to every advance in biological science. Science must fail for ID to succeed. In the famous "explanatory filter" of William A. Dembski, one finds "design" by eliminating "law" and "chance" as explanations. This, in effect, makes it a zero-sum game between God and nature. What nature does and science can explain is crossed off the list, and what remains is the evidence for God. This conception of design plays right into the hands of atheists, whose caricature of religion has always been that it is a substitute for the scientific understanding of nature.[9]

Now the way some ID proponents write can lead to misunderstandings, if read with a jaundiced eye. When discussing aspects of biology, ID theorists frequently contrast the role of natural laws like gravity with the role of intelligent design. They may speak of natural selection as a "mindless" or "brute" or even "purposeless" process, and contrast such processes with intelligent design. For instance, in their lucid introduction to ID, William Dembski and Jonathan Witt ask: "Are things in nature the product of mindless forces alone, or did creative reason play a role?"[10]

When interpreted theologically, it can *sound* (contrary to the intentions of the author) as if ID implies that God only acts apart from these natural forces, that he is merely an artificer who rearranges pre-existing material under the control of mindless forces. God's action, then, would be set up against nature, and left to fill the "gaps" that nature leaves empty.

These ideas would certainly create troublesome zero-sum games, as Barr charges, if anyone actually advanced them; but no theistic ID proponent ever has. Remember the distinctions from the introduction to this volume. No theist worth his salt believes that God is aloof from the world except when he acts directly in nature. For theists, God transcends the world, is free to act directly in it, and always remains intimately involved with it. God can act both primarily and directly as well as through so-called "secondary causes." These include natural processes and laws that he has established, such as the electromagnetic force.[11] An event might be an expression of both a physical law and the purposes of God. It's not as if atheists appeal to gravity while theists appeal to miracles. Gravity is as consistent with theism as are miracles. It's just that most theists and atheists agree on gravity but not on miracles.

In most contexts, though, ID arguments are meant to be metaphysically minimal. That is, they are modest attempts to introduce purpose, intelligent agency, and the like as a legitimate way of explaining at least some things in nature, to the open-minded skeptic, with a minimum of metaphysical baggage. ID theorists don't want to smuggle theism, for instance, into their design arguments (even if those arguments have theistic implications). To argue that design is at least sometimes detectible or detected in nature, however, doesn't imply that *only* that part of nature is designed.

There's also a reason that ID theorists sometimes speak of the Darwinian "mechanism" (but not ordinary physical laws) as a "mindless process." As we noted earlier, the Darwinian "mechanism" is qualitatively different from ordinary physical mechanisms like gravity. You can express this law in mathematical form. Newton's law is $F = \mathbf{G}m_1m_2/r^2$. If you know it and a few other things, you can predict where Jupiter is going to be in the sky on July 12, 2035. You can't use it to describe everything, of course, but the law explains a certain set of facts, and very reliably.

The Darwinian selection-variation "mechanism" isn't like that. Darwin simply proposed a designer-substitute by extrapolating from a somewhat trivial, known set of facts. In the twentieth century, this variation came to be identified with genetic mutations. And yet, despite some trivial examples, the power of this "mechanism" to explain as much as Darwin intended has not been demonstrated. If anything, the evidence suggests that its powers are quite limited. We don't even have reason to believe that natural selection and *directed* genetic mutations can do much creative work.[12]

Of course, while Darwin proposed variation and natural selection as a "mindless" substitute for design, it doesn't follow that these processes could not be features of the world God has created. To some degree, they obviously are. We have several good though modest examples of natural selection preserving survival-enhancing variations. Still, we need to recognize Darwin's mechanism for what it is, and for what is was intended to be. We need to take a hard look at the evidence and not just assume it can really fulfill Darwin's aspirations. Moreover, if God is in charge, then even where natural selection is working, no variations will be literally random in the sense that most Darwinists understand the word. They won't be purposeless.

When an ID theorist is also a theist (as is often though not always the case), these distinctions are always in the background, even if they don't show up in every argument. That's because ID arguments often focus on isolated, empirical evidence of design in nature—that is, with "design" insofar as it is detected and tractable. The theological implications, such as there are, can be treated separately.

Consider Behe again. When he is discussing the bacterial flagellum, he is evaluating the powers and limits of regular, repetitive physical laws (or, as I would say, of matter insofar as it acts in the God-given, normative ways that we refer to as physical laws), and of the Darwinian "mechanism." He concludes that these processes, which are not intelligent agents, probably don't have the power, by themselves, to produce the bacterial flagellum. That's because the locomotive function of the flagellum is inaccessible to the cumulative power of natural selection acting on random mutations. It is, as Behe says, "irreducibly complex." It needs many separate parts working together before it gets the survival-benefitting function. That's the negative part of his argument.

To get a working flagellum, Behe argues, you almost surely need foresight—the exclusive jurisdiction of intelligent agents. That's the positive part of his argument. An agent can produce a system for a future purpose, for an end. It's the obvious purpose of the flagellum, along with the fact that it is almost surely inaccessible to Darwinian selection—not merely the fact that it's really complicated—that justifies his conclusion that the bacterial flagellum is better explained by intelligent design than by repetitive natural laws or the Darwinian mechanism alone.

But it's a misunderstanding to construe Behe's arguments as complete descriptions of what *God* is doing. Behe is talking about isolated, miniscule bits of one organism in a subfield of biology, in which the limits of mutation and natural selection can be discerned. In the case of the bacterial flagellum, intelligent design goes beyond what known, repetitive, natural processes, as well as selection and mutation would do if left to their ordinary capacities. So we invoke intelligent design rather than repetitive physical processes alone *in this case*. But this isn't how every design argument, focusing on every feature of nature, works.

Ordinarily, when a scientist invokes a physical law, he intends to appeal to some repetitive feature of the physical world. A ball falls to the ground when dropped from the Leaning Tower of Pisa "because" of gravity. So a scientist can say that gravity "causes" the ball to fall (once dropped). Since it's constant—it always does the same, mathematically describable thing—and isn't an intelligent agent, gravity is seen as an impersonal property of matter. The very fact that it's so general suggests that it is great for explaining some general truths about nature, but not nearly everything.

The ultimate origin of such laws is a separate question, so the scientist need not be intending to exclude God's role in some broader sense, or that God is so excluded whatever the scientist intends. Similarly, Behe's argument does not imply that nature is a self-contained entity going on its merry way except when God decides to jump in to build a bacterial flagellum. Nor does it imply that natural laws or so-called impersonal processes are outside God's purposes or control, or can explain themselves. He's simply talking about one way of *detecting* design in tiny domains of biology that we understand well, and treating impersonal constants and mechanisms, such as gravity and the Darwinian mechanism, as givens.[13]

ID Arguments Are Different

Stephen Barr, like Ed Feser earlier, offers another representative complaint, which is that ID is quite different from the design argument made by the Church Fathers and scholastics: "The emphasis in early Christian writings was not on complexity, irreducible or otherwise, but on the beauty, order, lawfulness, and harmony found in the world that God had made."[14]

The obvious response is, So what? Neither Behe's nor any other ID argument implies that there are no other good design arguments. If you're thinking in broadly Aristotelian terms, for instance, then Thomas' argument based on a sophisticated use of Aristotle's ideas on causation is just fine. And if you're thinking in terms of physical laws and constants, then the "fine tuning" of those constants is, I would argue, evidence for intelligent design.

There's no reason to concede the laws of physics to materialists, for whom the very idea of "laws" ought to be perplexing.

It's just that we might only notice the "designedness" of physical laws or constants when we attend to them directly. In biology, physical laws are often part of the background. Think of it this way. Imagine an expedition of future astronauts is exploring a distant planet. They find something that looks like a stack of thin tablets, which look like they contain written text. The language is unique; no one has ever seen it before; and the "book" is quite exotic. So it's given to a crack team of exoplanet cryptographers, who eventually determine that it is in a fact a book—a cookbook. Now they know that the text was written by intelligent beings. They would not have been able to determine that without focusing on the text, and treating the tablets, binding, and cover as background contrasts to the text—like pages.

But once they decrypt the text, they could then turn their attention to the book, that is, the tablets and cover. Noting their unusual chemical composition, artful design, and user-friendly layout, they would then realize that the book itself, and not just the text, is the product of intelligence.

In the same way, one can make a design argument based on some narrow feature in the biological world without denying that the background media, considered on their own, are also evidence of design. One can focus on parts without denying the whole. ID theorists do that all the time, pointing to evidence from physics, cosmology, and astronomy, *and* appealing to beauty and rationality in nature, as Benjamin Wiker and Jonathan Witt have done in *A Meaningful World*.[15]

But like the case of the book and text, even if fine-tuning in physics is evidence of design, it doesn't follow that the fine-tuning of, say, gravity and electromagnetism can give rise to reproducing cells or other things we discover in nature. One can argue that one set of conditions, which is evidence of design, is necessary for life, but not sufficient for it, and that another set of conditions that goes beyond the first set is also evidence for design.

In fact, we may detect design at the cellular level in part by *contrasting* it with the background regularities of physics. Such evidence would be an example of what philosopher Del Ratzsch calls "counterflow":[16] it exhibits features we have good reason to doubt can be produced by ordinary, repetitive physical processes, but that could be produced by an intelligent agent. Pointing to evidence of counterflow is no insult to physics or to God, anymore than saying that gravity alone couldn't produce Shakespeare's sonnets is an insult to gravity—though it might be an insult to Shakespeare to say otherwise. The properties of nature that we refer to as physical constants are highly abstract, isolated regularities. They explain some things, not everything. And we have no reason to assume that the Darwinian "mechanism" can explain everything else. We should determine the explanatory power and limits of all these means by empirical observation.

As we saw previously, the Church already infallibly has declared at least one place in nature where we should expect evidence of counterflow: the origin of human beings. God immediately creates human souls (at least), and these cannot be reduced to pre-existing material constituents or mechanisms.[17] Human beings are body/soul unities, not merely souls inhabiting bodies, so our souls must have effects in the physical world. That means there can be no Catholic objection, *in principle*, to the existence of counterflow within the natural world. At the very least, we should be open to discovering it elsewhere.

Of course, scientific study is not the only way to experience or gain knowledge of nature. And this leads us to the next issue.

Why Claim that ID is Science?

Though science is not the only way to learn about the world, many ID proponents have argued that ID arguments should be considered within the purview of natural science. I generally find these turf discussions, where we try to demarcate philosophy, science, theology, and so forth, tedious and fruitless. It seems absurd to expect that nature and reality should conform neatly to the departments of a modern state university. But the discussion is necessary because many people, including some Catholic intellectuals, misunderstand why many ID proponents claim that ID is science. They accuse ID theorists of accepting the same "scientistic" assumptions they otherwise oppose, by implying that only scientific explanations are legitimate. Moreover, and sometimes contradictorily, they insist that ID is at best philosophy or "religion,"[18] not science.[19]

Methodological Naturalism

In objecting to ID, some Catholic critics invoke "methodological naturalism," the idea that science by its very nature must be limited to naturalistic, that is, non-intelligent, explanations like chance and natural law.

So Catholic philosopher Ernan McMullin says:

> But, of course, methodological naturalism does not restrict our study of nature; it just lays down which sort of study qualifies as scientific. If someone wants to pursue another approach to nature—and there are many others—the methodological naturalist has no reason to object. Scientists have to proceed in this way; the methodology of natural science gives no purchase on the claim that a particular event or type of event is to be explained by invoking God's creative action directly.[20]

A great deal has been written responding this type of objection, which I don't think survives scrutiny.[21] But in any case, why would orthodox Catholics accept this limit on scientific inquiry in the first place?

To make things really complicated, this objection is often framed (as we will see) as if it were based on traditional Catholic categories. This allows the critic to pit Catholic tradition against ID. The result is quite bizarre. A certain kind of Catholic critic will accuse ID theorists of limiting God's action to a few discrete interventions. Then they argue, notwithstanding, that ID is bad science because it invokes divine causality and does not accept methodological naturalism.

There are several problems and confusions here. Let's consider each in turn.

A Hybrid of Traditional and Modern

Thomas and other scholastics bequeathed to Roman Catholics a positive view of reason that encompasses our knowledge of nature, morality, God, and other persons. We can have scientific, metaphysical, and moral knowledge. This is much broader than the modern view, which often limits our reason to knowledge drawn from natural science. The traditional view is often related to Aristotle's four causes. But some modern Catholics unwittingly impose Francis Bacon's anti-Aristotelian distinctions between science and philosophy onto Aristotle's four causes, producing a hybrid view that is part traditional/part modern. So they argue along these lines: Philosophy and theology differ from science because they focus on different causes, whether material, efficient, formal, or final. As one encyclopedist explains it, "these causes clarify the relations between philosophy, faith, and theology."[22]

Science, so the argument goes, deals only with efficient and material causes, but ignores other causes. That's okay, as long as the intentionally limited nature of the scientific enterprise is kept in mind. This way of framing science becomes a problem only when science becomes arrogant or forgetful, that is, when it either tries to explain everything in terms of efficient and material ("mechanistic") causes, or when it denies the reality of anything outside its domain—such as free will or morality. When let loose to thus ravage the countryside, it becomes *scientism*.[23]

There's much to be said for this way of thinking. For instance, it offers the various disciplines some autonomy. It's possible to do chemistry, for instance, without adjudicating the dispute between Aristotelians, Platonists and nominalists on the nature of universals, or passing judgment on the *filioque* clause in the Nicene Creed. That means that people of different philosophical and religious bents can practice science together. It's a public enterprise based on public evidence that doesn't depend on special revelation or philosophical conformity.

Moreover, this approach is non-reductionist. It doesn't try to make all of reality conform to the microscope, the telescope, or the calculus. It reminds us that natural science, however it's defined, does not exhaust either reality or human reason, or even our knowledge of nature.

The classical view also reminds us that our knowledge drawn from science depends on more basic, philosophical knowledge of truth, logic, and so forth. "Metaphysics" means, quite literally, "beyond physics." The traditional category of "natural philosophy" captured the breadth of reality and reason quite nicely.

So the classical view dignifies reason in its fullness. Quite contrary to being anti-science, it defends reason against the skeptics and post-modernists that haunt not just the halls of the humanities, but of current philosophy of science as well.

Now, if you're thinking both classically but also accepting modernist boundaries for and definitions of science and philosophy, it would make sense to say that intelligent design arguments are philosophical rather than scientific, since they seem to be appealing to something like formal and final causes. They often speak of "information," for instance, and appeal to purposes and ends. ID arguments could be rational, compelling, true, essential for completely explaining nature, and even more certain than anything derived from science. But they would still be philosophy, not science.

No Thick, Black Border: Problems with Bright-line Distinctions between Science and Philosophy

The Rhetorical Setting

Despite its value, however, there's a serious rhetorical problem in fusing a classical understanding of reason with a Baconian demarcation of the disciplines. The assumption that reason is broader than science is easily lost on most of our contemporaries, who know little about Aristotle's four causes and have otherwise not thought much about the subject. Nowadays, science is often taken as the only valid form of objective, public knowledge. Philosophy is commonly regarded as something that people do when they have too much time on their hands, since philosophers never resolve anything.

This way of speaking of "science" and "philosophy" would have been alien to Thomas and other scholastics. For them, these words had different meanings than they do to most moderns. In particular, Thomas didn't limit "science" to the quantitative of study of nature. On the contrary, he applied the word, as Thomist Edward Feser correctly explains, to "an organized body of knowledge of both the facts about some area of study and of their causes or explanations." So, for Thomas, metaphysics, ethics, and theology were sciences as well.[24] He wouldn't have said ID is philosophy *rather than* science, though he might have asked in which *science* it properly belonged.

Whether we like it or not, however, when natural science (understood along Baconian lines) retained the word "science" (from *scientia*, which means knowledge) while the other disciplines lost it, everything other than natural science was demoted. "Natural science" is now assumed to be the only source of public knowledge. Everything else is seen as speculation or opinion.

Given this context, to say that an argument is philosophy and not science, will be understood to mean that the argument might be an intellectual curiosity, but it is basically a private opinion. (Theology fares even worse than philosophy in this regard, since it is presumably based on revelation rather than reason.) That's unjust and confused, but it's our modern rhetorical reality. To pretend otherwise is to court injustice and confusion.

You can recognize this rhetorical reality without conceding the mistaken assumptions behind it. That is precisely the position of ID theorists. One can think an argument is scientific without thinking that natural science is the only source of knowledge.

An Uneasy Hybrid

Besides the rhetorical problem, there's also a philosophical problem. As we saw above, some traditional Catholics use Aristotle's categories but unwittingly combine them with a sharp distinction between philosophy and natural science. But that black line is of very recent vintage. Even many nineteenth century figures that we would call "scientists" would have seen themselves as "natural philosophers." Ironically, we owe the convention of calling natural science "science" to a man sometimes called "the last great natural philosopher": William Whewell.[25]

So when most English speakers use the word "science," they're using a word coined not in the Middle Ages, but in the mid-nineteenth century. Whewell's sense of the word "science," though now practically universal, is *not* based on a traditional Catholic understanding of reason. And yet, many such Catholics accept his definition, along with the boundaries laid down by their nemeses Bacon and Descartes. Thus, they assign efficient and material causes to natural science and formal and final causes to philosophy and theology.

For Aristotle and Thomas, however, this rigid segregation of the four causes into separate disciplines would have made *no sense* whatsoever. In fact, it would have seemed irrational. For them, the formal cause is prior to the material cause, and the final cause is prior to the efficient. As a result, the material cause is *unintelligible* without the formal, and the efficient cause is unintelligible without the final. The final cause is the "cause of all causes," and so gives sense to the other three.[26] Why would they recommend that natural scientists apply a method that makes the object under study unintelligible?

In truth, the modern scientific concept of cause (insofar as there is one) is not really the same as Aristotle's concepts of efficient and material causes. The four causes derive from a metaphysical apparatus alien to the modern scientific mental environment. The definitions of the four causes are interdependent. As a result, it's very hard to translate them individually into quite different mental environments. So it's misleading at best to say that modern science simply and self-consciously limits itself to "efficient" and "material" causes.

Catholic Compartmentalism

Unfortunately, this way of delineating science and philosophy can become a rigid form of Catholic compartmentalism that fails to capture modern science in its diversity in favor of an "ideal" science that is not practiced in the real world. The goal may be to immunize philosophy or metaphysics from the corrosive effects of scientism, or to preserve the autonomy of natural science; but the effect is to concede the realm of public debate and evidence to materialists. It is a potentially fatal, and unnecessary capitulation to modernism.

One popular claim is that science and philosophy, or science and religion, are two complementary but non-overlapping ways of describing the world. While there's something to that claim, it's far too simplistic. It also has the inevitable result of downgrading philosophy and religion in practical affairs.

A similar trope is to say that science asks "how" questions, whereas philosophy and theology ask "why" or "purpose" questions.[27] The problem with this formula is that it's simply not true. Scientists continually ask "why" questions: Why do objects fall to Earth when dropped? Why do peacocks have bright plumage? What is the purpose of the heart? Why do giraffes have long necks? Why does matter have the properties it has? Why is the night sky dark? Scientists have been asking such questions for as long as there's been something called science, and there's no reason to think they'll stop doing it anytime soon.

This is especially true of Darwinian Theory,[28] but cosmologists and physicists also often speak of their work as searching for the "meaning of it all," and of trying to determine "where we came from."[29] Even when asking "how" questions, scientists invariably get into origins. The three "basic questions" that scientists ask are: "What's there?" "How does it work?" And "How did it come to be this way?"[30]

We can complain that this is scientism, but science is, at least in part, what scientists do. Should Catholics be in the business of telling scientists what questions they should and shouldn't ask? Isn't it better to show, if possible, that science, when it refuses to consider the possibility of purpose, fails to explain the empirical realities of nature fully and accurately, and that, if purpose is a real part of nature, then any science that seeks an accurate understanding of nature should be open to that possibility? If narrowly materialistic approaches to nature are wrong or profoundly incomplete, wouldn't it be better to show *that*, rather than merely to complain that these approaches are scientistic and should leave room for "other ways of knowing"?

New Developments and Real Diversity in Natural Science

An overly hasty segregation of science and philosophy also fails to accommodate the many developments in logic and philosophy of science since Aristotle—and since Bacon for that matter. The Philosopher tended to think of science as a deductive enterprise. In his *Posterior Analytics*, Bk. I, 1–2, 4, for instance, Aristotle claims that *scientia* is a matter of seeing what necessarily follows from what one sees to be necessarily true. This may be true for some natural science—such as when physicists discover universal constants or "laws"—though these never involve logical necessity. As a result, Catholic scholars often assume that science is always and everywhere concerned with discovering laws.[31] Unfortunately, this definition would leave out much natural science, which rarely provides conclusions that are necessarily true (either physically or logically), and does not always appeal to natural law. It would also disqualify many of the intellectual tools that modern scientists actually use.[32]

In reality, natural science is a diverse enterprise. We don't have at hand hard and fast criteria for defining "science"; as a result, it's rarely easy to demarcate "science" from other intellectual activities like "philosophy." Modes of reasoning, and degrees of certainty, vary from discipline to discipline. Science includes not just lab physics and chemistry but molecular biology, cosmology, and astronomy.[33] And with the emergence of various multiple universe theories in cosmology, it's clear that materialists don't recognize any bright line between "science" and "philosophy." Catholics and other Christians who continue to do so are essentially burdening themselves with a self-imposed handicap.

And this is just in the so-called natural sciences. There are also human sciences such as economics, sociology, and psychology, not to mention archaeology, forensic science, and the Search for Extraterrestrial Intelligence (SETI), all of which presuppose a concept of intelligent agency that can be studied. You can't put agents under a microscope, of course, or reduce them to mathematical equations (though some may try). Still, agents often leave empirical marks of their activity behind. And we have increasingly sophisticated ways of isolating, tracing, and studying these marks—in information and communications theory. Some of those tools have been developed by ID theorists such as William Dembski.[34]

Even within natural science, there's a difference between experimental sciences like lab chemistry and physics, which allow repeatable experiments, observational sciences like astronomy, which often don't, and historical or origins sciences like cosmology, origin-of-life research, and evolutionary biology, that depend heavily on comparing competing hypotheses. All these sciences use empirical observation, and all admit of confirmation; but the way you confirm a hypothesis in chemistry differs from how you confirm a hypothesis about the origin of matter or the origin of life (though even here, there's no black line between the two enterprises).

The historical sciences generally follow what has been called abduction or "inference to the best explanation," in which multiple competing hypotheses are compared with regard to their causal adequacy.[35] With this method, the most causally adequate hypothesis is the most probable, and so it wins out, even if we can never be sure that it's correct. A great deal of science, including the science most likely to have significant metaphysical implications, proceeds along these lines.[36]

The distinction between experimental and historical or origins science is especially important when dealing with intelligent design and evolution. Evolutionary theory is largely historical and abductive, and most contemporary design arguments focus likewise on the historical sciences in cosmology, origin-of-life research and biological evolution.

Ironically, given the quasi-traditional but really modern way of distinguishing science and philosophy, some Catholics conclude that Darwinism is wrong in proposing a "mechanistic" explanation for teleology, but still qualifies as scientific, while teleological explanations are correct but are not scientific.[37] This is perplexing since, in at least some cases, the Darwinian and teleological explanations are *competitors* in trying to explain the same empirical reality. We have a right to suspect something isn't quite right if one of these explanations gets tagged as "scientific" but the other one as "philosophical."[38] It amounts to a pre-emptive surrender to materialism.

Materialist Infection of Science

Finally, the classical understanding, when combined with a modernist definition of science, can lead to naïveté about the role of philosophical assumptions within science itself, especially in those sciences removed from lab chemistry and physics. The problem here isn't merely that science has transgressed its proper boundaries, as one often hears. The problem is that materialism has entered into the foundations and content of the science itself.

This materialist infection of science is distinct from scientism and requires a different response. Failure to tread carefully here leads many well meaning Catholics to take the scientific claims of Darwinists for granted, and then to find a theological excuse for not looking at the actual scientific evidence. The recent book *Faith, Science, & Reason* by Christopher Bagley is a poignant example of this tendency. It is in many ways an excellent book, designed to help faithful Catholics deal with the claims of both faith and science with integrity. And yet when it comes to Darwinism, Baglow is shockingly uncritical. Rather than looking carefully at the evidence for the creative role of random genetic mutations, he spends time explaining how "randomness" or "chance" is not incompatible with divine providence, as long as we view them as different levels of explanation.[39] An event may be random from our perspective or a scientific perspective, but ultimately the work of providence.

As a logical point, that's certainly correct: looking random and being random are two different things. However, the effect of Bagley's argument on the uncritical reader is, first, to prevent that reader from ever spending any time looking at the actual evidence for key Darwinian claims, second, to leave the reader wondering how this mere logical distinction preserves the Church's teaching that God's design is clearly manifest in nature, and third, to misconstrue what Darwinian Theory means to the vast majority of its proponents, who consistently define "random" as "purposeless."

To redefine the word "random" privately in our heads to mean "apparently random at the level of our observation but ultimately the work of providence" is no response to the Darwinist denial of teleology. The truly trenchant response, when accompanied by a careful presentation of the evidence, is: "No. You're mistaken. We know on empirical grounds that natural selection and random variation explain very little. Moreover, nature, from cosmology to physics to biology, has observable features that are better explained by intelligent design than by the Darwinian and other materialist explanations."

Why Argue that ID is Science?

So then, at last, is ID scientific? Why do ID theorists claim that it is? They do so for three good reasons. The first is that any definition of science broad enough to encompass the diversity of science will also encompass ID arguments. (Whether those arguments are right or wrong is a separate issue.) Darwinism has always presupposed that design arguments in biology are coherent, specific, and falsified. So they are already part of the content of science. The special creation of individual species is the constant foil in Darwin's *Origin of Species*, but it isn't the only possible alternative to Darwinism. Rather, design broadly construed is the alternative to which the mind naturally inclines if the Darwinian explanation is defeated. Even arch-Darwinist and atheist Richard Dawkins has admitted this.

Second, ID arguments are based, not on religious texts or even broad philosophical questions, but on publicly available, empirical evidence drawn from nature. ID doesn't focus on agents themselves, but on the empirically detectible traces of intelligence in nature. As a result, ID arguments offer answers to questions that scientists ask *as scientists*. While a complete list of criteria for what constitutes science remains elusive, it's generally agreed that scientific arguments ought to be based on public evidence from nature, and ought, as philosopher Del Ratzsch has said, "to be put in empirical harm's way." Normally that means that they can be verified, falsified, or tested against competing hypotheses. Given these reasonable criteria, a number of contemporary ID arguments qualify.

Third, ID follows argumentative forms well attested in natural science. For instance, Steve Meyer's argument for intelligent design as the best explanation for biological information follows the same canons of reasoning that Charles Darwin followed in the *Origin of Species*.[40]

So there are perfectly sound reasons for claiming that ID be considered within science itself, even if, like so many scientific ideas including Darwinism, it has larger philosophical implications.

Conclusion

To summarize the conclusions of the last three chapters, the difficulty certain traditional Catholics have in negotiating the difficult debate involving Darwinism and intelligent design derives from: (1) using terms like "evolution" and "mechanism" ambiguously, (2) misinterpreting the arguments of ID proponents, (3) taking the scientific claims of Darwinists for granted, (4) sitting too tightly and uncritically on anti-Catholic aspects of Aristotle's philosophy and (5) overlaying modernist distinctions of academic disciplines on a classical understanding of reason. Together these inclinations constitute an unwitting appeasement, then a capitulation, to the forces of modernism, materialism, and secularism.

Once we've worked through these complicated issues properly, however, I'm convinced that orthodox Catholics not only will see the problems with Darwinism, but will find a lot to like about intelligent design. Properly understood, ID theorists are defending reinvigorated concepts of formal and final causation as legitimate explanations of nature, categories that were supposedly banished in the early modern era. We Catholics should be on the front lines of this effort to liberate science and culture from the grip of materialism, not looking for quasi-Catholic ways to maintain a creaky and materialist status quo.

IV.

Jews and Evolution

14. The Maimonides Myth and the Great Heretic

Can a Jew Be a Darwinist? Part 1

David Klinghoffer

The Jewish people, whether considered historically or religiously, are set apart from others. "Behold!" the Torah says. "It is a nation that will dwell in solitude and not be reckoned among the nations" (Numbers 23:9). There is a certain way that religious Jews have of being slightly abstracted from the world—apparently as intended by the Torah's own laws, which have been responsible for Jewish survival as a distinct people for millennia. Keeping kosher and observing the Sabbath, to cite two important examples, have preserved a unique Jewish identity, even as they make total social integration with the wider society difficult. The flipside of this otherwise desirable characteristic might be that religious Jews are slow to pick up on legitimate concerns grasped more readily by some members of other faiths.

Maybe it should be no surprise, then, that religiously committed Jews have been less ready than similarly committed Christians to recognize the threat to the integrity of their beliefs posed by Darwinian thought. Secular and liberal Jews have every reason to embrace Darwinism precisely to the extent their secular peers of other ethnicities and comparable socio-economic status do. Orthodox and otherwise tradition-minded Jews, on the other hand, are committed to a religious worldview that, looked at objectively, stands at loggerheads with Darwinian evolution. Yet the Orthodox community has hardly grappled with the evolution debate.

Among the ultra-Orthodox, a common view would be some variation on Young Earth Creationism, but without the attempt to justify the view in the scientific or pseudo-scientific language that you find among similarly inclined Christians. In fact, ultra-Orthodox education models don't emphasize science—nor math, English, history, or anything other than intensive study of traditional religious texts. The black-hatted and black-suited Charedim generally see science as beside the point.

More puzzling is the indifference to the Darwin debate among centrist and modern Orthodox Jews, who value secular education. Representing mainstream Orthodoxy, this community doesn't appear to have given the problem much sustained thought. As a result, you'll often hear modern Orthodox Jews dismiss the evolution controversy with a verbal shrug. They will tell you, "Darwin is a problem for Christians. Not for us." The reality is they haven't looked carefully at either Darwinism or at what our own Jewish sources say.

In this and the following essay I propose to survey some of those sources. For a believing Jew, one who takes Jewish tradition seriously, Darwinism should be a virtually impossible thought—impossible, that is, if Jewish faith is to be maintained. On the contrary, evidence of intelligent design in nature is precisely what Judaism prompts Jews to expect that science will find. Nor is this merely a scientific or theological issue. It goes to the heart of the relationship God seeks with Jews in particular—as a people in intimate contact with him, and one charged with carrying moral illumination to the world. Before examining the relevant classical sources, however, I want to acknowledge that most Jews initially will find my case counterintuitive.

The *Wall Street Journal* has promoted as a representative Jewish view that of Yeshiva University biologist Carl Feit,

> who is an ordained rabbi and Talmudic scholar.... Professor Feit says that in nearly a quarter-century of teaching introductory biology, he has always taught evolution—supported by traditional Jewish source material—and that 'there has never been a blip on the radar here.' His assessment echoes the official line of the Modern Orthodox rabbinical association, which states that evolution is entirely consistent with Judaism.[1]

In contrast, a popular 2008 theatrical documentary film featured three Jewish skeptics of Darwinism. *Expelled: No Intelligence Allowed* was a film about the suppression of intelligent design advocates and Darwin skeptics in the modern academy. It was narrated by Jewish lawyer and comedian Ben

Stein. In one memorable scene Stein, mathematician David Berlinski, and Orthodox Israeli physicist Gerald Schroeder (with a yarmulke) were shown touring the ruins of the Berlin Wall, a symbol of Darwinism's authoritarian tyranny in the academic world. They were the three Darwin-doubting Jewish musketeers. After seeing the movie, a colleague of mine half-joked, "I didn't realize intelligent design was *such* a Jewish enterprise." I savored the comment.

When you get out into the wider world, however, Jewish hostility or indifference to arguments for intelligent design remains very much the rule. Most conservative Christians understand very well what is at stake in the Darwin debate. They appreciate and support the scientists and writers in the intelligent design movement. Jews, from Reform to Orthodox, do not.

A few years ago, the Orthodox Union's magazine, *Jewish Action*, included a package of three articles on intelligent design. The first in the lineup was by Bar-Ilan University physicist Nathan Aviezer. He claimed to discover "a striking similarity between ID and the ideas that underlie idolatry." What ideas? Well, the ancient pagans "observed phenomena of nature that seemed completely inexplicable, and then postulated supernatural beings (analogous to today's 'intelligent designers') to explain phenomena."[2]

Actually, design theorists argue not from what is "inexplicable" in nature but rather from what is known today about, for example, the software in the cell (including DNA), which presents positive indications of design just like the software that runs in your computer. ID, an increasingly confident minority view among scientists, thrives because modern science understands more than Darwin did when he published *The Origin of Species* in 1859.

Another scientific writer in the *Jewish Action* ID symposium, Arnold Slyper, is a pediatric endocrinologist at Loyola University Medical Center in Illinois. Dr. Slyper grasped what Dr. Aviezer does not, that ID "is Jewish to the core and one of our fundamental beliefs."

But he presumably would be classed as a heretic by Orthodox rabbi Natan Slifkin, who once gave a lecture at Yeshiva University's Stern College for Women, titled "The Heresy of Intelligent Design." "I find it [i.e., intelligent design] theologically offensive," said Slifkin,[3] who oddly is the same famous fellow who protested when his own his books were banned for supposed heresy by some Charedi rabbis who objected to his views on the (old) age of the earth.

But the situation with the Reform and Conservative movements is also surprising. Two prominent non-Orthodox rabbis of the previous generation were actually proto-ID-advocates. That's one of the fascinating points made clear in a recent book, *Jewish Tradition and the Challenge of Darwinism*.[4] It is a collection of essays, one of which describes the thought of Roland Gittelsohn and Robert Gordis, representing respectively the Reform and Conservative rabbinates, who challenged natural selection and random variation as a sufficient explanation of how life got to be the way it is.

If it's not apparent by now that Darwinism does not divide Jews along an easily defined and expected religious divide, I could cite the late neoconservative icon Irving Kristol, who was not known as a traditionally observant Jew but wrote critically of Neo-Darwinism in a 1986 *New York Times* op-ed piece, "Though this theory is usually taught as an established scientific truth, it is nothing of the sort. It has too many lacunae."[5] Or I could cite the essayist Joseph Epstein, Jewish but of no publicly declared religious commitment, who commented on the *Wall Street Journal* op-ed page:

> Not only have the past 50 or so years been largely bereft of grand ideas, but much of the best intellectual work of the period has been devoted to eliminating the major ideas, or idea systems, of the previous 100 or so years: notably, Marxism and Freudianism, with Darwinism perhaps next to tumble.[6]

Rabbi Slifkin and those who think like him, people like Christian genome scientist Francis Collins, believe you can coherently affirm God and Darwinism. Their strategies for reconciling the two ideas are diverse and Protean, some clever, some not. But all insist that whatever role God played in life's development, it is undetectable, thus unfalsifiable.

At the outset, then, we may note that one difference between the reconcilers and intelligent design advocates is that the latter at least take the risk that their particular arguments for design may be shot down by scientific counterevidence. What's wrong with shying away from that risk? For one thing, Judaism sees God as asking us to be his witnesses in the world. That is why we stand up to say the benediction over wine, *Kiddush*, on the Sabbath eve night that commemorates Creation, to simulate giving court testimony about God as Creator. In Jewish legal proceedings, testimony is always given while standing. In a legal context, one of the features of an acceptable witness is that his testimony must be potentially falsifiable. As the Talmud and common sense agree, if a witness's testimony cannot even in principle be knocked down, then that testimony has to be dismissed. Jewish Darwin apologists appear to understand Judaism as if Torah were capable of being reconciled with any putatively scientific idea imaginable, even if the idea happens to be wrong.

Waving Maimonides

Religiously informed Jews, when arguing among themselves, are always lobbing rabbinic citations back and forth and at one another. The argument typically comes down to a game of capture the flag, where the flags are famous rabbis and their opinions from the Talmudic, medieval, and modern periods. When the argument is about God and Darwin, the preeminent flag is Maimonides. Every self-respecting Jewish religious intellectual wants to identify himself with Maimonidean rationalism and respect for science. Can we have any hope of inferring what Maimonides' position would be? Leon Wieseltier, the celebrated literary editor at *The New Republic*, evidently thought so when he wrote an essay tarring Darwin-doubters as naïve creationists and Bible literalists.[7]

One thing ID advocates are not guilty of is reading the Bible simplistically—they overwhelmingly accept that the earth is about 4.5 billions years old and most don't worry much about common descent, the idea that organisms share common ancestors. They are concerned with the *mechanism* of evolution, not the fact of change over vast stretches of time. But never mind that. Wielseltier recalls that when he was young and foolish, he was troubled by apparent conflicts between science and religion. That is, until he read Maimonides' *Guide of the Perplexed*. For in that book, the great sage dispels any compulsion the religious reader may feel to read the Bible only "literally," including Genesis 1, which describes the earth's creation.

Now, Maimonides was writing specifically about a scientific controversy, hot in his day, regarding the eternity of the world. Strict Aristotelians said the world had always existed. As Maimonides wrote:

> Know that our shunning the affirmation of the eternity of the world is not due to a text figuring in the Torah according to which the world has been produced in time. For the texts indicating that the world has been produced in time are not more numerous than those indicating that the deity is a body. Nor are the gates of figurative interpretation shut in our faces.[8]

Since the Bible can indeed be interpreted figuratively when appropriate, a faithful Jew surely may without qualms embrace Darwinism. Right?

Not so fast. Seen through the eyes of our rabbinic arch-rationalist, the answer isn't nearly so simple. In the very same chapter of the *Guide* quoted above (II:25), Maimonides goes on to say something that Leon Wieseltier missed or ignored. The sage writes that he rejects the eternity of the world for two reasons. First, because it "has not been demonstrated." Second, because it makes nonsense of Judaism: "If the philosophers would succeed in demonstrating eternity as Aristotle understands it, the Torah as a whole would become void, and a shift to other opinions would take place. I have thus explained to you that everything is bound up with this problem."

He was saying that though parts of the Bible's text may indeed be interpreted in diverse ways, theologically, an eternal universe is still incompatible with the God of the Torah. Simply put, Aristotle makes God's role in the world, as a creator and guide, superfluous.

Darwinism does the very same thing, ascribing all creation to blind material processes, as Darwin himself said: "I would give absolutely nothing for the theory of natural selection if it requires miraculous additions at any one stage of descent."[9] There is no God in that picture.

Maimonides would ask if Darwinism nevertheless has been "demonstrated." Well, Darwin's followers reached a high point of self-confidence in 1959 with the Centennial Celebration held at the University of Chicago to mark the hundredth-year anniversary of the publication of *The Origin of Species*. The event was notable for the total conviction on the part of many speakers that any debate about Darwin was over and done.

But since then, the trend has changed directions and the doubts continue to mount. The Discovery Institute, which drives much of the debate about Darwinism and intelligent design, has compiled a list of professed Darwin-doubting scientists who have signed a statement to that effect. The list currently stands at more than 800 signers, including researchers at UCLA, Princeton and MIT.

That doesn't sound like a theory that has been unambiguously "demonstrated." Nor is it one that may be comfortably reconciled with Torah. Maimonides, I suspect, would tell us there is a choice every Jew must make: between God and Darwin. As anyone who takes ideas seriously needs to recognize, you can have one or the other, but not both.

Yet the clichés continue unabated. One such cliché goes approximately like this: "Maimonides was a physician. A physician is a kind of scientist. Maimonides was therefore a religious scientist. Consequently any attempt to merge any science-flavored idea, such as Darwinism, with Judaism would meet with Maimonides' approval." This really is how many Jewish people think. And in their own respective context, so do many Christians. Thus Francis Collins, currently director of the National Institutes of Health, spoke at a Pew Forum conference and reaffirmed the Maimonides Myth:

> Basically, if you look at Judaism and Islam, you will find a range of views about origin. Certainly in Judaism, conservative and reform Jews are generally accepting of evolution, and a lot of Orthodox Jews are as well. Maimonides is often cited here as a reason to assume that if you have a conflict between science and the Torah, there's been an error and a misinterpretation, not that science is evil.[10]

There you have it. According to Maimonides, our understanding of Torah must yield before anything scientists happen to say at a given moment.

Since such glib reconciliations are so common, what a breath of fresh air it was to open the relatively new and admirably lucid Maimonides biography by Joel L. Kraemer of the University of Chicago, *Maimonides: The Life and World of One of Civilization's Greatest Minds*. Professor Kraemer asks what a student of Maimonides like himself should reply if asked, "What is the most important idea taught by Maimonides in his scientific and philosophic writings?" Answers Kraemer: "A good answer would be that it is the idea of an orderly universe governed by laws of a cosmic intelligence." Contemporary relevance, please? Replies Kraemer:

> Maimonides grasped the great divide between monotheists, who believe that an intelligence guides the universe, and Epicureans, who believe that everything happens by chance. The argument continues nowadays between intelligent adherents of intelligent design and Darwinian atheists who believe in chance mutation.[11]

Maimonides was fighting the good intellectual fight for intelligent design almost eight hundred years ago. It was *his top philosophical and scientific concern*. Should anyone need any further proof that ID, right or wrong, is at any rate an authentically Jewish cause?

Epicurus, the Great Heretic

In fact, traditional Jewish sources go further than merely placing a kosher certification on intelligent design. Kraemer's reference to Epicureans is very much on point, for the Greek philosopher Epicurus gave his name to the Talmud's term for a heretic.

As I mentioned earlier, fervently religious Jews often seem to assume that since we already know God made the world, there is, in a scientific vein, hardly anything else worth saying on the matter. We have the Torah. Why do we need science? We need it because Jewish tradition itself tells us that this is a subject of which we can't afford to be ignorant. The Mishnah's tractate *Pirke Avot* instructs us to "Know how to answer an *apikoros*" (2:19)—a heretic, or literally, an Epicurean. The Mishnah uses the word without explanation, for a category of persons who have no share in the World to Come. The Talmud links it with insolence either to the face of the Sages or in their presence.[12] Maimonides finds an etymological connection to an Aramaic word for "disparagement."

But what of the idea content of the term? In the Mishnah's context, it is linked with other heretical ideas. The *apikoros* is listed alongside other heretics, those who say the resurrection of the dead has no support in the Torah and those who deny the Torah's divine origins. In a Hebrew dictionary, the word is defined as an "atheist, freethinker, heretic."

Rabbi Joseph Albo, a medieval luminary, explains *apikoros* as deriving from the name of Epicurus (born c. 342 BCE) and his school (*Sefer ha-Ikkarim* 1:10). In Hebrew, Epicurus is "*Epikoros*." *Apikoros* and *Epikoros* are spelled the same way, though the Talmudic pronunciation, unlike modern Hebrew, gives the initial vowel sound as an "a" rather than an "e." In popular English usage today, an "Epicurean" means someone who seeks pleasure in fine food or wine, but that is not what Epicurus himself was. Epicurean thought does stress the pursuit of pleasure but not the short-term kind. Rather, it urges us to avoid pain and think in terms of longer term, though not eternal, happiness. Among other things, to escape emotional pain, Epicureanism advocated masturbation over sexual relationships.[13]

Part of Epicurus' program was to eliminate fear of divine justice. The gods, he explained, were off in their distant celestial realm, indifferent to our world. In line with this, the philosopher taught that human life is a purely material affair. Even the soul is made of matter. There is nothing to fear from the gods in part because once you're dead, you're dead. There is no afterlife. This is understood to be a comfort.

Reality, Epicurus taught, is all composed of "atoms." The universe came into being through the unguided colliding of these atoms. "The world is, therefore, due to mechanical causes and there is no need to postulate teleology"—purpose or design—summarizes Frederick Copleston in *A History of Philosophy*.[14] For the rabbis, this last point, which achieved its modern biological expression in Darwin's theory, is the key to what's wrong with Epicureanism.

In the classic medieval philosophical work *Kuzari*, which takes the form of a dialogue between a rabbi and the king of the Khazars, Rabbi Yehudah HaLevi makes this point explicit. In the Fifth Essay (5:9, 5:20), the rabbinic protagonist teaches,

> We perceive [divine] wisdom in many creations, and the necessary purposes they serve.... This wisdom was alluded to by King David in the Psalm [104], "How great are Your deeds, God." He wrote it to refute the arguments of Epicurus the Greek, who believed that the universe came about incidentally [*b'mikreh*, lit. by chance].

HaLevi explains:

> All phenomena are traceable back to the Primary Cause in one of two ways: either directly from God's will, or through intermediaries. An example of the first way is the order and assembly that is *evident* in living creatures, plant life, and celestial spheres. No intellectual person can attribute this to happenstance. It is rather attributable [directly] to the *design* of the wise Maker. [emphasis added].

Let's not miss what the *Kuzari* has just said. No intelligent person would deny the *evidence of design* in living creatures. That is, nobody would do so unless he was an Epicurean, against which teaching the Jews are called to stand as a witness:

> The Jewish people provided every nation with a refutation against the Epicureans, who follow the beliefs of Epicurus the Greek. He said that all things happen incidentally, and that nothing in this world shows evidence of intent from a [higher] sentient being. His colleagues were called hedonists because they believed that pleasure is the ultimate objective and the principal good.

I have strictly adhered to the excellent new translation of the *Kuzari* by Rabbi N. Daniel Korobkin.[15] But if you look at the Hebrew (itself a translation from Arabic), you will see that the Hebrew word that Rabbi Korobkin translates as "design" (*kavanah*) in the second passage quoted above, he translates as "intent" in the third. You could just as well translate that penultimate sentence as, "[Epicurus] said that all things happen incidentally, and that nothing in this world shows evidence of design from a [higher] sentient being."

Rabbi Korobkin clarifies in a footnote: "Because of his radical beliefs and the wanton behavior of some of his followers, Epicurus was viewed by the Sages as one of the most morally destructive of all Greek philosophers."[16]

Epicureanism, which scholars like my colleague Benjamin Wiker have connected by a clear intellectual genealogy to Darwinism, should be regarded as anathema to the Jewish mind. The Jewish sages fully expected the world to show "evidence of design from a [higher] sentient being." Abraham ibn Ezra (1089–1164) is one of a handful of the most important classical commentators on the Torah. Explaining the first of the Ten Commandments, "I am the Lord your God Who took you out of Egypt" (Exodus 20:1), Ibn Ezra asks why God commands our belief using this formulation with its two clauses. The sage argues that religious believers whose faith is gained through tradition alone will be in good shape

> until a heretic begins to argue with them that there is no God, [then] they put their hands over their mouths, because they don't know what to respond. On the other hand, if a person applies himself to the study of the sciences, which are like steps to help a person reach his desired destination, he will be able to discern the handiwork of God in metals, plants, living creatures, and in the human body itself. ... From God's ways [in nature] the discerning person comes to know God. That is why Scripture writes, "I am the Lord your God" [—the God Whom you discern in nature]. But this can only be appreciated by someone who is extremely wise.... As for the miracles done in Egypt ... everyone saw this, both the wise and the unwise, both adults and children.... Therefore, Scripture first writes, "I am the Lord," for the discerning individual. It then writes, "Who took you out," so that even the non-discerning individual may understand.

For the simple, in other words, simple faith. For the wise, intelligent design. Incidentally, when Ibn Ezra speaks above of a "heretic" who challenges the simple in faith about God's existence, but who gets an adequate response only from the discerning Jew who contemplates nature's intelligent design, the word that he uses for "heretic" is *apikoros*.

Another classical source for the expectation of evidence for design is Rabbi Moshe Chaim Luzzatto (1707–1746). In *The Way of God*, he affirmed that the Creator's existence can indeed be "demonstrated from what we observe in nature and its phenomena. Through such scientific disciplines as physics and astronomy, certain basic principles can be derived, and on the basis of these, clear evidence for these concepts deduced" (1:1:2).

A revered modern expositor of Jewish tradition, Rabbi Joseph B. Soloveitchik, wrote of the tension between God's hiding from us and his revealing himself, a dance expressed in romantic terms in the Biblical Song of Songs. Rav Soloveitchik was not one who believed that science somehow "proves" God—a caricature often hung on intelligent design advocates by their Darwinist critics. However, Soloveitchik was in no doubt about the revelatory quality in nature. We formally affirm God's design on the Sabbath. Rav Soloveitchik taught that God's self-revelation in the natural world is the theme of Friday night, the Sabbath eve, whereas on Saturday morning the liturgical focus is on revelation in the transcendent realm. Soloveitchik cited Nachmanides (Rabbi Moses ben Nachman, 1194–1270) on Deuteronomy 12:15—"You shall seek His Presence, and you shall come there." By seeking God's Presence, the *Shechinah*, in nature, we come closer to appreciating His transcendent reality outside nature. This is the spiritual journey we reenact every *Shabbat*, in the movement from Sabbath eve to Sabbath morning.[17]

The expectation of evidence for design in living creatures may be more specific even than this broadly shared assumption that God is revealed in nature. Life's code, DNA, acts like computer software, or like a language consisting of letters and words; it is arranged in specific sequences to accomplish a specific task or convey a specific meaning. The letters of this genetic "alphabet" in the correct combinations code for functional proteins. Kabbala too speaks of such an alphabet, comprised of the letters of the Hebrew alphabet, with which God continually speaks the world into existence. Different combinations of letters produce different creatures. A century and a half before Watson and Crick, Rabbi Schneur Zalman of Liadi sought to make Kabbala accessible to ordinary readers. In the *Tanya* (1796), he writes of how "the creatures are divided into categories [both] general and particular by changes in the combinations, substitutions and transpositions [of the letters]" (*Shaar Hayichud Vehaemunah,* Chapter 12).

I don't mean to attribute more prescience to the rabbis about developments in biology than they actually possessed. Yet their words remain strikingly relevant today to the debate about Darwinian evolution. Thus a standard weapon in the Darwinist's arsenal is the argument from apparently poor, botched, or suboptimal design. As Darwinian biologist and atheist Richard Dawkins writes in *The Greatest Show on Earth,* regarding the extravagantly lengthy and circuitous recurrent laryngeal nerve of the giraffe, "Any intelligent designer would have hived off the laryngeal nerve on its way down, replacing a journey of many meters by one of a few centimeters."[18] Darwinists think they have discovered a devastating "Ah hah! Gotcha!" sort of a response to religious believers who, it's assumed, never realized that nature has a certain painful lack of perfection built into it. Yet the Hebrew Bible alerts us early on that creation is afflicted with a "lack" or "deficiency" (*chesron*), as Jewish philosophy terms it. In the sixteenth century, the Maharal of Prague discussed this theme in his book on Chanukah, *Ner Mitzvah*.

The Maharal finds evidence that deficiency in creation was not only intended and foreseen by God but is a necessary feature of it, alluded to in the opening verses of Genesis: "In the beginning God created the heaven and the earth, and the earth was without form, and void; and darkness was upon the face of the deep."

One particular "deficiency" that was tinkered with and corrected was the initial solitude of Adam, the first man, depicted as lonely and single: "And the Lord God said, It is not good that the man should be alone; I will make him an help meet for him" (Genesis 2:18). Read that again carefully. God's creative activity produced something that was "not good." That it was fixed later through the creation of Eve doesn't take away from the startling admission by the Bible itself. Dawkins again: "This pattern of major design flaws, compensated for by subsequent tinkering, is exactly what we should *not* expect if there really were a designer at work."[19] The Hebrew Bible's reply would be, "Oh, really?"

In the context of Chanukah, with its theme of the wicked Greek kingdom's oppression of the Jews in their land and the subsequent civil war pitting religiously loyal Jews against secularist Greek-loving Jews, the theme emerges a little differently. In the Biblical scheme of history, four ancient kingdoms arose and sequentially divested God's presence in the world of some of its splendor. Each did so by depriving the Jews of sovereignty in their land, where Israel was intended to carry out her spiritual mission to the fullest extent possible. One kingdom was Greece. Another was Rome, in whose exilic shadow we still live. The Maharal finds all four alluded to in the second verse in Genesis. It was foreseen—not a matter of chance—as part of the pattern that God knew full well would unfold.

Woven into creation from the start was a very painful thread of "deficiency," playing out on the historical stage. Why not, too, in nature? It could hardly be otherwise. If a trivial example like my habitually sore knee is "bad design" and a point scored for Darwinism, then any trivial lack of perfection in created reality is enough to trigger the atheist response. *Any* evil in nature, *any* suffering.

The world can be rough and it is obviously not all a matter of people freely choosing evil. The verse in Isaiah (45:7) says so directly: "I form the light, and create darkness: I make peace, and create evil: I the Lord do all these things." Well, consider the alternative. A world without evil. What would that be like? It would be the perfect hamster cage or turtle terrarium, where all our needs are provided, there are no predators, no contagious disease, no confusion, no loneliness, no sin, no particular purpose, no growth, no freedom; just spinning aimlessly on our exercise wheel or swimming idly in our calm, algaed paddling pool.

Creatures that could never grow or change spiritually because they were unchallenged and therefore totally uninteresting? What's the point? Once we admit that some "lack" in creation, at least as we perceive it, was inevitable if there was to *be* a creation with freedom and growth, then what extent of deficiency was going to be enough? Maybe a little, maybe a lot. You will have to ask God when you meet him. Taking it for granted that part of his purpose in creating us was to relate to us as free beings, surely it makes sense that he would want to relate to something more than a hamster in a cage.

Tinkering with Nature

THE RABBIS ALSO SEEMED to anticipate objections from Darwin apologists that it would be somehow beneath God's dignity to "tinker" with nature, involving himself with its development over the course of billions of years. After all, doesn't it blow a huge hole in the idea of a Designer at work shaping nature if this Designer feels the need to interfere so frequently with the products of his creativity—whether to perform a miracle like splitting the Red Sea, or to shape, somehow, a species?

In his book *Finding Darwin's God*, Brown University biologist and theistic evolutionist Kenneth Miller asks sarcastically:

> Is the designer being deceptive? Is there a reason why he can't get it right the first time? Is the designer, despite all his powers, a slow learner? He must be clever enough to design an African elephant, but apparently not so clever that he can do it the first time.[20]

It is like a car that needs to go into the shop all the time. We assume it was poorly designed. The question asked by Miller is fair.

This peculiarity of nature was not discovered recently. Almost a thousand years ago, in the classic work of moral reflection *Duties of the Heart*, the Spanish sage Rabbi Bachya ibn Paquda explained the verse in Psalms (135:6): "Whatever God willed, He did—in heaven and on earth." Rabbeinu Bachya, as he's called, emphasized the importance of reflecting on the signs of God's wisdom in nature. Creation, he writes, was not all accomplished through one law. If it had been, that would have left a deceptive impression:

> When a thing always acts in a certain way, this indicates that its actions are not the expression of its own will but only of the nature imposed upon it; that there is some force that compels it to act in this fashion.... One who acts out of free will, however, acts in various ways at various times.

God wished to "indicate his oneness and free will in all his actions." So he "created diverse things, according to the dictates and timing of his wisdom" (*Gate of Reflection*, Chapter 2). A creation that was entirely ruled by law, smoothly and gradually building up complexity, would have precluded this. It would have seemed that this natural law was the source of "creativity," rather than God's being so. Instead, the Designer's work went forward in fits and starts, with relatively sudden bursts and radiations of creativity—"tinkering." Paleontology shows exactly this pattern in the fossil record.

What positive good is served by God's having set nature up this way? The ancient tradition that explains the Hebrew Bible has an answer. God is exemplifying his own freedom as a moral example to us—a point to which we'll return in the next chapter, when we consider the moral implications of Darwinian thinking.

15. God's Image, Our Mission

Can a Jew Be a Darwinist? Part 2

David Klinghoffer

Something is out there beyond nature, guiding the destinies of living creatures. Such an understanding, of nature driven by a force outside nature, was dominant in biology before Darwin. Baron Georges Cuvier (1769–1832), director of Paris' Musee d'Histoire Naturelle, held that there was an unknown biological "formative impulse," an organizational principle of some kind that directed the formation of diverse kinds of life. The concept goes still further back. Much further.

What Cuvier called the "formative impulse" was called God's "wisdom" by the rabbis. The Bible teaches, "The Lord founded the earth with wisdom" (Proverbs 3:19). Similarly, an ancient Aramaic translation of the Torah, *Targum Yerushalmi*, renders the first verse in Genesis as, "With *wisdom* God created the heavens and the earth." Whether we think of that impulse as indeed coming from God or from some other unknown agent makes a big difference. But the progress of science from imagining existence as a purely material affair, without purpose, as Darwinian evolution consistently and insistently portrays the matter, to the more advanced description toward which biology increasingly points, is a major step in the right direction. That direction is a Jewish one—for the theological reasons we saw in the previous chapter, but also for more broadly spiritual and moral ones.

Charles Darwin's wife, Emma, herself a passionate theist, unlike her husband, said that his theory of selection and random variation "put God further off." She said it well. Even believers in theistic evolution—the idea that while God exists, he created the biological world by Darwinian means—are necessarily committed to seeing God as greatly abstracted from his alleged work as life's creator. God's nearness to us, reflected in his image (*tzelem*) that we somehow bear upon ourselves, is no mere theological debating point. It is the necessary condition of his having a relationship with us all as the Bible envisions. But Darwinism, whether in its atheistic or theistic evolutionary form, makes the idea of God's *tzelem* in us, as many Jewish sources understand the idea, incomprehensible.

When you look at your face in the mirror, is what you see the stamp of God's own image—not his face, because he doesn't have a face or a body or any physical aspect, but his spiritual image? Or is your face the mere product of an inherently unguided evolutionary process, a configuration of physiognomic features reflecting God's intentions partially at best, probably not at all?

This is one of those points where many believers in theistic evolution ask religious believers to pare back theological beliefs to suit Darwinian doctrine. You know the key verses from Genesis (1:26–27):

> And God said, "Let us make man in our image, after our likeness: and let them have dominion over the fish of the sea, and over the fowl of the air, and over the cattle, and over all the earth, and over every creeping thing that creepeth upon the earth." So God created man in his own image, in the image of God created he him; male and female created he them.

On its pro-theistic evolution website, the BioLogos Foundation cites Cambridge University's Simon Conway Morris on "convergence." That is the hopeful idea, one that is very far from being accepted by most Darwinian biologists, that even unguided evolution would eventually "converge" on a creature *somewhat* like us. Thus, even without God's involvement as life's designer, we could still expect

> many of the traits that are particularly relevant for human-like beings. These examples include basic senses like balance, hearing and vision, as well as highly advanced features like the human brain.... Characteristics such as a large brain capable of consciousness, language and complex thought would inevitably have to emerge from the evolutionary process.
>
> The exact anatomical features of this ultimate sentient being might not be precisely specified by the evolutionary process, however. This thought can be unsettling to anyone who imagines our particular body plan is part of the *imago Dei*, or image of God.[1]

What BioLogos means is, say goodbye to the idea that your face and body necessarily represent God spiritually in any meaningful way. Theistic Darwinist Kenneth Miller has been blunter about this. He asks what you would get if, in Stephen Jay Gould's famous thought experiment, the videotape of life's history were re-run from the beginning with, as in the first go around, no purposeful design (emphasis added):

> [E]ventually I think you would also get a large, intelligent, reflective, self-aware organism with a highly developed nervous system. *Now it might be a big-brained dinosaur, or it might be a mollusk with exceptional mental capabilities....* [M]y point is that I think eventually under the conditions that we have in this universe you would get an intelligent, self-aware and reflective organism, which is to say you'd get something like us. It might not come out of the primates, it might come from somewhere else.[2]

The brainiest of mollusks are squid, octopus and cuttlefish. So if you are willing to believe the face or body of one of those can reflect God's image as well as ours, then you may be comfortable with Miller's brand of theistic evolution. But in that case, you had better also be comfortable with abandoning the clear meaning of the verses from Genesis. The relevant Hebrew words, *tzelem* (image) and *demut* (likeness), mean respectively "appearance" and "similarity in form or deed." These are the definitions given by the classical Spanish medieval commentator and mystic Nachmanides, based on an analysis of how the words are used elsewhere in the Hebrew Bible. Our being created in God's image, he writes, is meant to "stress the remarkable phenomenon that distinguished [man] from [all] the rest of the creations." This includes "[man's] facial expression, [which is an expression of] wisdom and knowledge and perfection of deed." This is God's image sealed in our own faces.

The Zohar (1:191a), relating the mystical interpretation of the Torah, says this:

> [W]hen the blessed Holy One created the world, he fashioned every single creature of the world in its own fitting image, and afterward he created the human being in a supernal image [i.e., God's image corresponding to the divine emanations, or *sefirot*, depicted in a configu-

> ration of ten reflecting the shape of a man], granting him dominion over them all through this image. For as long as a human exists in the world, all those creatures of the world raise their heads and gaze upon the supernal image of the human being; then they all fear and tremble before him, as is said: "Fear and dread of you shall be upon every living thing of the earth" [Genesis 9:2].[3]

Don't worry if you do not understand exactly what that means. The key point to take away is that even animals somehow perceive the Godly image in man. They would not perceive it an octopus, however intelligent.

Nor is this some kind of exclusively mystical insight. Rabbi Samson Raphael Hirsch, Darwin's contemporary, emphasizes the same point in his own classic Torah commentary, and he disclaimed any kabbalistic influence. Instead, he emphasized the practical worldview emanating from the text—conveyed, most characteristically, by an exquisitely careful examination of the Hebrew language in which the Biblical tradition is transmitted.

His approach to Torah was scientific, in the sense that he let the words, the data, say what they say rather than fitting them to an *a priori* idea. Based on an etymological analysis, he too concludes (on Genesis 1:26) that "image" (*tzelem*) "only means the outer covering, the bodily form." So: "The bodily form of man proclaims him as the representative of God, as the divine on earth, ... such as complies with, is adequate to, a being having the calling of being 'godlike.'" Clearly, not just any bodily form would serve the purpose.

Finally, you can't get any more fundamental understanding of what the Torah means than from the supreme classical commentator, Rashi. He cites a parable from the Talmud (Sanhedrin 46b) in explanation of a verse in Deuteronomy (21:23). If a man is convicted of a capital crime and then executed and hung, care must be taken that his body not be displayed overnight: "for a hanging person is an insult to God."

Why so?

> It is a degradation of the King, for man is made in the likeness of his image, and Israel are his sons. This can be compared to two twin brothers who resembled each other. One became a king, while one became ensnared in banditry, and was hung. Whoever would see him [hanging] would say, "The King is hanging!" Any instance

of *k'lalah* (insult) in Scripture means treating lightly and in a demeaning fashion.

God is demeaned by the person in His image being hung overnight. There are other ways to understand this remarkable parable, but the obvious one, given what we have said so far, is that seeing a degraded human body also degrades God, since "man is made in the likeness of his image."

I have not tried to suggest how the human face, or hands and feet, reflect this image, because I'm not sure how that works. It is much easier to say how our posture, unique in the world of creatures, bears witness to our human mission on earth.

The Maharal defines the Godly "image," at least in part, as residing in our standing and walking erect. This is the same Rabbi Judah Lowe of Prague (Maharal is an acronym) credited in Jewish lore with creating the Golem. He discusses the question in his book *Be'er ha'Golah,* among other places, explaining that it is because of man's having been granted kingship over this lower world in which we live that animals walk crouched over, to one degree or another, while humans stand up straight:

> Man stands upright, straight, like a pillar that is upright, which is not the case with any other being, as none of them stand up straight but rather all walk hunched over. And this is an indication of the [exalted] level of man, for man is king over the lower plane of existence, and all serve him.... This is called the "image of God."

Needless to say, neither an octopus nor a dinosaur, however brainy, would fit this criterion.

If you were Francis Collins you might ask why all this matters. Are we insisting on a woodenly literal reading of Scripture? Is this all about the dreaded phantom menace of literalist creationism? As we saw in the previous chapter, Maimonides (who incidentally takes a different, more intellectualizing view on the meaning of *tzelem*) writes in the *Guide of the Perplexed* that where no larger philosophical or moral issue is at stake, and where science goes against the literal meaning, figurative interpretations of the Scriptural text can be an option. But where such an interpretation would throw an authentic religious worldview into chaos, and where in any event the scientific evidence doesn't compel it, then certainly we should resist abandoning the plain meaning.

Hirsch, as always, clarifies the relevant worldview implications. It matters urgently that we not entirely spiritualize the meaning of our being imprinted with God's image. That way lies moral catastrophe, with our bodily acts being relegated in importance to a mere afterthought.

As Hirsch writes in his Torah commentary, the verses in Genesis teach (emphasis added)

> the godlike dignity of the human body. *Indeed the whole Torah rests primarily on making the body holy.* The entire morality of human beings rests on the fact that the human body, with all its urges, forces and organs, was formed commensurately with the godly calling of man, and is to be kept holy and dedicated exclusively to that godly calling. *Nothing digs the grave of the moral calling of man more effectively than the erroneous conception which cleaves asunder the nature of man.* Only recognizing godlike dignity in the spirit, it directs the spirit to elevate itself to the heights, and in mind and thought to soar upwards to a higher sphere, but leaves the body to unbridled license, animal-like, nay lower than animal.

When you hear someone say that our spirit may come from God, but our bodies reflect his will in only the vaguest possible way, that is, in other words, a prescription for moral disaster. Animalism, I mean, of the kind we see around us today.

The Moral Message of Darwinism

Religious people around the world look to their spiritual guides for courage, displayed in a direct confrontation with corrosive modern error. Whether considered from a scientific or a religious vantage point, theistic evolution is an empty vessel. We may adapt the famous saying in praise of the Torah in *Pirke Avot*: Turn it over and over, *nothing* is in it.[4]

It must be emphasized, however, that Darwinism has more than just theological or spiritual implications. Its moral message also bears very directly on the Jewish stake in the evolution debate. Over the past century and a half, Darwinian evolutionary theory has cast a shadow over European and American history, inspiring eugenics and Nazi race theory, encouraging the dehumanizing tendency in modern culture that finds no more value in a man, woman, or unborn baby than in any other "animal." Darwinism, springing directly from Darwinian theory, gives comfort and support to those who make the case for abortion and euthanasia. If we are the product of design, then the designer can set the moral order in which we may operate. If we were cast up on the cosmic shore by a purposeless, unguided process, then perhaps, if man has any sort of freedom, then he can determine his own values. More than that, maybe every individual can decide for himself what is right and wrong.

Darwin watered the seeds of contemporary nihilism and relativism, as, indeed, evolutionary theory's early critics recognized. Samson Raphael Hirsch, for one, commented acidly on this point in the context of the idol Baal Peor, referred to in the Torah and worshipped in the most grotesquely animalistic fashion. Hirsch warned of "the kind of Darwinism that revels in the conception of man sinking to the level of beast and, stripping itself of its divine nobility, learns to consider itself just a 'higher' class of animal."[5]

We discussed in the last chapter God's "tinkering" or "interfering" with nature. Why does he act this way, causing confusion among the Kenneth Millers of this world? Because God sought to model freedom of will for us. So Rabbi Hirsch, writing at the same time that Darwin's *Descent of Man* was published, explains a verse in the song the Jews sang at the splitting of the Red Sea on their way out of Egypt: "Who is like unto thee, O God, amongst the gods! Who like thee is uniquely powerful in holiness!" (Exodus 15:11).

To be "holy" doesn't mean to be pious. In a human, holiness "is the highest possible degree of moral freedom, in which the will to morality has no longer any resistance to overcome, but is absolutely ready to carry out the will of God." As the verse from Exodus says, God in his own supreme freedom and independence is differentiated from false "gods"—"all other powers and forces that are deified by men," forces, as those in nature, to which men ascribe supreme power. God alone performs "miracles" or "wonders" (in Hebrew, *feleh*), actions that occur, according to Hirsch "purely and absolutely by the free and untrammeled will of the doer, independent of, and mostly in contrast to, the existing laws of Nature."

God's first "wonder" was the creation of those laws themselves. His "further interference" with them "remains a *feleh,* a wonder." In an etymological note that confirms this, Hirsch points out that *feleh* is also the Hebrew word for a vow (Numbers 6:2), an act in which a person takes on himself an obligation independently of moral law. The purpose is all for us, that we should be "holy," recognizing our moral freedom from nature: "Ye shall be holy: for I the Lord your God am holy" (Leviticus 19:1).

Whether we are free, or entirely products of a blind and deterministic nature, is a question of deep moral importance and one that divides Jews from believers in many other thought systems, including Nazism. In *Mein Kampf,* Hitler used Darwinian language to make his case for racial war against the Jews. He rallied the millions of Germans who bought his bestselling book with an appeal to biology, which, as he argued, revealed certain iron laws of Nature—principally the struggle for supremacy pitting the superior races against the inferior.

Defy Nature, he wrote, and then the "whole work of higher breeding, over perhaps hundreds of thousands of years, might be ruined with one blow." The major Hitler biographers—Toland, Fest, Kershaw, Bullock—all agree on Hitler's debt to Darwinism, however debased. A gentle soul, Darwin himself never advocated genocide. But in *The Descent of Man,* he predicted that the logic of natural selection made inevitable something like what Hitler attempted against the Jews: "At some future period, not very distant as measured by centuries, the civilized races of man will almost certainly exterminate and replace throughout the world the savage races."[6]

What you would not readily foresee from reading Darwin's writings is that the race requiring extermination would turn out to be us Jews. But Hitler perceived an inner logic in Darwinism that even Charles Darwin did not. In the same chapter of *Mein Kampf* where the Darwinist flavor is most pronounced—Chapter XI, "Nation and Race"—Hitler comments that while his philosophical outlook is based on respecting Nature's laws, the Jews with their "effrontery" say the opposite: that "Man's role is to overcome Nature!"

Hitler notes with disgust that "[m]illions thoughtlessly parrot this Jewish nonsense and end up by really imagining that they themselves represent a kind of conqueror of Nature."[7]

There is, in other words, a Darwinian case for seeing the Jews as an arch-enemy. Darwin's portrait of reality in his books is one where "Nature" determines all. In *The Descent of Man*, he explains that even our morality, like everything else about us, is a product of natural selection. The Jews, Hitler wrote, defy nature and call others to do so. This is the characteristic "Jewish nonsense." Hitler in fact had put his finger on a profound theme in rabbinic literature. The greatest sages of the Jewish past—from the Maharal of Prague to Moshe Chaim Luzzatto to Samson Raphael Hirsch—taught that overcoming nature, and teaching others likewise, is nothing less than the Jewish mission to the world.

This starts with seeking to overcome our own "nature," bending it to God's will. As the Maharal and others explained, the symbol of this unique Jewish mission is circumcision, a most unnatural thing to do. We perform the *bris* specifically on the eighth day of an infant boy's life. That is because in the system of Jewish number symbolism, seven signifies the natural order of the world, which in the biblical narrative was created in seven days. Transcending this natural order is represented by seven plus one, or eight.

The *bris* on the male organ became, then, a most logical symbol of Jews and Judaism. A remarkable rabbinic image in the ancient midrashic work *Tanchuma* tells how the archetypal enemy of the Jews in Scripture, the wickedly nihilistic tribe of Amalek, abused the bodies of slain Jewish males. They would "cut off the circumcised organs and fling them upward," a sign of contempt for Heaven.[8]

Comparing the Nazis with Amalek is common in modern Jewish discourse, but some Nazis too saw themselves that way. When Julius Streicher was hung, his last words were to cry out bitterly, "Purim Festival 1946!" It was a reference to the Jewish holiday commemorating the events recounted in the book of Esther. In the story, a minister in the Persian royal court, Haman, descendant of the Amalekite king Agag, seeks to exterminate the Jews but is executed himself in the end, by hanging. As historian Robert Conot writes in *Justice at Nuremberg*, this demonstrates Streicher's "fascination with and knowledge of Judaism."[9] We could say the same of Hitler.

Truth and Consequences

It can't be said often enough that Darwinism's intellectual fruits, including, indirectly, Nazism, do not of course by themselves invalidate Darwin's theory of evolution. But it is a very good reason to think twice about whether the idea is true, to ask yourself if you have really examined the evidence, independently and critically, on all sides. Many theistic evolutionists, wishing to avoid joining the argument against Darwinism, may go on constructing their ingenious, ever shifting rationales, like clouds that take the fanciful shapes of a rabbit or an elephant on a breezy day but soon disperse and blow away. Meanwhile in the real world, the consequences of Darwin's idea remain as they always have been, contributing to the alienation of men and women from their Creator, reinforcing a vulgar, degraded public life where "evolution made me do it" is a reasonable excuse for all manner of immorality. Clever cosmologists and philosophers console themselves in the high tower of the academy. Down below on the street, Darwinism's effect has been to make it harder than ever for human beings to muster the personal fortitude to be good.

To say, in Maimonides' phrase, that everything is bound up with this problem is no exaggeration. More than any medieval sage ever knew, it is the reality with which Jews, other religious people and friends of religion are now confronted daily. Will we capitulate, surrendering without even bothering to consider the issue, all the while pretending to have done no such thing? Or will we resist?

Conclusion

Jay W. Richards

The debate about God and evolution bridges academic disciplines as diverse as physics and theology, draws on the latest scientific evidence, engenders controversy in virtually every setting and bears on significant and perennial questions: Does the universe point beyond itself, or does it, so far as we can tell, appear to stand alone, without need of explanation? Is it a creature or a brute fact? Does the universe exist for a purpose? Does it bear evidence of the same? Can we explain everything we see around us in nature by impersonal, even apparently purposeless processes? Should everything be explained thus?

The debate also raises questions that religious believers must ask in every generation: What truths should we accommodate, and what truth claims should we challenge? And how do we tell the difference between truths and mere truth claims, especially when there is social pressure to conform to prevailing opinions?

The ancients Hebrews often struggled to resist the claims of the Canaanites, Babylonians, and Greeks. Early Christians had to learn to defend their young faith against what St. Ireneus called the "foreign erudition" of pagan philosophers, and the Church Fathers had to fight Arian, Gnostic, and other heresies.

It's child's play to detect the intellectual compromises of believers in earlier generations. We are troubled, for instance, that many Christians once thought they had the right to own their fellow human beings. And we are amused that many Christians once identified the Aristotelian-Ptolemaic model of the universe with Christian theology. The real test of discernment, however, is to detect and resist false intellectual orthodoxies in one's own time.

If there is one recurring critique in this volume, it is that prominent theistic evolutionists have failed to pass this test. They have erred in trying to integrate Neo-Darwinism and naturalistic assumptions into their theology. While any responsible theology will accommodate whatever is truly known about the natural world, Neo-Darwinism writ large is based, not on a clear evaluation of the evidence of nature, but on social convention and an arbitrary requirement—called methodological naturalism—that science appeal only to impersonal, even purposeless processes. Social convention, however, is not scientific evidence. On simple logical grounds, if nature cannot be adequately explained without intelligence, then any rule that prevents scientists from exploring that possibility will lead them to false conclusions.

In this volume, we have seen what happens when theists try to reconcile their faith with Darwinism and methodological naturalism: they speak in vague and confusing generalities, use crucial terms ambiguously, and/or jettison essential parts of their theological traditions. If orthodox theism and orthodox Darwinism are incompatible creeds, this is just what we should expect.

With enough mental gymnastics and redefinition of terms, perhaps it is possible to concoct a hybrid of Darwinism and theism that is neither fish nor foul. But surely, before exploring such arcane possibilities, we should do our best to evaluate carefully the empirical evidence for Darwinian (and other materialist) claims. Instead, too many theistic evolutionists have put accommodation ahead of evaluation, showing remarkable naïveté in their evaluation of the relevant evidence. The recurring appeal to "junk DNA" is only one example. Over a century and a half after Darwin first proposed his theory, all we have are trivial examples that no one disputes—antibiotic resistance in bacteria, fluctuating beak size in finches, and so forth—and little or no evidence that life otherwise submits to Darwinian explanation. On the contrary, we have strong evidence that such explanations are extremely limited in scope.

Nevertheless, theistic evolutionists such as Francis Collins, Ken Miller, Denis Lamoureux and others insist that believers accommodate the Darwinian story. In some cases, the accommodation on offer is so comprehensive that we are told that Darwinism or methodological naturalism arises not from biology but from theology. Some, like Michael Tkacz, claim that the doctrine of creation requires that every organism have an explanation within nature, and seem more troubled by intelligent design than by Darwinism. Others, like Howard Van Till, discover in the writings of the Church Fathers, not intelligent design, but rather a "Robust Formational Economy Principle," in which nature does its own creating. Presumably the fact that these putatively theological arguments fit like a hand in a glove with methodological naturalism is just a coincidence.

Others, such as Karl Giberson, argue that, rather than challenging theology, Darwinism is actually a welcome partner. If life is largely the result of selection and random variation, so the argument goes, then God is less implicated in natural evil. But as we've seen, such arguments either fail to solve the dilemma confronting them, or *dissolve* the dilemma by compromising one if its premises. If God has delegated his creative activity to the functional equivalent of the Gnostic Demiurge, then he has merely passed the buck. There is no payoff in this scenario unless God fails to know or control the future. In that case, God may be less implicated in the natural evil described in the Darwinian narrative, but he is no longer providentially guiding his creation in any traditional sense. Such problems, it seems, are intrinsic rather than incidental to the project of reconciling traditional theism and Darwinism.

What about intelligent design? As we have seen, ID is often misrepresented, not least by prominent theistic evolutionists, who may identify it with young earth creationism, special creationism, or even heresy. In truth, intelligent design is compatible with many meanings of the word "evolution" and with non-Darwinian accounts of biological evolution. Intelligent design goes far beyond a critique of Darwinism and other materialistic theories, of course. ID proponents provide positive arguments for design based on public evidence from cosmology to biology. Believers have every right to be cautious about ID arguments. But it should be obvious that, all things being equal, intelligent design is much more congenial to theistic religion than is the theory that atheist Richard Dawkins observed "made it possible to be an intellectually fulfilled atheist."[1] As the arguments in this volume make clear, to the degree that theistic evolution is theistic, it will not be fully Darwinian. And to the degree that it is Darwinian, it will fail fully to preserve traditional theism.

Discussion Questions

Introduction: Squaring the Circle

Jay W. Richards

1. What popular stories are frequently used to illustrate the supposed "warfare" between science and religion? How have recent historians of science responded to this warfare thesis?

2. How would you answer the question: Are science and religion in conflict?

3. Why does the word "creationism" carry so much baggage? Can you describe the different "creationist" views? What is intelligent design and how does it differ from creationism?

4. Is it possible to believe in both "theistic evolution" and "intelligent design"? Why or why not?

5. What are the main differences between theism, deism, pantheism, and panentheism?

6. What do theologians mean when they say that God can act both primarily and through "secondary" causes?

7. Can you define the different senses of the word "evolution"? Why is it important to distinguish these senses when considering God and evolution?

8. What is the central dilemma in reconciling theism with Darwinian evolution?

1. Nothing New under the Sun

by John West

1. Has the doctrine of creation been important historically in Christian theology?

2. Who were the Epicureans and what did they believe? How did early Christians respond to them?

3. What are two common beliefs held by the Gnostics in the early centuries of Christianity?

4. What current debate does John West compare to the debate between the early Christians and the Epicureans? What current debate does he compare to the debate between the early Christians and Gnostics? How do the contemporary debates differ from these early debates?

5. How do most theistic evolutionists today differ from prominent theistic evolutionists who were contemporaries of Darwin, such as Alfred Wallace and Asa Gray?

6. Why does West compare natural selection to the Gnostic idea of a Demiurge?

7. Why does West suggest that many theistic evolutionists depict God as a "cosmic trickster"?

8. How important is the doctrine of a historical fall to Christian theology?

2. Having a Real Debate

John West

1. Does John West think that theistic evolution can win over much of the scientific community? Why or why not?

2. Why, according to West, is there a resurgence of interest in theistic evolution?

3. What do you think of Karl Giberson's "practical" reasons for remaining a Christian?

4. West mentions scientific critiques of Neo-Darwinism. What are they? In your opinion, why do many theistic evolutionists avoid this evidence?

5. West argues: "In their zeal for promoting Darwin, some theistic evolutionists exhibit an almost _____ ______ to the authority of scientists to set the agenda for the rest of the culture." Why does he see this as a troubling tendency?

6. How does West think Christians should respond to claims of scientific "consensus" on controversial issues?

3. Smelling Blood in the Water

Casey Luskin

1. What are the two camps of Neo-Darwinism?

2. Casey Luskin argues that the National Center for Science Education's (NCSE) pro-religion stance may be a "posture." What evidence does he provide for that charge?

3. Biologist Francisco Ayala is a former priest who now refuses to state his religious views. What reason does Luskin imply that Ayala might have for doing so?

4. What is the "big debate" involving new atheists, the evolution defense lobby, and theistic evolutionists?

5. What does NOMA (Non-Overlapping Magisteria) refer to? Why does Luskin think the idea fails?

6. What elements in Darwin's theory of evolution (as most textbooks define it) bring it into conflict with theistic religion?

7. Why does Luskin think it is foolish for theistic evolutionists to ally themselves with atheists?

4. Death and the Fall

William Dembski

1. Why does William Dembski think that anyone who accepts "evolution" as Neo-Darwinists understand it will have a hard time holding onto the concept of *imago dei*?

2. What is theistic evolutionist Karl Giberson's view of the fall, and how does it differ from the traditional orthodox account?

3. Is the fall an important doctrine in Christian theology? Why or why not?

4. Former Catholic priest Francisco Ayala has argued: "A major burden was removed from the shoulders of believers when convincing evidence was advanced that the design of organisms need not be attributed to the immediate agency of the Creator, but rather is an outcome of natural processes." What burden does Ayala think has been removed?

5. Why does Dembski think that Darwinism is no help to theism in resolving the problem of evil?

6. Theistic evolutionist Karl Giberson has argued: "[T]he gift of creativity that God bestowed on the creation is theologically analogous to the gift of freedom God bestowed on us. Both we and the creation have freedom." What does Dembski say about the claim that impersonal nature has "freedom"? Do you agree with Giberson or Dembski? Why?

7. Why does Dembski think that Darwinism, if true, would make the problem of evil more rather than less acute? Do you agree with him? Why or why not?

5. Random Acts of Design

Jonathan Witt

1. What evidence of intelligent design does Francis Collins affirm?

2. How does Collins characterize ID arguments in biology?

3. Can you summarize Michael Behe's argument that certain "molecular machines" are irreducibly complex?

4. Some critics have raised an objection to Behe's argument that the bacterial flagellum is irreducibly complex. What is it? What are the responses to this objection? Does Francis Collins appear to know the details of this debate, including the responses Witt mentions?

5. What is methodological naturalism (or methodological materialism)? Does Collins adhere to this principle consistently? Why or why not?

6. What dilemma does Witt argue plagues Francis Collins's view of providence?

6. Darwin of the Gaps

Jonathan Wells

1. According to Jonathan Wells, it "is not evolution in general, but _____________ that intelligent design proponents reject." Why is this distinction important in the question of God and evolution?

2. According to Francis Collins, what evidence "provides 'powerful support for Darwin's theory of evolution. . . .'"? What are the two parts of Darwin's theory? How does Jonathan Wells respond to Collins's claim?

3. What is the difference between macroevolution and microevolution? Collins claims that the distinction is "increasingly seen to be artificial." How does Wells respond to this claim?

4. According to Wells, what is the evidence against the claim that study of DNA provides overwhelming support for universal common ancestry?

5. What is "junk DNA"? Why does Francis Collins argue that it provides evidence against intelligent design? What is wrong with Collins's argument, according to Wells?

6. What does Wells mean when he calls Collins's argument a "Darwin of the gaps" argument?

7. Making a Virtue of Necessity

Jay W. Richards

1. What is Howard Van Till's "Robust Formational Economy Principle"?

2. According to Jay Richards, what is the chief advantage of the Principle?

3. Why does Richards think the Principle has closer affinities to naturalism than to theism?

4. According to Richards, what unstated theological principle lies behind the Robust Formational Economy Principle?

5. What does Richards argue is "perhaps the most obvious objection to Van Till's theological aesthetic, at least for Christians"?

6. What, according to Richards, is the "fundamental theological question for justifying one's belief that the Principle holds in our world"?

7. What reasons does Richards give for his charge that Howard Van Till is making a virtue of necessity?

8. The Difference It Doesn't Make

Stephen C. Meyer

1. Can you summarize Denis Lamoureux's view of theistic evolution, or what he calls teleological evolution" or "evolutionary creation"?

2. Why does Lamoureux object to the phrase "theistic evolution"?

3. What is Lamoureux's objection to intelligent design?

4. Does intelligent design rely on a specific interpretation of Scripture?

5. Why does Meyer argue that modern ID arguments are not God-of-the-gaps arguments?

6. Why does Meyer argue that repetitive physical laws cannot produce the information we see in the biological world?

7. What are the two ways to understand Lamoureaux's argument concerning the origin of biological information, according to Stephen Meyer?

8. What dilemma does Meyer detect in Lamoureux's use of the word "teleological"?

9. Everything Old is New Again

Denyse O'Leary

1. What was the older Catholic apologists' "main point" against Darwinism?

2. Why, according to Denyse O'Leary, was Wallace neglected and ridiculed, but Darwin lionized?

3. Did Belloc object to "evolution" generally? What did he say about Darwin's theory?

4. What was Chesterton's view of Darwin's theory? What did Chesterton predict would happen to the theory? Was he right?

5. What was Mivart's critique of natural selection? Do you think it is still relevant today? Why does O'Leary refer to Mivart as a "tragic figure"?

6. Why does O'Leary think that Darwinism continues to prosper despite the lack of evidence in its favor?

7. How does O'Leary depict the Church's response to Darwin's theory?

10. Can a Thomist Be a Darwinist?

Logan Gage

1. What is an "essence" in Aristotelian and Thomistic thought? According to Logan Gage, how does this concept conflict with Darwinism?

2. What is transformism? Why does Gage think that "those defending the tradition of natural philosophy found in Aristotle and St. Thomas simply cannot accept" it? What does he think the Thomist would need to change in Darwinian Theory to accept transformation of one species into another?

3. What is occasionalism? How did St. Thomas' view differ from occasionalism?

4. What is an exemplar cause? Why should the concept matter to Thomists who wish to affirm Darwinism?

5. What are the four causes, according to Aristotle?

6. What criticisms have some Thomists raised against ID? How does Gage respond to these criticisms?

7. How do modern ID arguments differ from St. Thomas' arguments for the existence of God? Must one choose one or the other?

11. Is There a Catholic View of Creation and Evolution?

Jay W. Richards

1. According to Catholic teaching, on what basis can we know that God exists?

2. How would you explain the "official" Catholic position on evolution, as Richards describes it? Do you think this position is widely understood by Catholics?

3. What senses of the word "evolution" can make it hard for Catholics to understand the Church's objections to Darwinian evolution?

4. What is creation *ex nihilo*?

5. What types of ways does God act in the world?

6. Richards argues that the claims of Michael Tkacz "look like deductions from naturalism, rigid Aristotelianism, or a hybrid of the two, not like implications of Thomism." What does he mean? Why does Richards make this charge?

7. What types of ID critics does Richards describe? Does he think that ID is well understood by most of these critics?

12. Separating the Chaff from the Wheat

Jay W. Richards

1. What is the difference between "mechanism" and "reductionism"? Why does Richards argue that Catholics should be careful to distinguish the two concepts?

2. What is "teleo-mechanism"? Why, according to Richards, is this a better term to apply to thinkers such as Isaac Newton and William Paley?

3. Which of Aristotle's beliefs, according to Richards, is most at odds with Christian theology?

4. What is "immanent teleology?" How does St. Thomas' view of teleology differ from Aristotle's?

5. How does Thomas' concept of exemplar cause "split the difference between Platonic ideas or forms ... and Aristotelian forms"?

6. Why did Robert Boyle object to Aristotelian philosophy? Richards argues that "held together, the contrasting views of nature in Boyle and Aristotle provide a valuable lesson." What is that lesson?

13. Understanding Intelligent Design

Jay W. Richards

1. Jay Richards argues that ID is a tertium quid. What does he mean?

2. According to Richards, "Earlier design arguments tended to work with only two categories: order and chance/chaos." How do modern ID arguments go beyond these earlier design arguments?

3. Why do ID theorists sometimes contrast intelligent design with natural laws and constants? Does it follow that, according to ID, laws are not the result of design?

4. How does the Darwinian "mechanism" differ from other physical mechanisms?

5. Why does Richards argue that many Catholics unwittingly adopt a view of science that is a hybrid of traditional and modern ideas?

6. What problems are there with distinguishing strongly between natural science and other disciplines, such as philosophy? How would St. Thomas have distinguished different areas of inquiry?

7. What is scientism? How are scientism and materialism different?

8. Why do ID proponents often maintain that ID is science?

14. The Maimonides Myth and the Great Heretic

David Klinghoffer

1. According to David Klinghoffer, how have many Orthodox Jews responded to Darwinism? Does he think this response divides Jews "along an easily defined and expected religious divide"?

2. Does Klinghoffer think Maimonides would have sought to reconcile God and Darwin? Why or why not?

3. Was intelligent design a concern to Maimonides?

4. Who is the great heretic? Why does Klinghoffer connect Epicurus and Darwin?

5. Klinghoffer summarizes Ibn Ezra's teaching in this way: "For the simple, ... simple faith. For the wise, intelligent design." What does Klinghoffer mean?

6. What "deficiency" or "lack" in the creation does Scripture describe? Why, according to Rabbinical interpretation, would this need to be part of the world God has created?

7. Why, according to Rabbeinu Bachya, did God not accomplish all of creation through one law?

15. God's Image, Our Mission

David Klinghoffer

1. What do you think it means to be made in the image of God?

2. Why does David Klinghoffer argue that Darwinism makes the biblical idea that we are created in God's image "incomprehensible"?

3. Before reading Klinghoffer's essay, did you know that the image of God described in the Bible refers not just to our spiritual aspect, but to our physical aspect as well? Why is this important?

4. What is "the moral message of Darwinism"?

5. What does "holiness" mean, according to Klinghoffer?

6. According to Klinghoffer, what "inner logic" did Hitler perceive in "Darwinism that even Charles Darwin did not"?

Endnotes

Introduction by Jay W. Richards

1. This is a popular way of speaking. See for instance "Creation Through Evolutionary Means" at: http://science.drvinson.net/creationviaevolution-online. In the last several years, there have been a number of books that have defended some version of this thesis. See, for example, Denis Alexander, *Creation or Evolution: Do We Have to Choose?* (Oxford: Monarch Books, 2008); Ted Peters and Martinez Hewlett, *Can You Believe in God And Evolution? A Guide for the Perplexed* (Nashville: Abingdon Press, 2006); Keith Miller, editor, *Perspectives on an Evolving Creation* (Grand Rapids: Eerdmans, 2003); Denis O. Lamoureux, *Evolutionary Creation: An Evangelical Approach to Evolution* (Eugene: Wipf & Stock, 2008); and Karl W. Giberson, *Saving Darwin: How to Be a Christian and Believe in Evolution* (New York: HarperOne, 2008).

   Jeremy Manier reports on a number of the recent thinkers who try to reconcile God and evolution:

   > The religious response is simple, some believers say: God used evolution to create us. But invoking God's direct guidance raises daunting scientific hurdles. Ever since the publication in 1859 of Darwin's "Origin of Species," religious writers have tried to cram the idea of design back into evolution, often without success.

   In "The New Theology," *Chicago Tribune* (Jan. 19, 2008).

2. From J. R. Lucas, "Wilberforce and Huxley: A Legendary Encounter," *The Historical Journal* 22, no. 2 (1979): pp. 313–330, available online at: http://users.ox.ac.uk/~jrlucas/legend.html. Lucas goes on to debunk the "legendary" aspects of the story.
3. The Wikipedia entry "1860 Oxford evolution debate" characteristically perpetuates this mythology. See: http://en.wikipedia.org/wiki/1860_Oxford_evolution_debate.
4. I discuss Warfield's views in detail in "The Evolving Debate Over Origins," *Review of B. B. Warfield: Evolution, Science and Scripture, Selected Writings* edited by Mark A. Noll and David N. Livingstone, *Touchstone* (October 2001), at: http://www.touchstonemag.com/archives/article.php?id=14-08-044-b.
5. For a brief summary, see Guillermo Gonzalez and Jay Richards, *The Privileged Planet: How Our Place in the Cosmos is Designed for Discovery* (Washington, DC: Regnery, 2004), chapter 11. See also Rodney Stark, *For the Glory of God: How Monotheism Led to Witch Hunts, Science, and the End of Slavery* (Princeton: Princeton University Press, 2004), and Alvin Plantinga, "Religion and Science," *Stanford Encyclopedia of Philosophy*, at: http://plato.stanford.edu/entries/religion-science/.
6. Aristotle distinguished between four "causes," that is, factors that explain or are responsible for something else. To put it simply, the material cause explains what something is made of; the formal cause explains what something is; the final cause explains the ultimate purpose toward which something tends, and the efficient cause explains where something came from, or what or who produced it.
7. "Intelligent Design: A Brief Introduction," in *Evidence for God*, edited by William A. Dembski and Michael R. Licona (Grand Rapids: Baker Books, 2010), p. 104.
8. Daniel R. Brooks and E. O. Wiley, *Evolution as Entropy* (Chicago: University of Chicago Press, 1986), p. xi, quoted in M. A. Corey, *Back to Darwin: The Scientific Case for Deistic Evolution* (Lanham: University Press of America, 1994), p. 3.

9. G. G. Simpson, *The Meaning of Evolution: A Study of the History of Life and of Its Significance for Man*, revised edition (New Haven: Yale University Press, 1967), p. 345.
10. Francis Darwin, *Life and Letters of Charles Darwin*, volume 1 (New York: Appleton, 1887), pp. 280, 283–284, 278–279.
11. For a discussion of the relationship between creation and conservation, see Jonathan Kvanvig, "Creation and Conservation," *Stanford Encyclopedia of Philosophy*, at: http://plato.stanford.edu/entries/creation-conservation/.
12. Here's how Merriam-Webster defines theism: "belief in the existence of a god or gods; specifically: belief in the existence of one God viewed as the creative source of the human race and the world who transcends yet is immanent in the world," at: http://www.merriam-webster.com/dictionary/theist.
13. This seems to be the position of theologian John Haught, for instance, who follows the thought of Teilhard de Chardin. See Haught's *Making Sense of Evolution: Darwin, God, and the Drama of Life* (Philadelphia: Westminster/John Knox Press, 2010), and *God After Darwin: A Theology of Evolution* (Boulder: Westview Press, 2007).
14. Beyond this point, the details get complicated. There is an extensive literature dating back to the medieval Catholic scholastics, who discussed the various modes of divine action, and how God's action relates to natural causation. The three main views are "occasionalism," "mere conservationism," and "concurrentism." In occasionalism, God alone brings about every effect in nature. In mere conservationism, according to Alfred Freddoso, "God contributes to the ordinary course of nature solely by creating and conserving natural substances along with their active and passive causal powers or capacities." In concurrentism,

    > a natural effect is produced immediately by both God and created substances, so that, contrary to occasionalism, secondary agents make a genuine causal contribution to the effect and in some sense determine its specific character by virtue of their own intrinsic properties, whereas, contrary to mere conservationism, they do so only if God cooperates with them contemporaneously as an immediate cause in a certain "general" way which goes beyond the conservation of the relevant agents, patients, and powers, and which renders the resulting effect the immediate effect of both God and the secondary causes.

    Alfred J. Freddoso, "God's General Concurrence with Secondary Causes: Pitfalls and Prospects," at: http://www.nd.edu/~afreddos/papers/pitfall.htm. See also Alfred J. Freddoso, "God's General Concurrence with Secondary Causes: Why Conservation is Not Enough," *Philosophical Perspectives* 5 (1991): pp. 553–585, available online at: http://www.nd.edu/~afreddos/papers/conserv.htm. Among the Abrahamic faiths, some version of concurrentism (though not necessarily designated as such) is by the far the most common position. However, many modern theistic evolutionists seem inclined toward "mere conservationism." In that sense, these traditional debates are relevant to our current discussion. For our purposes, however, the crucial point is that God in the theistic view is free to act in a variety of ways both within the created order and as Creator of that order.
15. In *Darwinism, Design, and Public Education*, edited by John Angus Campbell and Stephen C. Meyer (Lansing: Michigan State University Press, 2004).
16. See the explanation for the meaning of "evolution" from the BioLogos Foundation, which seeks to give a Christian interpretation and defense of evolution. The explanation begins with "change over time," then goes on to fill out the definition with common descent and the Darwinian mechanism. But it quickly slips from defining the term to presenting the details as if they were uncontested facts. At: http://biologos.org/questions/what-is-evolution/.

17. See discussion of this point in the comments of Thomas Cudworth, "Olive Branch from Karl Giberson," *Uncommon Descent* (April 15, 2010), at: http://www.uncommondescent.com/intelligent-design/olive-branch-from-karl-giberson/#more-13010.
18. (New York: Penguin Books, 2008).
19. I am not referring to Darwin's personal beliefs, which seem to have varied over time. I am referring to the actual arguments Darwin made in the *Origin*, which use deistic rather than atheistic or materialistic premises. Whether this reflected Darwin's personal beliefs or merely a rhetorical strategy is a separate and difficult question. See, for instance, Neil C. Gillespie, *Charles Darwin and the Problem of Creation* (Chicago: University of Chicago Press, 1979), and M. A. Corey, *Back to Darwin: The Scientific Case for Deistic Evolution*, pp. 6–25.
20. Michael Ruse, *Darwinism Defended: A Guide to the Evolution Controversy*, with a foreword by Ernst Mayr (Reading, Mass.: Addison-Wesley, 1982), xi–xii. Quoted in Ibid.
21. William Dembski, "Converting Matter into Mind: Alchemy and the Philosopher's Stone in *Cognitive Science*," *Perspectives on Science & the Christian Faith* 12 (1990), at: http://www.asa3.org/ASA/topics/PsychologyNeuroscience/PSCF12-1990Dembski.html#7.
22. Richard Dawkins, *The Blind Watchmaker: Why the Evidence of Evolution Reveals a Universe Without Design* (New York: W. W. Norton and Co., 1996), p. 6.
23. Timothy Keller, *The Reason for God: Belief in God in an Age of Skepticism* (New York: Dutton, 2008), p. 94.
24. Ibid., p. 87.
25. Keller is quoting from David Atkinson, *The Message of Genesis 1–11* (Downers Grove: InterVarsity Press, 1990), p. 31.
26. This is in a white paper written for the BioLogos Foundation, "Christian, Evolution, and Christian Laypeople," available online at: http://www.biologos.org/uploads/projects/Keller_white_paper.pdf.
27. The first six meanings are drawn from the essay by Stephen C. Meyer and Michael N. Keas, "The Meanings of Evolution."

1. Nothing New Under the Sun by John West

1. It should be emphasized that the Christian doctrine of "creation" is not the same thing as what is commonly referred to as "creationism." The Christian doctrine of creation focuses on the foundational truth that the universe (and everything within it) was created intentionally by a knowing and loving God rather than by a blind and undirected process or by multiple deities. A doctrine of creation need not specify the particular method God used to create the universe; nor does it require a particular reading of the Genesis account of creation. By contrast, "creationism" as it is commonly understood today seeks to use the Genesis account of creation to defend a specific account of the development of the universe and life, usually involving a young earth and creation in six 24-hour days rather than over billions of years.
2. Karl Giberson, *Saving Darwin: How to Be a Christian and Believe in Evolution* (HarperOne, 2008), p. 10, emphasis in the original.
3. Irenaeus, *Against Heresies*, Book II, chapter 1, in Alexander Roberts and James Donaldson, editors, *The Ante-Nicene Fathers*, vol. 1: *The Apostolic Fathers, Justin Martyr, Irenaeus* (1885), available online at: http://www.ccel.org/ccel/schaff/anf01.ix.iii.ii.html.
4. Philip Schaff, *History of the Christian Church*, vol. III (Grand Rapids: Eerdmans Publishing, 1980), p. 668.

5. See discussion in Schaff, *History of the Christian Church*, vol. II, pp. 538–541; also see the comparative table of creedal statements on pp. 536–537.
6. Ibid., p. 540.
7. See discussion in John G. West, *Darwin Day in America: How Our Politics and Culture Have Been Dehumanized in the Name of Science* (Wilmington: ISI Books, 2007), pp. 5–10.
8. Cited in West, *Darwin Day in America*, p. 8.
9. Theophilus, *Theophilus to Autolycus*, Book I, Chapter V, available in *Design in the Bible and the Early Church Fathers* (Discovery Institute, 2009), p. 11, available online at: http://www.discovery.org/scripts/viewDB/filesDB-download.php?command=download&id=4431.
10. Ibid.
11. Dionysius, *The Books on Nature*, Part II, in *Design in the Bible and the Early Church Fathers*, p. 13.
12. Ibid., p. 14.
13. Lactantius, *The Divine Institutes*, Book II, Chapter IX, in *Design in the Bible and the Early Church Fathers*, p. 19.
14. For further examples, see *Design in the Bible and the Early Church Fathers*, pp. 11–22; also see William A. Dembski, Wayne J. Downs, and Fr. Justin B. A. Frederick, editors, *The Patristic Understanding of Creation: An Anthology of Writings from the Church Fathers on Creation and Design* (Riesel, TX: Erasmus Press, 2008).
15. See Richard Dawkins, *The God Delusion* (New York: Mariner Books, 2008); Sam Harris, *Letter to a Christian Nation* (New York: Alfred Knopf, 2006); Daniel Dennett, *Darwin's Dangerous Idea: Evolution and the Meanings of Life* (New York: Touchstone, 1996).
16. Richard Dawkins, *River Out of Eden: A Darwinian View of Life* (New York: Basic Books, 1995), p. 133.
17. For good introductions to the modern theory of intelligent design, see William Dembski, *Intelligent Design: The Bridge Between Science and Theology* (Downers Grove: InterVarsity Press, 1999); William Dembski and James Kushiner, eds., *Signs of Intelligence: Understanding Intelligent Design* (Grand Rapids: Brazos Press, 2001); and Guillermo Gonzalez and Jay W. Richards, *The Privileged Planet: How Our Place in the Cosmos is Designed for Discovery* (Washington, DC: Regnery, 2004).
18. Hippolytus, *The Refutation of All Heresies*, Book VI, Chapter XXVIII, in Alexander Roberts and James Donaldson, editors, *The Ante-Nicene Fathers*, vol. 5: *Hippolytus, Cyprian, Caius, Novatian, Appendix*, available online at: http://www.ccel.org/ccel/schaff/anf05.iii.iii.iv.xxix.html.
19. Irenaeus, *Against Heresies*, Book I, Chapter XXVI, in *The Ante-Nicene Fathers*, vol. 1: *The Apostolic Fathers, Justin Martyr, Irenaeus*, available online at http://www.ccel.org/ccel/schaff/anf01.ix.ii.xxvii.html.
20. Peter Bowler, *Darwinism* (New York: Twayne Publishers,1993), p. 6.
21. Charles Darwin, *The Variation of Animals and Plants under Domestication*, second edition (London: John Murray, 1875), vol. I, p. 6, emphasis added; also see Charles Darwin, *On the Origin of Species by Means of Natural Selection, or the Preservation of Favoured Races in the Struggle for Life*, third edition (London: John Murray, 1861), pp. 84–85.
22. Charles Darwin, *The Autobiography of Charles Darwin and Selected Letters*, edited by F. Darwin (New York: Dover Publications, 1958 reprint of 1892 ed.), p. 63.
23. West, *Darwin Day in America*, pp. 23–24.

24. Alfred Russel Wallace, *An Anthology of His Shorter Writings*, ed. by Charles H. Smith (New York: Oxford University Press, 1991), p. 33.
25. Asa Gray to James Dwight Dana, June 22, 1872, in Daniel C. Gilman, *The Life of James Dwight Dana* (New York: Harper and Brothers, 1899), p. 302.
26. See Edward Larson, *Summer for the Gods: The Scopes Trial and America's Continuing Debate Over Science and Religion* (New York: Basic Books, 1997), pp. 22–26.
27. John Polkinghorne, *Quarks, Chaos, and Christianity* (New York: Crossroad Publishing Company, 2005), p. 113.
28. George V. Coyne, S.J., "The Dance of the Fertile Universe," p. 7, available at: http://www.aei.org/docLib/20051027_HandoutCoyne.pdf.
29. Kenneth R. Miller, *Finding Darwin's God: A Scientist's Search for Common Ground Between God and Evolution* (New York: HarperCollins, 1999), p. 244.
30. Ibid., p. 272.
31. Ibid., pp. 238–239.
32. Kenneth Miller, comments during "Evolution and Intelligent Design: An Exchange," March 24, 2007, at the "Shifting Ground: Religion and Civic Life in America," conference, Bedford, New Hampshire, sponsored by New Hampshire Humanities Council.
33. See Giberson, *Saving Darwin*, pp. 12–13 and Miller, *Finding Darwin's God*, pp. 273–274.
34. See Proverbs 8: 1–3, 22–35 on the role of wisdom in God's creation of the world, esp. v. 30: "Then I was beside Him as a master craftsman." (NKJV) In the Christian tradition, the character of "wisdom" in these verses has been identified with Christ. More generally, on the artistic genius in creation, see Benjamin Wiker and Jonathan Witt, *A Meaningful World: How the Arts and Sciences Reveal the Genius of Nature* (Downers Grove, IL: InterVarsity, 2006). Note that I am not claiming here that God is merely a master artist or craftsman, only that these terms accurately describe one facet of God's role as creator in the Christian tradition.
35. Giberson, *Saving Darwin*, p. 213, emphasis in the original.
36. Giberson, *Saving Darwin*, p. 210. In a footnote to the statement cited here, Giberson inexplicably assures readers that "[t]his view of God's creative involvement does nothing, in principle, to compromise the traditional biblical affirmation that God 'numbers the hairs on our heads' or 'cares about the sparrow's fall.' There is no reason why God cannot be intimately concerned about such details, just because God is the [sic] not the immediate creative agent behind those details." Aside from sheer assertion, it is unclear how "rejecting the idea that God is responsible for the details" leads one to maintain that God is "intimately concerned about such details."
37. The quotes that follow come from the "Can a Christian Be a Darwinist?" event with Karl Giberson and John West at Biola University, Feb. 5, 2009, transcribed from the audio available at: http://www.faithandevolution.org/debates/can-a-christian-be-a-darwinist.php. For a description of this event, see http://www.biola.edu/news/articles/2009/090223_darwinist.cfm.
38. For a description of the "Open Theology and Science Conference," see: http://www.enc.edu/history/ot/open_theo.html. The conference was funded by the John Templeton Foundation. For a classic exposition of open theism by one of its proponents, see Clark Pinnock, *Most Moved Mover: A Theology of God's Openness* (Grand Rapids: Baker Academic, 2001). For a critique of open theism, see Bruce Ware, *God's Lesser Glory: The Diminished God of Open Theism* (Wheaton, IL: Crossway, 2000).

39. Francis S. Collins, *The Language of God: A Scientist Presents Evidence for Belief* (New York: Free Press, 2006), p. 136. For a critique of this view, see biologist Jonathan Wells's "Darwin of the Gaps" essay in this volume, and his book *The Myth of Junk DNA* (Discovery Institute Press, 2011).
40. Collins, *The Language of God*, p. 205.
41. Richard Dawkins, *The Blind Watchmaker: Why the Evidence of Evolution Reveals a World without Design* (New York: W.W. Norton, 1996), p. 1, emphasis added.
42. See *Design in the Bible* and the *Early Church Fathers*.
43. COMPASS Interview of John Shelby Spong by Geraldine Doogue , July 8, 2001, transcript available at: http://www.abc.net.au/compass/intervs/spong-01pv.htm.
44. Giberson, *Saving Darwin*, p. 12.
45. Ibid., p. 12.
46. Transcribed from the audio for the "Can a Christian Be a Darwinist?" event.
47. Collins, *The Language of God*, pp. 208–209. The passage cited by Collins can be found in C. S. Lewis, *The Problem of Pain* (New York: Collier Books, Macmillan Publishing, 1962), pp. 79–85.
48. See Mike Perry, article on "Evolution" in *The C. S. Lewis Readers' Encyclopedia*, edited by Jeffrey D. Schultz and John G. West (Grand Rapids: Zondervan, 1998), pp. 158–159. In 1951, Lewis wrote Bernard Acworth of the Evolution Protest Movement that he was now inclined "to think that you may be right in regarding it [evolution] as the central and radical lie in the whole web of falsehood that now governs our lives." Lewis to Bernard Acworth, Sept. 13, 1951, in *The Collected Letters of C. S. Lewis*, edited by Walter Hooper (San Francisco: HarperSanFrancisco, 2007), vol. III, p. 138. Also see Lewis's parody of unguided evolution in his letter to Dorothy Sayers, March 4, 1954, ibid., pp. 434–437.
49. Lewis, *The Problem of Pain*, pp. 78–79.
50. Ibid., p. 88, emphasis added.
51. Transcribed from the audio for the "Can a Christian Be a Darwinist?" event.
52. "Reason or faith? Darwin expert reflects," interview with Ron Numbers (Feb. 3, 2009), at: http://www.news.wisc.edu/16176, emphasis added.
53. "The 'Evidence for Belief': An Interview with Francis Collins" (April 17, 2008), at: http://pewforum.org/events/?EventID=178.
54. Giberson, *Saving Darwin*, p. 215.

2. Having a Real Debate by John West

1. The comment comes from Dawkins's interview by Ben Stein in the documentary *Expelled* (2008); it is quoted in David Klinghoffer, "Richard Dawkins: A Biography," at: http://www.discovery.org/a/10291.
2. See John West, "Meet the Materialists, part 1: Eugenie Scott, 'Evolution Evangelist,'" *Evolution News and Views* (October 9, 2007), at: http://www.evolutionnews.org/2007/10/meet_the_materialists_part_1_e004326.html.
3. See "Misconception: Evolution and religion are incompatible," at http://evolution.berkeley.edu/evosite/misconceps/IVAandreligion.shtml; *Statements from Religious Organizations* (Oakland, California: National Center for Science Education), http://www.ncseweb.org/resources/articles/5025_statements_from_religious_orga_12_19_2002.asp#home.
4. Eugenie Scott, "Dealing with Antievolutionism," at: http://www.ucmp.berkeley.edu/fosrec/Scott2.html.

5. In her tips for activists who want to support evolution before their school board, Eugenie Scott advises: "Call on the clergy. Pro-evolution clergy are essential to refuting the idea that evolution is incompatible with faith.... If no member of the clergy is available to testify, be sure to have someone do so—the religious issue must be addressed in order to resolve the controversy successfully." Eugenie C. Scott, "12 Tips for Testifying at School Board Meetings," at: http://www.ncseweb.org/resources/articles/7956_12_tips_for_testifying_at_scho_3_19_2001.asp (accessed July 16, 2005, but since removed), emphasis in original.
6. *Congregational Study Guide for Evolution* (Oakland: National Center for Science Education, 2001), at: http://ncse.com/religion/companion-guide-to-pbs-evolution-series.
7. Phina Borgeson, "Introduction to the Congregational Study Guide for Evolution," at: http://ncse.com/webfm_send/138.
8. For information about these initiatives, see: http://blue.butler.edu/~mzimmerm/rel_evol_sun.htm.
9. Richard Dawkins, "Is Science a Religion?" at: http://www.thehumanist.org/humanist/articles/dawkins.html.
10. Dawkins, quoted in Klinghoffer, "Richard Dawkins: A Biography."
11. "Humanism and Its Aspirations: Humanist Manifesto III." (Washington, DC: American Humanist Association), at: http://www.americanhumanist.org/who_we_are/about_humanism/Humanist_Manifesto_III.
12. Michael Shermer, "Darwin on the Right: Why Christians and conservatives should accept evolution," *Scientific American* (October 2006), at: http://www.scientificamerican.com/article.cfm?id=darwin-on-the-right.
13. Michael Shermer, "Science Is My Savior," at: http://www.science-spirit.org/article_detail.php?article_id=520 (accessed Feb. 23, 2007, but as of August 2010 this article no longer appears to be available on the web).
14. Philip Kitcher, *Living with Darwin* (New York: Oxford University Press, 2007), p. 152.
15. Neil Gross and Solon Simmons, "How Religious are America's College and University Professors?" (Feb. 6, 2007), available at: http://religion.ssrc.org/reforum/Gross_Simmons.pdf.
16. See Larry Witham, *Where Darwin Meets the Bible: Creationists and Evolutionists in America* (New York: Oxford University Press, 2002), pp. 271–273.
17. Gregory W. Graffin and William B. Provine, "Evolution, Religion and Free Will," *American Scientist* 95 (July–August 2007): pp. 294–297; results of Cornell Evolution Project survey, at: http://www.polypterus.com/results.pdf.
18. Interview with Francis Collins for PBS's "The Question of God" (2004), at: http://www.pbs.org/wgbh/questionofgod/voices/collins.html; see also Collins, *Language of God*, pp. 21–31.
19. For a discussion of Darwin's theory of morality presented in *The Descent of Man*, see West, *Darwin Day in America*, pp. 25–37.
20. Ibid., pp. 23–24.
21. Michael Shermer, comments made at the Origin of Life Debate sponsored by the American Freedom Alliance, Nov. 30, 2009, in Beverly Hills, California. The audio from which these comments were transcribed can be found at: http://www.youtube.com/watch?v=0tHW0nBskM4.
22. See West, *Darwin Day in America* for a detailed account.
23. Giberson, *Saving Darwin*, pp. 155–156, emphasis in the original.

24. See "From Calvinism to Freethought: The Road Less Traveled," Freethought Association of West Michigan, at: http://www.freethoughtassociation.org/minutes/2006/May24-2006.htm.

25. Data compiled from Templeton Foundation website, http://www.templeton.org/what-we-fund/grant-search/. Grants include $876,639 for The Adaptive Logic of Religious Belief and Behavior (2008), $2,028,238 for The Language of God: BioLogos Website and Workshop (2008), $248,779 for Dialogue, Research, and Public Broadcast at the Intersection of Darwinism and Religion (2007), $295,800 for the Venice School of Science and Religion (2007), $74,800 for *Saving Darwin: How to Be a Christian and Believe in Evolution* (2007), $2,000,000 for the Faraday Institute for Science and Religion (2006), $1,999,124 for *The Evolution and Theology of Cooperation: The Emergence of Altruistic Behavior, Forgiveness and Unselfish Love* (2005), $6,187,971 for the Templeton-Cambridge Journalism Fellowships and Seminars in Science and Religion (2004), $1,002,414 for Science, Theology, and the Ontological Quest (2003), and $5,351,707 for the AAAS Dialogue on Science, Ethics, and Religion: Promoting a Public Conversation (1996). Additional grants went for projects relating to evolution and the evolution of religious belief.

26. "About the BioLogos Foundation," at: http://biologos.org/about.

27. Interview with Francis Collins, at: http://www.pbs.org/faithandreason/transcript/coll-body.html.

28. "The Language of God: Biologos Website and Workshop," at: http://www.templeton.org/what-we-fund/grants/the-language-of-god-biologos-website-and-workshop.

29. See West, *Darwin Day in America*.

30. Ibid., pp. 123–162.

31. See, for example, Michael Denton, *Evolution: A Theory in Crisis* (Chevy Chase: Adler and Adler, 1986); Michael Behe, *Darwin's Black Box: The Biochemical Challenge to Evolution* (New York: Free Press, 1996) and *The Edge of Evolution* (New York: Free Press, 2007); William Dembski, *No Free Lunch: Why Specified Complexity Cannot Be Purchased without Intelligence* (Lanham: Rowman and Littlefield, 2002); John Angus Campbell and Stephen C. Meyer, eds., *Darwinism, Design, and Public Education* (East Lansing: Michigan State University Press, 2003); William A. Dembski and Michael Ruse, eds., *Debating Design: From Darwin to DNA* (New York: Cambridge University Press, 2004); Stephen C. Meyer, Scott Minnich, et. al., *Explore Evolution: The Arguments for and Against Neo-Darwinism* (London: Hill House Publishers, 2007); William Dembski and Jonathan Wells, *The Design of Life* (Dallas: Foundation for Thought and Ethics, 2008); David Berlinski, *The Deniable Darwin and Other Essays* (Seattle: Discovery Institute Press, 2009). Also see the "Peer-Reviewed and Peer-Edited Scientific Publications Supporting the Theory of Intelligent Design (Annotated)," Aug. 26, 2010, at: http://www.discovery.org/a/2640; the "Further Debate" section of the *Explore Evolution* website, at: http://www.exploreevolution.com/; and the scientific journal *BIO-Complexity*, at http://bio-complexity.org/.

32. John West, "Scientist Says His Peer-Reviewed Research in the Journal of Molecular Biology 'Adds to the Case for Intelligent Design,'" *Evolution News & Views* (Jan. 10, 2007), at: http://www.evolutionnews.org/2007/01/journal_of_molecular_biology_a.html. Also see Douglas D. Axe, "Estimating the Prevalence of Protein Sequences Adopting Functional Enzyme Folds," *Journal of Molecular Biology* 341, no. 5 (August 27, 2004): pp. 1295–1315.

33. "Response from Ralph Seelke to David Hillis Regarding Testimony on Bacterial Evolution before Texas State Board of Education, January 21, 2009" (March 23, 2009), at: http://www.discovery.org/a/9951; see also A. K. Gauger, S. Ebnet, P. F. Fahey, R. Seelke, "Re-

ductive evolution can prevent populations from taking simple adaptive paths to high fitness," *BIO-Complexity* 2 (2010): pp. 1–9. doi: 10.5048/BIO-C.2010.2.

34. The complete list of signers of the statement can be viewed at: http://www.dissentfromdarwin.org.
35. Quoted in Darry Madden, "UMass scientist to lead debate on evolutionary theory," *Brattleboro Reformer* (Feb. 2, 2006).
36. Phillip Skell, "Open Letter to the South Carolina Education Oversight Committee" (January 23, 2006), at: http://www.discovery.org/a/3174.
37. For a response to Collins, see Jonathan Wells's essay "Darwin of the Gaps" in this volume.
38. Stephen Meyer, *Signature in the Cell: DNA and the Evidence for Intelligent Design* (New York: HarperOne, 2009); Francisco Ayala, "On Reading the Cell's Signature," Jan. 7, 2010, at: http://biologos.org/blog/on-reading-the-cells-signature/.
39. For responses to Ayala's review and an account of what happened, see David Klinghoffer, editor, *Signature of Controversy* (Seattle: Discovery Institute Press, 2010), pp. 17–35, 57–58.
40. For a list of relevant endorsements of the book, see: http://www.signatureinthecell.com/quotes.php.
41. See John West, "Where's the Dialogue? Alas, Colleague of Francis Collins at 'Biologos' Doesn't Offer Any" (June 3, 2009), *Evolution News and Views*, at: http://www.evolutionnews.org/2009/06/wheres_the_dialogue_alas_colle021051.html.
42. John West, "Broadening the Faith and Evolution Debate," *Washington Post*'s On Faith website (June 2, 2009), available at: http://www.discovery.org/a/11081.
43. See John West, "Clarity and Confusion: Stephen Barr Answers My Questions," *Evolution News and Views* July 8, 2009), at: http://www.evolutionnews.org/2009/07/clarity_and_confusion_stephen022721.html.
44. G. Sermonti, "Darwin is a Prime Number," *Rivista di Biologia* 95 (2002): p. 10.
45. For information on those discriminated against or persecuted for their beliefs about Darwin or intelligent design, see West, *Darwin Day in America*, pp. 231–241; for responses to those who deny the reality of such discrimination, see *NCSE Exposed*, at: http://www.ncseexposed.org/.
46. For a discussion of some of Collins's bioethics views, including his support for embryonic stem cell research, some forms of human cloning, and his ambivalence on eugenic abortions, see Wesley J. Smith, "Collins heads to NIH," *To the Source* (July 30, 2009), at: http://www.tothesource.org/7_30_2009/7_30_2009.htm; and David Klinghoffer, "Francis Collins on Abortion: Obama's Pick for NIH and His 'Devout' Views on Terminating Down Syndrome Children," *Kingdom of Priests Blog, Beliefnet* (July 8, 2009), at: http://blog.beliefnet.com/kingdomofpriests/2009/07/francis-collins-on-abortion.html#ixzz0vbKsK6jG. Collins raises doubts about the "insistence that the spiritual nature of a person is uniquely defined at the very moment of conception" in the appendix to *The Language of God*, p. 251.
47. John West, "The Abolition of Man? How Politics and Culture Have Been Dehumanized in the Name of Science" (February 8, 2008), at: http://www.heritage.org/Research/Lecture/The-Abolition-of-Man-How-Politics-and-Culture-Have-Been-Dehumanized-in-the-Name-of-Science.
48. Kenneth P. Green and Hiwa Alaghebandian, "Science Turns Authoritarian," *The American* (July 27, 2010), at http://www.american.com/archive/2010/july/science-turns-authoritarian.

49. See Richard Weaver, *Ideas Have Consequences* (Chicago: University of Chicago Press, 1948), pp. 1–17; Richard Weaver, "Humanism in an Age of Science," "Social Science in Excelsis," and "Concealed Rhetoric in Scientistic Sociology," in Ted J. Smith III, editor, In *Defense of Tradition: Collected Shorter Writings of Richard M. Weaver*, 1929–1963 (Indianapolis: Liberty Fund, 2000), pp. 61–72, 141–143, 317–331; Eric Voegelin, "The Origins of Scientism," in Ellis Sandoz, editor, *The Collected Works of Eric Voegelin*, vol. X (Columbia: University of Missouri Press, 1989), pp. 168–196; G. K. Chesterton, *The Everlasting Man* (San Francisco: Ignatius Press, 1993), pp. 23–29 and *Orthodoxy* in *Collected Works of G. K. Chesterton*, vol. I (San Francisco: Ignatius Press, 1986); C. S. Lewis, *The Abolition of Man* (New York: Macmillan Publishing, 1947) and *That Hideous Strength* (New York: Macmillan Publishing, 1946). For a discussion of scientism, see Michael D. Aeschliman, *The Restitution of Man: C. S. Lewis and the Case Against Scientism* (Grand Rapids: Eerdmans, 1998).
50. Charles Darwin, *The Origin of Species* (New York: The Modern Library, 1993), p. 19.

3. Smelling Blood in the Water by Casey Luskin

1. William B. Provine, "Abstract of Will Provine's 1998 Darwin Day Keynote Address, Evolution: Free will and punishment and meaning in life," at: http://fp.bio.utk.edu/darwin/1998/provine_abstract.html.
2. Daniel C. Dennett, *Darwin's Dangerous Idea: Evolution and the Meanings of Life* (New York: Simon & Schuster, 1995), p. 63.
3. See Gary Wolf, "The Church of the Non-Believing," *Wired* (November 2006): pp. 182–193.
4. Richard Dawkins, quoted in *Expelled: No Intelligence Allowed* (Premise Media, 2008), Chapter 11: Watchdogs.
5. Eugenie Scott, quoted in *Expelled: No Intelligence Allowed* (Premise Media, 2008), Chapter 11: Watchdogs.
6. See "Notable Signers," at: http://www.americanhumanist.org/Who_We_Are/About_Humanism/Humanist_Manifesto_III/Notable_Signers.
7. Eugenie Scott, "Dealing with Antievolutionism," at: http://www.ucmp.berkeley.edu/fosrec/Scott2.html (emphasis added).
8. See Jon D. Miller, Eugenie C. Scott, Shinji Okamoto, "Public Acceptance of Evolution," Science 313 (August 18, 2006): pp. 765–766.
9. President Obama used such words in his inaugural address, as he promised: "We'll restore science to its rightful place...." See "President Barack Obama's Inaugural Address" (January 21, 2009), at: http://www.whitehouse.gov/blog/inaugural-address/.
10. Eugenie C. Scott, "Monkey Business," *The Sciences* 36 (January/February 1996): pp. 20–25.
11. Jerry Coyne, "Truckling to the Faithful: A Spoonful of Jesus Helps Darwin Go Down," *Why Evolution Is True Blog* (April 22, 2009), at: http://whyevolutionistrue.wordpress.com/2009/04/22/truckling-to-the-faithful-a-spoonful-of-jesus-helps-darwin-go-down/.
12. Eugenie Scott, quoted in *More* magazine (August 2005), as quoted by William Dembski, "'Evolution Evangelist,'" *Uncommon Descent* (August 24, 2005), at: http://www.uncommondescent.com/evolution/evolution-evangelist/.
13. Interview with Eugenie Scott on Minnesota Atheists, *Time Index* 17:20, at: http://mnatheists.org/atheist_talk/atheists_talk_068_05_03_2009.mp3.

14. Ibid., at *Time Index* 17:10. See also "Frequently Asked Questions about NCSE," at: http://ncse.com/about/faq.
15. Eugenie C. Scott, "Review of Johnson (1992)," in *Darwin*, ed. Philip Appleman (New York: W.W. Norton, 2001), p. 592.
16. Ibid.
17. Minnesota Atheists Interview with Eugenie Scott, at *Time Index* 48:05–48:50.
18. Steven Newton, "Preparatory Materials for Speakers at the 21 January 2009 Texas SBOE Meeting," 32, 44, originally at: http://skepchick.org/blog/wp-content/uploads/2009/01/prep_materials_21janmeeting1.pdf (accessed January 27, 2009), but removed immediately after the talking points were exposed as opposing the supernatural. Document now available at: http://www.discovery.org/scripts/viewDB/filesDB-download.php?command=download&id=4411.
19. Ibid.
20. Frank Press, "Preface," in *Science and Creationism: A View from the National Academy of Sciences*, first edition (Washington, DC: National Academy Press, 1984), p. 6.
21. Ralph J. Cicerone, Harvey Fineberg, Francisco J. Ayala, "Preface," Science, Evolution, and *Creationism: A View from the National Academy of Sciences* (Washington, DC: National Academy Press, 2008), p. xiii (emphasis added).
22. Edward J. Larson and Larry Witham, "Leading scientists still reject God," *Nature* 394 (July 23, 1998): p. 313.
23. Ibid. See also Edward J. Larson and Larry Witham, "Scientists and Religion in America," *Scientific American* 281 (September, 1999): pp. 88–93.
24. Ibid. (emphasis added).
25. Michael Shermer, *Why Darwin Matters: The Case Against Intelligent Design* (New York, NY: Henry Holt and Co., 2006), p. 114.
26. Michael Shermer, "Science is my Savior," *Science & Spirit*, at: http://www.science-spirit.org/printerfriendly.php?article_id=520.
27. Michael Shermer, "Michael Shermer," in *What We Believe but Cannot Prove: Today's Leading Thinkers on Science in the Age of Certainty*, ed. John Brockman (New York: Harper-Perennial, 2006), p. 38.
28. Michael Shermer, "Science is my Savior."
29. Shermer, *Why Darwin Matters*, p. 122.
30. Ibid., p. 115.
31. James Robert Brown, "Fundamentally Mistaken," review of *Why Darwin Matters: The Case Against Intelligent Design*, *American Scientist* 95 (March–April 2007), at: http://www.americanscientist.org/bookshelf/pub/fundamentally-mistaken.
32. Ibid.
33. Alan D. Attie, Elliot Sober, Ronald L. Numbers, Richard M. Amasino, Beth Cox, Terese Berceau, Thomas Powell, and Michael M. Cox, "Defending science education against intelligent design: a call to action," *Journal of Clinical Investigation* 116 (2006): pp. 1134–1138.
34. Steve Paulson, "Seeing the light—of science," *Salon* (January 2, 2007), at: http://www.salon.com/books/int/2007/01/02/numbers/.
35. Ibid.
36. Ibid.
37. Michael Ruse, "Creationists Correct?: Darwinians wrongly mix science with morality, politics," *National Post* (May 13, 2000).

38. Tamlar Sommer, "Interview with Michael Ruse," reprinted from *The Believer*, at: http://evans-experientialism.freewebspace.com/ruse.htm.
39. Michael Ruse, *Can a Darwinian be a Christian?: The Relationship between Science and Religion* (New York: Cambridge University Press, 2000), p. 217.
40. This enlightening exchange was leaked to William Dembski who posted it as "Remarkable exchange between Michael Ruse and Daniel Dennett" (February 21, 2006), at: http://www.uncommondescent.com/intelligent-design/the-ruse-dennett-briefwechsel-the-clash-between-evolution-and-evolutionism/.
41. See Claudia Dreifus, "A Conversation with Francisco J. Ayala; Ex-Priest Takes the Blasphemy Out of Evolution," *New York Times* (April 27, 1999), at: http://www.nytimes.com/1999/04/27/science/conversation-with-francisco-j-ayala-ex-priest-takes-blasphemy-evolution.html.
42. "Francisco and Hana Ayala: 'We Both Dare to Think Out of the Box,'" at: http://www.aaas.org/news/releases/2006/0308ayala.shtml and "Francisco J. Ayala," at: http://archives.aaas.org/people.php?p_id=198 and "Biologist Francisco J. Ayala Wins National Medal of Science" (May 9, 2002), at: http://www.universityofcalifornia.edu/news/article/4322.
43. See Cornelia Dean, "Roving Defender of Evolution, and of Room for God," *The New York Times* (April 29, 2008), at: http://www.nytimes.com/2008/04/29/science/29prof.html.
44. See Claudia Dreifus, "A Conversation with: Francisco J. Ayala; Ex-Priest Takes the Blasphemy Out of Evolution," *New York Times* (April 27, 1999), at: http://www.nytimes.com/1999/04/27/science/conversation-with-francisco-j-ayala-ex-priest-takes-blasphemy-evolution.html.
45. Ralph J. Cicerone, Harvey Fineberg, Francisco J. Ayala, "Preface," Science, Evolution, and *Creationism: A View from the National Academy of Sciences* xiii (Washington, DC: National Academy Press, 2008).
46. Francisco J. Ayala, *Darwin's Gift to Science and Religion*, (Washington, DC: Joseph Henry Press, 2007), p. 6.
47. Emphasis added. Ayala statements in "Beyond Belief" atheist conference at the Salk Institute on November 5, 2006 as seen at http://thesciencenetwork.org/programs/beyond-belief-science-religion-reason-and-survival/session-3-3 [Time index 46:45.]
48. Deposition of Francisco J. Ayala, Association of Christian Schools International et al. vs. Romans Stearns (June 12, 2007), p. 71.
49. Cornelia Dean, "Roving Defender of Evolution, and of Room for God," *The New York Times* (April 29, 2008).
50. Chris Mooney and Sheril Kirshenbaum, *Unscientific America: How Scientific Illiteracy Threatens Our Future* (New York, NY: Basic Books, 2009), pp. 97–98.
51. Chris Mooney, "Darwin's Sanitized Idea," *Slate* (September 25, 2001), at: http://bbs.slate.com/id/115965/.
52. Ibid.
53. Ibid.
54. *Yale Bulletin and Calendar*, at: http://www.yale.edu/opa/ybc/campusnotes.html.
55. Ibid.
56. Chris Mooney, "How Christian is gay bashing?" *Yale Herald* (October 23, 1998), at: http://www.yaleherald.com/archive/xxvi/10.23.98/opinion/gay.html.
57. Chris Mooney, "Humanists offer an alternative for agnostics," *Yale Herald* (April 17, 1998), at: http://www.yaleherald.com/archive/xxv/4.17.98/opinion/humanists.html.

58. "Chris Mooney," *Skeptical Inquirer*, at: http://www.csicop.org/author/christophermooney.
59. Chris Mooney, "Global Warming and the Posture of Skepticism," *DeSmogBlog* (February 28, 2008), at: http://www.desmogblog.com/global-warming-and-the-posture-of-skepticism-bjorn-lomborg.
60. "Chris Mooney Wins American Meteorological Society Book Award," *Science Progress* (January 14, 2009), at: http://www.scienceprogress.org/2009/01/chris-mooney-amsoc-battan-award/.
61. Chris Mooney, "Do I contradict myself? Very well then I contradict myself," *Discover Magazine Blogs The Intersection* (June 3, 2009), at: http://blogs.discovermagazine.com/intersection/2009/06/03/do-i-contradict-myself-very-well-then-i-contradict-myself/.
62. See Mooney and Kirshenbaum, *Unscientific America*, p. 18.
63. Ibid., p. 131.
64. Ibid.
65. Ibid., p. 132.
66. Ibid., p. 97.
67. Chris Mooney, "Why We Celebrate Darwin," *Discover Magazine Blogs The Intersection* (November 2, 2009), at: http://blogs.discovermagazine.com/intersection/2009/11/02/why-we-celebrate-darwin/.
68. Mooney, "Do I contradict myself?" (emphasis added).
69. Ibid.
70. Chris Mooney, "Why Evolution is True, But Coyne is Wrong About Religion, Part II: Lessons of Dover," *Discover Magazine Blogs The Intersection* (June 5th, 2009), at: http://blogs.discovermagazine.com/intersection/2009/06/05/why-evolution-is-true-but-coyne-is-wrong-about-religion-part-ii-lessons-of-dover/.
71. Mooney may have been enlightened on this strategy by the NCSE. A 2006 directive given by then-NCSE spokesperson Nicholas Matzke on the pro-Darwin blog *Panda's Thumb* warns: "We don't need the anti-creationists going and mixing their views on religion into their science" because "this is probably the surest path to disaster politically and in the courts." Matzke concludes: "Anyone who wants to do this has the right to do it, but it ain't helpful or particularly smart." See "Ron Numbers interview and article," *Panda's Thumb Blog* (June 24, 2006), at: http://www.pandasthumb.org/archives/2006/06/ron_numbers_int.html#comment-107918.
72. "Blasphemy Challenge targets youngsters," WSLS NewsChannel 10 (January 5, 2007) (accessed March 20, 2007).
73. Richard Dawkins, "Why There Almost Certainly Is No God," *The Huffington Post* (October 23, 2006), at: http://www.huffingtonpost.com/richard-dawkins/why-there-almost-certainl_b_32164.html.
74. Ibid.
75. Nicholas Matzke, on *The Inoculated Mind Podcast* (March 21, 2007), Time Index Approximately 1:19:00–1:20:00, at: http://www.inoculatedmind.com/?p=168.
76. Jerry Coyne, "Mooney and Kirshenbaum self-destruct at last" (August 11, 2009), at: http://whyevolutionistrue.wordpress.com/2009/08/11/mooney-and-kirshenbaum-self-destruct-at-last/.
77. See Jerry Coyne, "Seeing and Believing," *The New Republic* (February 4, 2009), at: http://www.tnr.com/article/books/seeing-and-believing?id=1e3851a3-bdf7-438a-ac2a-a5e381a70472.

78. Ibid.
79. Coyne, "Truckling to the Faithful."
80. Ibid.
81. Coyne, "Mooney and Kirshenbaum self-destruct at last."
82. Ibid.
83. Ibid. (emphasis in original).
84. Jerry Coyne, quoted in "Pope Enters the Debate on Intelligent Design" (September 1, 2006), at: http://video.msn.com/v/us/msnbc.htm?g=04456350-90D7-4D37-BC00-C70117F888E2&f=00&fg=copy.
85. Coyne, "Mooney and Kirshenbaum self-destruct at last" (emphasis added).
86. Larry Moran, "The Trouble with NCSE," *Sandwalk Blog* (April 22, 2009), at: http://sandwalk.blogspot.com/2009/04/trouble-with-ncse.html.
87. Sean M. Carroll, "Science and Religion are Not Compatible," *Discover Magazine Blogs Cosmic Variance* (June 23, 2009), at: http://blogs.discovermagazine.com/cosmicvariance/2009/06/23/science-and-religion-are-not-compatible/.
88. Richard Dawkins, Comment #368197 (April 22, 2009), at: http://richarddawkins.net/articles/3767#368197.
89. Ibid.
90. Ibid.
91. P. Z. Myers, Interview at: http://www.infidelguy.com/ (on file with author).
92. P. Z. Myers, "Chat with PZ Myers HERE!" at: http://ravingatheists.com/forum/showthread.php?t=13602&page=2.
93. P. Z. Myers, Interview at: http://www.infidelguy.com/.
94. P. Z. Myers, "Chat with PZ Myers HERE!," at: http://ravingatheists.com/forum/showthread.php?s=e3189a3730ec9110272298cc6cdf86bd&t=13602.
95. P. Z. Myers, "Chat with PZ Myers HERE!," at: http://ravingatheists.com/forum/showthread.php?t=13602&page=3.
96. P. Z. Myers, "Chat with PZ Myers HERE!," at: http://ravingatheists.com/forum/showthread.php?s=e3189a3730ec9110272298cc6cdf86bd&t=13602.
97. P. Z. Myers, "Chat with PZ Myers HERE!," at: http://ravingatheists.com/forum/showthread.php?t=13602&page=2.
98. P. Z. Myers, "Jerry Coyne lobs another bomb at the accommodationists ... to the barricades!" (April 22, 2009), at: http://scienceblogs.com/pharyngula/2009/04/jerry_coyne_lobs_another_bomb.php.
99. Ibid.
100. Ibid.
101. P.Z. Myers, "Is Religion Rational?," *Pharyngula Blog* (September 17, 2007), at: http://scienceblogs.com/pharyngula/2007/09/is_religion_rational.php.
102. Jason Rosenhouse, "Mooney on Dover," *EvolutionBlog* (June 6, 2009), at: http://scienceblogs.com/evolutionblog/2009/06/mooney_on_dover.php.
103. Ibid.
104. Jason Rosenhouse, "Miller Joins the Party," *EvolutionBlog* (June 11, 2009), at: http://scienceblogs.com/evolutionblog/2009/06/miller_joins_the_party.php.
105. Ibid.
106. Rosenhouse, "Mooney on Dover."

107. Rosenhouse, "Miller Joins the Party."

108. Jerry Coyne, "Seeing and Believing," *The New Republic* (February 4, 2009), at: http://www.tnr.com/article/books/seeing-and-believing?id=1e3851a3-bdf7-438a-ac2a-a5e381a70472.

109. Coyne, "Truckling to the Faithful."

110. Ibid.

111. Daniel Dennett quoted in "The Kitzmiller Decision," *Butterflies and Wheels Blog* (January 25, 2008), at: http://www.butterfliesandwheels.com/articleprint.php?num=162.

112. Dennett, *Darwin's Dangerous Idea*, p. 310.

113. Lori Arnold, "'Expelled' documentary explores Darwin, Intelligent Design, religion debate," *Christian Examiner* (April 2008), at: http://www.christianexaminer.com/Articles/Articles%20Apr08/Art_Apr08_03.html.

114. Peter Atkins, "Will science ever fail?," *New Scientist* 1833 (August 8, 1992): pp. 32–35.

115. Peter Atkins, "Awesome Versus Adipose: Who Really Works Hardest to Banish Ignorance?," *Free Inquiry* 18 (Spring 1998), at: http://www.secularhumanism.org/index.php?section=library&page=atkins_18_2.

116. Atkins, "Will science ever fail?"

117. Peter W. Atkins, *Creation Revisited: The Origin of Space, Time and the Universe* (London, UK: Penguin Books, 1994), p. 17.

118. Sam Harris, "Science Must Destroy Religion," in *What Is Your Dangerous Idea?: Today's Leading Thinkers On The Unthinkable*, ed. John Brockman (New York: Harper Perennial, 2007), p. 148.

119. Sam Harris, "Scientists should unite against threat from religion," *Nature* 448 (August 23, 2007): p. 864.

120. Ibid.

121. E. O. Wilson, *Consilience: The Unity of Knowledge* (New York: Vintage Books, 1998), p. 290.

122. E. O. Wilson quoted in Steven Benowitz, "Irreligious Researchers Differ In Their Views On Faith," *The Scientist* 9 (Apr. 17, 1995), at: http://www.the-scientist.com/article/display/16467/.

123. E. O. Wilson, *On Human Nature* (Cambridge: Harvard University Press, 1978), p. 1.

124. Jerry Coyne, quoted in Richard Dawkins, *The God Delusion* (New York: Bantam Press, 2006), p. 67.

125. Jerry Coyne, "Chris Mooney and Barbara Forrest love the faithful more than me," *Why Evolution Is True Blog* (June 2, 2009), at: http://whyevolutionistrue.wordpress.com/2009/06/02/chris-mooney-and-barbara-forrest-love-the-faithful-more-than-me/.

126. Jerry Coyne, "Rosenhouse vs. Mooney," *Why Evolution Is True Blog* (June 6, 2009), at: http://whyevolutionistrue.wordpress.com/2009/06/06/rosenhouse-vs-mooney/.

127. Francis Collins, *The Language of God: A Scientist Presents Evidence for Belief* (New York: Free Press, 2007), p. 6. Quoted by the NAS, "Excerpts of Statements by Scientists Who See No Conflict Between Their Faith and Science," at: http://www.nationalacademies.org/evolution/StatementsScience.html.

128. Richard Dawkins, "Why There Almost Certainly Is No God," *The Huffington Post* (October 23, 2006), at: http://www.huffingtonpost.com/richard-dawkins/why-there-almost-certainl_b_32164.html.

129. Daniel C. Dennett, *Breaking the Spell: Religion as a Natural Phenomenon*, (New York: Penguin, 2006), p. 30.
130. Massimo Pigliucci, "Personal Gods, Deism, & the Limits of Skepticism," *Skeptic* 8 (2000): pp. 38–45, at: http://psy.ucsd.edu/~eebbesen/Psych110/SciRelig.htm.
131. Wilson, *Consilience*, p. 286.
132. Ibid.
133. John Dupré, *Darwin's Legacy: What Evolution Means Today* (Oxford, UK: Oxford University Press, 2003), p. 58.
134. Ibid., p. 56.
135. Dawkins, *The God Delusion*, p. 91.
136. Ibid.
137. See Bora Zivkovic quoted in Anika Smith, "Lying in the Name of Indoctrination," *Evolution News & Views Blog* (August 27, 2008), at: http://www.evolutionnews.org/2008/08/lying_in_the_name_of_indoctrin.html.
138. In his atheist manifesto *Letter to a Christian Nation*, Sam Harris argues that the NAS's promotion of NOMA is contrived for political purposes:

    > [The NAS's] statement is stunning for its lack of candor. Of course, scientists live in perpetual fear of losing public funds, so the NAS may have merely been expressing raw terror of the taxpaying mob. The truth, however, is that the conflict between religion and science is unavoidable. The success of science often comes at the expense of religious dogma; the maintenance of religious dogma always comes at the expense of science. Our religions do not simply talk about "a purpose for human existence." Like science, every religion makes specific claims about the way the world is.
    >
    > From: Sam Harris, *Letter to a Christian Nation* (New York: Vintage Books, 2006), p. 63.

139. Kent Greenawalt, "Establishing Religious Ideas: Evolution, Creationism, and Intelligent Design," *Notre Dame Journal of Law, Ethics & Public Policy* 17 (2003): p. 334.
140. Stephen C. Meyer, "A Qualified Agreement Response," in *Science and Christianity: Four Views*, ed. Richard F. Carlson (Downers Grove: InterVarsity Press, 2000), pp. 112, 120.
141. David Hull, "The God of the Galápagos," *Nature* 352 (August 8, 1991): pp. 485–486.
142. Charles Darwin, *The Descent of Man and Selection in Relation to Sex* (1871) (New York: Barnes & Noble Books, 2004), p. 77.
143. Ibid., p. 78.
144. Julian Huxley, quoted in "Gloomy Debate," *Time* (June 19, 1944), at: http://www.time.com/time/printout/0,8816,711861,00.html.
145. Peter J. Bowler, *Evolution: The History of an Idea* (Berkeley & Los Angeles: University of California Press, 2003), p. 177.
146. Shermer, *What We Believe but Cannot Prove*, p. 39.
147. Gregory W. Graffin and William B. Provine, "Evolution, Religion and Free Will," *American Scientist* 95 (July/August, 2007): pp. 294–297, at: http://www.americanscientist.org/issues/id.3747,y.0,no.,content.true,page.2,css.print/issue.aspx.
148. Marc D. Hauser, "Our Universal Moral Grammar's Immunity to Religion," in *What Is Your Dangerous Idea?*, p. 60.
149. Ibid. at 60–61.
150. Michael Ruse, *The Darwinian Paradigm: Essays on its History, Philosophy, and Religious Implications* (London: Routledge, 1993), pp. 268–269 (citations omitted).

151. Michael Ruse and Edward O. Wilson, "The Evolution of Ethics," *New Scientist* 108 (October 17, 1985): pp. 50–52.

152. Ibid.

153. Dennett, *Breaking the Spell*, p. 93.

154. Ibid., pp. 210, 216.

155. Ibid., p. 53.

156. See "Expelled: No Intelligence Allowed—AAAS Response," from *AAAS YouTube*, Time Index 1:55–2:05, at: http://www.youtube.com/watch?v=58UDTq3kaZM.

157. Phillip Johnson, *The Wedge of Truth: Splitting the Foundations of Naturalism* (Downers Grove: InterVarsity Press, 2000), p. 99.

158. "The science of religion: Where angels no longer fear to tread," *The Economist* (March 19, 2008), at: http://www.economist.com/science/displaystory.cfm?story_id=10875666.

159. Open Letter to NABT, NCSE, and AAAS, at: http://www.metanexus.net/Magazine/ArticleDetail/tabid/68/id/2790/Default.aspx. Signatures on the open letter, at: http://fp.bio.utk.edu/darwin/Open%20letter/signatures.html (printed 6/17/02).

160. Ibid.

161. Ibid. (emphasis added).

162. William Corben, "The Nature of Science and the Role of Belief," *Science and Education* 9 (2000): pp. 219–246.

163. The Elie Wiesel Foundation for Humanity: Nobel Laureates Initiative (September 9, 2005), at: http://media.ljworld.com/pdf/2005/09/15/nobel_letter.pdf.

164. Kenneth R. Miller, *Kitzmiller v Dover Area School District*, Trial Transcript: Day 1 am Session (September 25, 2005), pp. 41, 44.

165. Kenneth R. Miller and Joseph S. Levine, *Biology* (Englewood Cliffs: Prentice Hall, 1991), p. 658 (emphasis in original); Kenneth R. Miller and Joseph S. Levine, *Biology* (Englewood Cliffs: Prentice Hall, 1993), p. 658 (emphasis in original); Kenneth R. Miller and Joseph S. Levine, *Biology* (Engelwood Cliffs: Prentice Hall, 1995), p. 658 (emphasis in original); Kenneth R. Miller and Joseph S. Levine, *Biology* (Upper Saddle River: Prentice Hall, 1998), p. 658 (emphasis in original); Kenneth R. Miller and Joseph S. Levine, *Biology* (Upper Saddle River: Prentice Hall, 2000), p. 658. (emphasis in original). For a detailed discussion of Miller's testimony on this topic, see Casey Luskin, "Ken Miller's 'Random and Undirected' Testimony," *Evolution News & Views Blog* (July 6, 2006), at: http://www.evolutionnews.org/2006/07/ken_millers_random_and_undirec.html, and Casey Luskin, "Truth or Dare with Dr. Ken Miller," p. 1, at: http://www.evolutionnews.org/KenMillerLectureGuide.pdf.

166. Joseph S. Levine and Kenneth R. Miller, *Biology: Discovering Life* (Lexington: D.C. Heath and Co. 1992), p. 152; Joseph S. Levine and Kenneth R. Miller, *Biology: Discovering Life* (Lexington: D.C. Heath and Co., 1994), p. 161 (emphasis in original).

167. Salvador E. Luria, Stephen Jay Gould, and Sam Singer, *A View of Life* (Menlo Park: Benjamin Cummings, 1981), p. 585.

168. Ibid., p. 586.

169. Burton S. Guttman, *Biology* (Boston: McGraw Hill, 1999), p. 37.

170. Douglas J. Futuyma, *Evolutionary Biology* (Sunderland: Sinauer Associates Inc. 1998), p. 5.

171. Ibid., p. 8.

172. Ibid., p. 5.

173. Monroe W. Strickberger, *Evolution* (London: Jones and Bartlett, 2000), p. 60.

174. Futuyma, *Evolutionary Biology*, p. 5.
175. Ibid.
176. Levine and Miller, *Biology: Discovering Life* (1992), p. 152; Levine and Miller, *Biology: Discovering Life* (1994), p. 161.
177. Miller and Levine, *Biology* (1991), p. 658; Miller and Levine, Biology (1993), p. 658; Miller and Levine, *Biology* (1995), p. 658; Miller and Levine, *Biology* (1998), p. 658; Miller and Levine, *Biology* (2000), p. 658.
178. Luria, Gould, and Singer, *A View of Life*, p. 574.
179. Guttman, *Biology*, p. 37.
180. Nicholas H. Barton, Derek E. G. Briggs, Jonathan A. Eisen, David B. Goldstein, and Nipam H. Patel, *Evolution* (Cold Spring Harbor: Cold Spring Harbor Laboratory Press, 2007), p. 435.
181. Miller and Levine, *Biology* (1991), p. 658; Miller and Levine, *Biology* (1993), p. 658; Miller and Levine, *Biology* (1995), p. 658; Miller and Levine, *Biology* (1998), p. 658; Miller and Levine, *Biology* (2000), p. 658.
182. Levine and Miller, *Biology: Discovering Life* (1992), p. 152; Levine and Miller, *Biology: Discovering Life* (1994), p. 161.
183. Helena Curtis and N. Sue Barnes, *Invitation to Biology* (New York: Worth Publishers, 1981), p. 475. The textbook teaches: "The real difficulty in accepting Darwin's theory has always been that it seems to diminish our significance. Earlier, astronomy had made it clear that the earth is not the center of the solar universe, or even of our own solar system. Now the new biology asked us to accept the proposition that, like all other organisms, we too are the products of a random process that, as far as science can show, we are not created for any special purpose or as part of any universal design."
184. Strickberger, *Evolution*, p. 70. The textbook teaches: "[T]he variability by which selection depends may be random, but adaptations are not; they arise because selection chooses and perfects only what is adaptive. In this scheme a god of design and purpose is not necessary."
185. Brian K. Hall and Benedikt Hallgrimsson, Strickberger's *Evolution: The Integration of Genes, Organisms, and Populations* (Sudbury: Jones and Bartlett, 2008), p. 659.
186. William B. Provine, "No free will," in *Catching up with the Vision*, ed. Margaret W. Rossiter (Chicago: University of Chicago Press, 1999), p. S123.

4. Death and the Fall by William Dembski

1. This chapter is adapted from *The End of Christianity: Finding a Good God in an Evil World* (Nashville: Broadman and Holman, 2009).
2. Philip Yancey and Paul Brand, *In His Image* (Grand Rapids: Zondervan, 1984), p. 22.
3. Denis Alexander writes, "Being an anatomically modern human was necessary but not sufficient for being spiritually alive." Quoted from *Creation or Evolution: Do We Have to Choose?* (Oxford: Monarch Books, 2008), p. 237.
4. Jerry Coyne, "Don't Know Much Biology," *Edge* June 6, 2007, available online at: http://www.edge.org/3rd_culture/coyne07/coyne07_index.html.
5. See the articles in Warren S. Brown, Nancey Murphy, and H. Newton Malony, *Whatever Happened to the Soul? Scientific and Theological Portraits of Human Nature* (Minneapolis: Augsburg Fortress, 1998). Compare Peter Van Inwagen, *Material Beings* (Ithaca: Cornell University Press, 1990).

6. Mario Beauregard and Denyse O'Leary, *The Spiritual Brain: A Neuroscientist's Case for the Existence of the Soul* (San Francisco: HarperOne, 2007). Jeffrey Schwartz and Sharon Begley, *The Mind and the Brain: Neuroplasticity and the Power of Mental Force* (New York: Regan Books, 2002).
7. See Nancey Murphy, "Nonreductive Physicalism: Philosophical Issues," in Brown *et al.*, *Whatever Happened to the Soul?* pp. 127–148.
8. "Humans are also unique in ways that defy evolutionary explanation and point to our spiritual nature. This includes the existence of the Moral Law (the knowledge of right and wrong) and the search for God that characterizes all human cultures throughout history." Francis Collins, *The Language of God: A Scientist Presents Evidence for Belief* (New York: Free Press, 2006), p. 200.
9. Charles Darwin, *The Descent of Man and Selection in Relation to Sex*, 2nd ed. (London: John Murray, 1882), p. 126.
10. Karl W. Giberson, *Saving Darwin: How to Be a Christian and Believe in Evolution* (San Francisco: HarperOne, 2008), p. 13. Emphasis in the original.
11. A. W. Montford, *The Hockey Stick Illusion: Climategate and the Corruption of Science* (London: Stacey, 2010).
12. William Dembski and Jonathan Wells, *The Design of Life: Discovering Signs of Intelligence in Biological Systems* (Dallas: Foundation for Thought and Ethics, 2008), ch. 1, which is on human origins.
13. This is, of course, a vast topic, and the references to Giberson's and my own work only scratch the surface. The evolutionary literature overwhelmingly argues against human uniqueness. For the other side, see Mortimer Adler's book cited in the next note as well as the following: Benjamin Wiker, *Moral Darwinism: How We Became Hedonists* (Downers Grove: InterVarsity, 2002); C. Stephen Evans, *Preserving the Person: A Look at Human Sciences* (Vancouver: Regent College Publishing, 1994); David Berlinski, *The Devil's Delusion: Atheism and Its Scientific Pretensions* (New York: Crown Forum, 2008).
14. For the distinction between a difference in kind and a difference in degree, especially as it applies to human uniqueness, see Mortimer Adler, *The Difference of Man and the Difference It Makes* (New York: Fordham University Press, 1993). This book, though originally published in 1967, is must-reading for anyone concerned with the problem of human uniqueness. It opens with a thought experiment about what would happen if it were possible to cross a human and an ape.

    Forty years later, Richard Dawkins proposed breaking the species barrier with "a successful hybridization between a human and a chimpanzee." See his brief article "Breaking the Species Barrier," *Edge* (January 2009), at: http://www.edge.org/q2009/q09_16.html. Dawkins continues, "Even if the hybrid were infertile like a mule, the shock waves that would be sent through society would be salutary. This is why a distinguished biologist described this possibility as the most immoral scientific experiment he could imagine: it would change everything!"

    Dawkins views such an experiment not as immoral but, if successful, as liberating: "Our ethics and our politics assume, largely without question or serious discussion, that the division between human and 'animal' is absolute." A "humanzee" would, for Dawkins, refute human uniqueness and thereby destroy the entire Judeo-Christian ethical system based on it—a prospect he relishes. Are theistic evolutionists like Karl Giberson prepared to follow Dawkins down this path? Absent human uniqueness, why not?
15. Giberson, *Saving Darwin*, p. 12. C. S. Lewis appreciated—and rejected—Giberson's point. According to Lewis, the skeptic of traditional theological doctrines "naturally listens with

impatience to our solutions of particular difficulties and our defences against particular objections. The more ingenious we are in such solutions and defences the more perverse we seem to him.... I have come to regard that attitude as a total misunderstanding." C. S. Lewis, *Miracles: A Preliminary Study*, rev. ed. (1960; reprinted San Francisco: Harper, 2001), pp. 109–110.

16. Giberson might demur that he is merely redefining or reconceptualizing the Fall in a way that preserves what historically has been most important about it. But Christian orthodoxy's understanding of the Fall is unrecognizable in his reconceptualization of it.
17. Giberson, *Saving Darwin*, p. 12.
18. For this line of reasoning taken to its logical conclusion, see Gaymon Bennett, Martinez J. Hewlett, Ted Peters, and Robert John Russell, eds., *The Evolution of Evil* (Göttingen, Germany: Vandenhoeck & Ruprecht, 2008).
19. Ayala seems concerned to preserve God's honor only when promoting Darwinian evolution. All quotes in this and the previous paragraph are from Francisco Ayala, *Darwin's Gift to Science and Religion* (Washington: Joseph Henry Press, 2007), pp. 159–160.
20. C. Robert Mesle, *Process Theology: A Basic Introduction* (St. Louis: Chalice Press, 1993), p. 62.
21. The subtitle of Giberson's *Saving Darwin* makes precisely this point: *How to Be a Christian and Believe in Evolution*.
22. Karl Giberson, "Evolution and the Problem of Evil," *Beliefnet* (September 28, 2009): available online at http://blog.beliefnet.com/scienceandthesacred/2009/09/evolution-and-the-problem-of-evil.html.
23. Clive Hayden, "Karl Giberson Responds to William Dembski," *Uncommon Descent* (September 29, 2009): available online at http://www.uncommondescent.com/evolution/karl-giberson-responds-to-dr-dembski.
24. John Polkinghorne, "God's Action in the World," 1990 J. K. Russell Fellowship Lecture, available online at: http://www.starcourse.org/jcp/action.html. Polkinghorne repeated this anecdote in the fall of 2002 at the Parchman Lectures, sponsored by Baylor University's Truett Seminary (September 30 and October 1, 2002).
25. Charles Darwin, *On the Origin of Species*, facsimile 1st ed. (1859; reprinted Cambridge: Harvard University Press, 1964), pp. 76, 79, 129, 490.
26. Ibid., p. 490.
27. Darwin, *Descent of Man*, p. 156. This passage is identical in both the 1871 first edition and the 1882 second edition of *The Descent of Man*.
28. Alexander, *Creation or Evolution*, p. 282.
29. Robert Wright, *Nonzero: The Logic of Human Destiny* (New York: Pantheon, 2000).
30. Cooperation can also be against the cruelties of inanimate nature, as when a fungus and an alga cooperate to produce lichen, the first stage of plant colonization of a rock face. See Lynn Margulis and Dorion Sagan, *Acquiring Genomes: A Theory of the Origins of Species* (New York: Basic Books, 2002), pp. 13–14.
31. Darwin, *Origin of Species*, p. 489.
32. Ibid., p. 490.
33. Alexander, *Creation or Evolution*, p. 282.
34. Unlike Darwin, who tried to minimize evolution's cruelty, some Darwinists positively revel in it. Take the annual Darwin Awards. According to Wendy Northcutt, author of the Darwin Award books, these are given posthumously to individuals who "ensure the long-

term survival of our species by removing themselves from the gene pool in a sublimely idiotic fashion." (Quoted from the front cover of her book *The Darwin Awards II: Unnatural Selection* [New York: Plume, 2003]). In these books, Northcutt details in case after case the misfortunes of people who died. Yes, the circumstances of their deaths are ridiculous. But I know of no other view, religious or secular, that inspires its adherents to celebrate the deaths of people they regard as stupid. Why is Darwin's name associated with these awards? Could it be because the name fits? Can anyone imagine Albert Einstein or Martin Luther King Jr. or Michelangelo lending their names to such an award?

Traditional religious believers are not alone in faulting Darwinian evolution for cruelty. Take New Age writer Lynne McTaggart:

> Our self-image grew even bleaker with the work of Charles Darwin. His theory of evolution—tweaked slightly now by the Neo-Darwinists—is of a life that is random, predatory, purposeless and solitary. Be the best or don't survive. You are no more than an evolutionary accident. The vast checkerboard biological heritage of your ancestors is stripped down to one central facet: survival. Eat or be eaten. The essence of your humanity is a genetic terrorist, efficiently disposing of any weaker links. Life is not about sharing and interdependence. Life is about winning, getting there first. And if you do manage to survive, you are on your own at the top of the evolutionary tree.

Quoted from: *The Field: The Quest for the Secret Force of the Universe*, updated edition (New York: HarperCollins 2008), pp. xxiv–xxv. Granted, this statement is a bit of a caricature. But why does Darwin's theory inspire such caricatures?

5. *Random Acts of Design by Jonathan Witt*

1. This comment can be found in the dust jacket endorsement by Collins of Kenneth Miller's book *Only a Theory: Evolution and the Battle for America's Soul* (New York: Viking, 2008).
2. Michael Denton, "Selected Excerpts: The Inverted Retina: Maladaptation or Pre-adaptation?" *Origins & Design* 19, no. 2, issue 37, available online at: www.arn.org/docs/odesign/od192/invertedretina192.htm.
3. See also Casey Luskin and Logan Paul Gage, "Appendix: A Reply to Francis Collins's Darwinian Arguments for Common Ancestry of Apes and Humans," *Intelligent Design 101: Leading Experts Explain the Key Issues*, edited by H. Wayne House (Grand Rapids: Kregel, 2008); James A. Shapiro and Richard Sternberg, "Why Repetitive DNA Is Essential to Genome Function," *Biological Review* 80 (2005): pp. 227–250; Sternberg, "On the Roles of Repetitive DNA Elements in the Context of a Unified Genomic-Epigenetic System," *Annals of the New York Academy of Sciences* 981 (2002): pp. 154–188; and Sternberg and James A. Shapiro, "How Repeated Retroelements Format Genome Function," *Cytogenetic and Genome Research* 110 (2005): pp. 108–116.
4. Scott Minnich and Stephen Meyer, "Genetic analysis of coordinate flagellar and type III regulatory circuits in pathogenic bacteria" in *Proceedings of the Second International Conference on Design and Nature* (September 1, 2004). See also William Dembski, "Still Spinning Just Fine: A Response to Kenneth Miller," (February 17, 2003), v.1, at: http://www.designinference.com/documents/2003.02.Miller_Response.htm.
5. "When Moses asks God who he should say sent him back to deliver the Israelites from Egypt, God replies: "This is what you are to say to the Israelites: 'I AM has sent me to you.'" (Ex. 3:14b)
6. Kenneth Miller quoted in Paul Nussbaum, "Evangelicals divided over evolution," *Philadelphia Inquirer* (May 30, 2005): p. A01.
7. See the discussion of Collins in the two chapters of this volume by John West.

6. Darwin of the Gaps by Jonathan Wells

1. "President Clinton Announces the Completion of the First Survey of the Entire Human Genome," White House Press Release (June 25, 2000), available online at: http://www.ornl.gov/sci/techresources/Human_Genome/project/clinton1.shtml.
2. "International Consortium Completes Human Genome Project," National Human Genome Research Institute (April 14, 2003), available online at: http://www.genome.gov/11006929.
3. Francis S. Collins, *The Language of God: A Scientist Presents Evidence for Belief* (New York: Free Press, 2006), pp. 2–3, 122.
4. Collins, *The Language of God*, pp. 107, 127, 141, 210.
5. Ibid., pp. 21–31, 67–78, 93.
6. Ibid., pp. 67, 140.
7. Ibid., pp. 183–186.
8. "What is the theory of intelligent design?" from Frequently Asked Questions at the web site of Discovery Institute's Center for Science and Culture, available at: http://www.discovery.org/csc/topQuestions.php.
9. Francis Darwin (editor), *The Life and Letters of Charles Darwin* (London: John Murray, 1887), volume 1, p. 309, available online at: http://darwin-online.org.uk/content/frameset?viewtype=side&itemID=F1452.1&pageseq=327.
10. Francis Darwin (editor), *The Life and Letters of Charles Darwin*, volume 2, p. 312, available online at: http://darwin-online.org.uk/content/frameset?viewtype=side&itemID=F1452.2&pageseq=328.
11. Michael J. Behe, *Darwin's Black Box: The Biochemical Challenge to Evolution* (New York: Free Press, 1996), p. 39.
12. Collins, *The Language of God*, p. 188.
13. Michael J. Behe, "In Defense of the Irreducible Complexity of the Blood Clotting Cascade: Response to Russell Doolittle, Ken Miller and Keith Robison," Discovery Institute (July 31, 2000), available online at: http://www.discovery.org/a/442.
14. Collins, *The Language of God*, p. 192.
15. Here Collins cites Kenneth R. Miller, "The Flagellum Unspun: The Collapse of Irreducible Complexity," pp. 81–97 in W. A. Dembski and M. Ruse (editors), *Debating Design: From Darwin to DNA* (Cambridge: Cambridge University Press, 2004), available online at: http://www.millerandlevine.com/km/evol/design2/article.html.
16. Michael J. Behe, "Afterword," *Darwin's Black Box 10th Anniversary Edition* (New York: Free Press, 2006), p. 268.
17. Scott A. Minnich and Stephen C. Meyer, "Genetic Analysis of Coordinate Flagellar and Type III Regulatory Circuits in Pathogenic Bacteria," Second International Conference on Design & Nature, Rhodes, Greece (September 1, 2004), available online at: http://www.discovery.org/a/2181.
18. Collins, *The Language of God*, p. 193.
19. David W. Snoke, "In Favor of God-of-the-Gaps Reasoning," *Perspectives on Science and Christian Faith* 53 (2001): pp. 152–158, available online at: http://www.cityreformed.org/snoke/gaps.pdf.
20. William A. Dembski, The Design Revolution: Answering the Toughest Questions About *Intelligent Design* (Downers Grove: InterVarsity Press, 2004), especially chapters 8–15.
21. Collins, *The Language of God*, pp. 127–131.

22. Theodosius Dobzhansky, *Genetics and the Origin of Species*, reprinted 1982 (New York: Columbia University Press, 1937), p. 12.
23. Collins, *The Language of God*, pp. 131–132.
24. Pamela F. Colosimo, Kim E. Hosemann, Sarita Balabhadra, Guadalupe Villarreal, Jr., Mark Dickson, Jane Grimwood, Jeremy Schmutz, Richard M. Myers, Dolph Schluter and David M. Kingsley, "Widespread Parallel Evolution in Sticklebacks by Repeated Fixation of Ectodysplasin Alleles," *Science* 307 (2005): 1928–1933. Abstract available online at: http://www.ncbi.nlm.nih.gov/pubmed/15790847?itool=EntrezSystem2.PEntrez.Pubmed.Pubmed_ResultsPanel.Pubmed_RVDocSum&ordinalpos=3&log$=free.
25. Collins, *The Language of God*, p. 132.
26. Ibid, pp. 126–129.
27. Ulfur Arnason, Joseph A. Adegoke, Kristina Bodin, Erik W. Born, Yuzine B. Esa, Anette Gullberg, Maria Nilsson, Roger V. Short, Xiufeng Xu and Axel Janke, "Mammalian mitogenomic relationships and the root of the eutherian tree," *Proceedings of the National Academy of Sciences USA* 99 (2002): pp. 8151–8156, available online at: http://www.pnas.org/cgi/content/full/99/12/8151.
28. Leigh Van Valen, "Deltatheridia, a New Order of Mammals," *Bulletin of the American Museum of Natural History* 132 (1966): pp. 1–126; Leigh Van Valen, "Monophyly or Diphyly in the Origin of Whales," *Evolution* 22 (1968): pp. 37–41.
29. Dennis Normile, "New Views of the Origins of Mammals," *Science* 281 (1998): pp. 774–775; Richard Monastersky, "The Whale's Tale: research on whale evolution," *Science News* (November 6, 1999), available online at: http://www.findarticles.com/p/articles/mi_m1200/is_19_156/ai_57828404.
30. Kenneth D. Rose, "The Ancestry of Whales," *Science* 293 (2001): pp. 2216–2217.
31. J. G. M. Thewissen, Lisa Noelle Cooper, Mark T. Clementz, Sunil Bajpai and B. N. Tiwari, "Whales originated from aquatic artiodactyls in the Eocene epoch of India," *Nature* 450 (2007): pp. 1190–1194. Abstract available online at: http://www.ncbi.nlm.nih.gov/pubmed/18097400?ordinalpos=1&itool=EntrezSystem2.PEntrez.Pubmed.Pubmed_ResultsPanel.Pubmed_RVDocSum.
32. Davide Pisani, Michael J. Benton and Mark Wilkinson, "Congruence of Morphological and Molecular Phylogenies," *Acta Biotheoretica* 55 (2007): pp. 269–281. Abstract available online at: http://www.springerlink.com/content/k452146636811u14/.
33. Collins, *The Language of God*, pp. 130–136.
34. Ibid, pp. 136–137.
35. Jonathan Wells, "Darwin's Straw God Argument," Discovery Institute (December 31, 2008), available online at: http://www.discovery.org/a/8101.
36. Douglas J. Futuyma, *Evolution* (Sunderland, MA: Sinauer Associates, 2005), p. 49.
37. Hidenori Nishihara, Arian F.A. Smit and Norihiro Okada, "Functional noncoding sequences derived from SINEs in the mammalian genome," *Genome Research* 16 (2006): pp. 864–874, available online at: http://www.genome.org/cgi/content/full/16/7/864.
38. Craig B. Lowe, Gill Bejerano and David Haussler, "Thousands of human mobile element fragments undergo strong purifying selection near developmental genes," *Proceedings of the National Academy of Sciences USA 104* (2007): pp. 8005–8010. Available online at: http://www.pnas.org/cgi/content/full/104/19/8005.
39. Michael Pheasant and John S. Mattick, "Raising the estimate of functional human sequences," *Genome Research* 17 (2007): pp. 1245–1253, available online at: http://www.genome.org/cgi/content/full/17/9/1245.

40. P. D. Mariner, R. D. Walters, C. A. Espinoza, L. F. Drullinger, S. D. Wagner, J. F. Kugel & J. A. Goodrich, "Human Alu RNA Is a Modular Transacting Repressor of mRNA Transcription during Heat Shock," *Molecular Cell* 29 (2008): pp. 499–509. Abstract available online at: http://www.ncbi.nlm.nih.gov/pubmed/18313387?ordinalpos=1&itool=EntrezSystem2.PEntrez.Pubmed.Pubmed_ResultsPanel.Pubmed_RVDocSum.
41. Collins, *The Language of God*, pp. 129–130.
42. Patricia Cortazzo, Carlos Cerveñansky, Mónica Marín, Claude Reiss, Ricardo Ehrlich and Atilio Deana, "Silent mutations affect in vivo protein folding in Escherichia coli," *Biochemical and Biophysical Research Communications* 293 (2002): pp. 537–541. Abstract available at: http://www.ncbi.nlm.nih.gov/pubmed/12054634?ordinalpos=2&itool=EntrezSystem2.PEntrez.Pubmed.Pubmed_ResultsPanel.Pubmed_RVDocSum.
43. Chava Kimchi-Sarfaty, Jung Mi Oh, In-Wha Kim, Zuben E. Sauna, Anna Maria Calcagno, Suresh V. Ambudkar and Michael M. Gottesman, "A 'Silent' Polymorphism in the MDR1 Gene Changes Substrate Specificity," *Science* 315 (2007): pp. 525–528. Abstract available online at: http://www.ncbi.nlm.nih.gov/pubmed/17185560?ordinalpos=5&itool=EntrezSystem2.PEntrez.Pubmed.Pubmed_ResultsPanel.Pubmed_RVDocSum.
44. For a more extensive critique of the junk DNA argument, see Jonathan Wells, *The Myth of Junk DNA* (Seattle: Discovery Institute Press, forthcoming 2011).

7. Making a Virtue of Necessity by Jay W. Richards

1. See his chapter, "The Fully Gifted Creation," in *Three Views on Creation and Evolution*, ed. J. P. Moreland and John Mark Reynolds (Grand Rapids: Zondervan, 1999), pp. 180–1, 198–215.
2. This is the term he used in "Science & Christian Theology as Partners in Theorizing," in *Science and Christianity: Four Views*, ed. Richard F. Carlson (Downers Grove: InterVarsity Press, 2000), pp. 195–24.
3. He seems to argue that detecting divine design is, as he says, "beyond my empirical grasp." See "The Creation: Intelligently Designed or Optimally Equipped?" *Theology Today* 55 (October 1998): p. 363. Van Till does not, however, confront arguments for the empirical detectability of design. Rather, he uses the Principle as a sort of regulative principle to sidestep these arguments. For one proposal for detecting intelligent design and distinguishing it from the "natural" causes of law and chance, see William A. Dembski, *The Design Inference* (Cambridge: Cambridge University Press, 1998). For arguments for design and against methodological naturalism in science, see William A. Dembski, *Intelligent Design: The Bridge between Science and Theology* (Downers Grove: InterVarsity Press, 1999); William A. Dembski, ed., *Mere Creation* (Downers Grove: InterVarsity, 1998); Michael J. Behe, *Darwin's Black Box* (New York: Free Press, 1996).

   For philosophical arguments against methodological naturalism, see Alvin Plantinga, "Methodological Naturalism," *Origins & Design* 18 no. 1 (Winter 1997): pp. 18–27, and "Methodological Naturalism, Part 2," *Origins & Design* 18, no. 2 (Fall 1997): pp. 22–33; Phillip E. Johnson, *Reason in the Balance* (Downers Grove: InterVarsity, 1994), pp. 205–18; J. P. Moreland, "Theistic Science & Methodological Naturalism," in *The Creation Hypothesis*, ed. J. P. Moreland (Downers Grove: InterVarsity, 1994).
4. He also, however, speaks of certain "marks" of design in the universe, implying that some design is detectable. This leads Paul Nelson to argue that Van Till has some ambivalence concerning the issue of detectability. Paul Nelson, "Is 'Intelligent Design' Unavoidable—Even by Howard Van Till? A Response," *Zygon* 34 (December 1999): pp. 677–82.

5. Van Till objects to this label (*Three Views*, p. 161, footnote), since he thinks many Christians use "theistic evolution" pejoratively.
6. Ibid., pp. 193. I will argue below that this way of formulating the question subtly biases the dispute in favor of Van Till's perspective.
7. Ibid., pp. 185–6.
8. Van Till posed the issue this way in a paper presented at "The Nature of Nature" Conference, Baylor University, April 12–15, 2000, entitled "Cosmic Evolution as the Manifestation of Divine Activity." The central question he addressed in this paper was: "Who owns the Robust Formational Economy Principle?" He answered that it was the legitimate intellectual property of theists.
9. At least since Galileo clashed with the Aristotelians, a central theme of modern science is that we should not merely try to deduce how nature must be from theological or philosophical first principles. As we will see below, on this point Van Till's reasoning has much in common with the Aristotelians.
10. Paul Nelson (in private correspondence) has pointed out, rightly, that for anyone to say he has "discovered" that the actual world manifests the Principle, he must be willing to allow that it might not do so: "No scientist can say that he has discovered that natural causes are sufficient if there was never any possibility that they were not."
11. See especially Stephen Meyer, *Signature in the Cell: DNA and the Evidence for Intelligent Design* (San Francisco: HarperOne, 2009); "Qualified Agreement: Modern Science & the Return of the 'God Hypothesis,'" in *Science and Christianity: Four Views*, pp. 127–74, and idem.
12. See, for example, Dembski, *Intelligent Design*, chapter 6.
13. *Three Views*, pp. 192–94.
14. In the introduction to *Three Views*, J. P. Moreland and John Mark Reynolds note: "On closer examination, the gaps argument turns out not to be an actual argument. It is more a bit of apologetic advice" (p. 22).
15. There is much more to be said on these points, but they have been discussed sufficiently elsewhere. For more on the God of the gaps argument, the historical theological basis for the concept of functional integrity, and Van Till's views on "special creationism," see his writings as follows, "Basil and Augustine Revisited: The Survival of Functional Integrity," in *Origins & Design* 19 no. 1 (Summer 1998); "Is Special Creationism a Heresy?" *Christian Scholar's Review* 22 (June 1993): pp. 380–95; "When Faith and Reason Meet," in *Man and Creation: Perspectives on Science and Theology* (Hillsdale: Hillsdale College Press, 1993), pp. 141–64; "God and Evolution: An Exchange (with Phillip E. Johnson)," *First Things* 34 (June/July 1993): pp. 32–41; "Basil, Augustine, and the Doctrine of Creation's Functional Integrity," *Science and Christian Belief* 8 (April 1996): pp. 21–3. For criticism of Van Till's claim that his view is rooted in the Christian tradition, see Jonathan Wells, "Abusing Theology: Howard Van Till's 'Forgotten Doctrine of Creation's Functional Integrity,'" *Origins & Design* 19, no. 1 (Summer 1998). For analysis of exegetical considerations, see the excellent book by C. John Collins, *The God of Miracles: An Exegetical Examination of God's Action in the World* (Wheaton: Crossway Books, 2000).
16. See Paul Nelson and John Mark Reynolds, pp. 64–7, and Walter Bradley, "Response to Howard J. Van Till," pp. 219–25, in *Three Views*. See also Collins, *The God of Miracles*, pp. 163–75.
17. Ibid., pp. 186–8. Van Till writes: "I find it theologically awkward to imagine God choosing at the beginning to withhold certain gifts from the creation, thereby introducing gaps into

the creation's formational history—gaps that would later, in the course of time, have to be bridged by acts of special creation" (p. 187).

18. Such theological arguments are common fare in the evolutionary biological literature, so Van Till is not alone in appealing to them. See Paul A. Nelson, "The Role of Theology in Current Evolutionary Reasoning," *Biology and Philosophy* 11 (1996): pp. 493–517. Incidentally, there is some dispute among Plato scholars about this portrayal of the Demiurge, but that's not important here.
19. *Three Views*, p. 188.
20. This metaphor might sound deistic. Van Till denies that his view is deistic, since he maintains that God not only creates but also actively sustains the creation. I am not contending that one should not use this metaphor because it is deistic, but simply because it does not adequately express God's relationship with creation if used uncritically.
21. In my original article, Van Till objected that I attributed to him the metaphor of creation as a "self-sustaining and even self-generating machine, like a watch." In any event, the argument is the same without the metaphor. Van Till's aesthetic judgment is that it is better for the creation to be able to generate all of its own complexity than for God to do so "directly." I do not see any basis for this judgment. At the very least, it is not nearly compelling enough to adopt it as an *a priori* principle.
22. Much more could be said on this point, but others have already done so adequately. For instance, see John Mark Reynolds, "Are There Gaps in the Gapless Economy? The Improbable Views of Howard J. Van Till," *Origins & Design* 19 no. 1 (Summer 1998): pp. 23–25.
23. Similarly, Van Till deflects the charge that his view is deistic by affirming that God sustains and upholds the universe moment by moment, even if these divine activities are empirically undetectable.
24. Robert Dehaan has noted (in private correspondence) that this strategy also contradicts a *prima facie* reading of "Psalm 65 and Job 38, and like passages, which depict God as deeply involved in the everyday economy of the natural world, quite apart from his involvement in 'salvation history.'"
25. *Three Views*, p. 187.
26. Ironically, perhaps, Van Till's distinction between salvation history and natural history will not prevent a conflict in biblical studies, since the requirements of methodological naturalism contradict certain historical claims intrinsic to Christian belief, such as the reality of Christ's incarnation and resurrection. So if a key motivation for proposing the Principle is to avoid conflict with methodological naturalism, it fails, at least in one discipline. For more, see my "Naturalism in Theology and Biblical Studies" in *Unapologetic Apologetics: Meeting the Challenges of Theological Studies*, ed. William Dembski and Jay Richards (Downers Grove: InterVarsity Press, 2001).
27. But what if we discovered that the Principle holds in our world? Would that mean that naturalism is true and theism or Christianity is false? Clearly, the answer is no, as Van Till has argued. The question is whether such a discovery would make naturalism or theism more epistemically probable. I do think that a Principle is more likely given naturalism than given theism, even if one can construct some *post hoc* theological arguments for it.
28. Jay Wesley Richards, "Howard Van Till's 'Robust Formational Economy Principle' as a Critique of Intelligent Design Theory," *Philosophia Christi* 4, no. 1 (2002): pp. 101–12; Howard J. Van Till, "Is the Creation's Formational Economy Complete?: A Response to Jay Wesley Richards," *Philosophia Christi* 4, no. 1 (2002): pp. 113–118; Jay Wesley Richards, "A Reply to Howard J. Van Till, *Philosophia Christi* 4, no. 1 (2002): pp. 119–23.
29. Van Till, "Is the Creation's Formational Economy Complete?," p. 113.

30. I made this same point in my initial essay. Since Van Till did not engage it in his response, however, I am reiterating it here.
31. One should note that the central claim of intelligent design theory, that design is empirically detectable in nature, does not require an "interventionist concept of divine creative action." In any event, Van Till uses this phrase in his response to avoid engaging my claim that there are a number of possible types of divine action that his Principle is unable to accommodate. By passing over this point, he seems to concede it.
32. Ibid., p. 114.
33. Ibid., p. 116.
34. He treats this subject in a paper he presented to the Freethought Association in Michigan in 2006, "From Calvinism to Free Thought: The Road Less Travelled. See the report on the event at: http://www.freethoughtassociation.org/minutes/2006/May24-2006.htm. Jeremy Manier says this about Van Till's current views:

    In some ways he's an agnostic, though he treasures the traditions he learned as a child. What keeps him from dropping his belief is an overpowering feeling that something he can't define has made this dazzling world possible, a limitless source of order and creation. For that, he feels he still must use the name of God, even if he no longer knows what it means. [In "The New Theology," *Chicago Tribune* (Jan. 19, 2008).]

35. Ibid., p. 115.
36. Note that this argument is not directed at the Robust Formational Economy Principle *per se*, but rather at holding it as a first principle in the way that Van Till does.
37. I note for the record that Van Till dislikes this term, and does not employ it himself. This has no bearing on my argument, however, since he clearly accepts what it denotes, even if he objects to what it often connotes in theological disputes. I prefer Phillip Johnson's term "epistemic naturalism" to "methodological naturalism," but since the latter is already well established in the literature, I have used it here.

8. *The Difference It Doesn't Make by Stephen C. Meyer*

1. This article is adapted from *Darwinism Defeated? The Johnson-Lamoureux Debate over Biological Origins* (Vancouver: Regent College Press, 1999).
2. George Gaylord Simpson, *The Meaning of Evolution*, revised edition (New Haven: Yale University Press, 1967), p. 345.
3. See Denis Lamoureux, "Evolutionary Creation: A Christian Approach to Evolution," available at: http://biologos.org/uploads/projects/Lamoureux_Scholarly_Essay.pdf. In the following, I will refer to this online essay. See also his recent book-length treatment, *Evolutionary Creation: A Christian Approach to Creation* (Eugene: Wipf & Stock, 2008).
4. Lamoureux, "Evolutionary Creation."
5. Quoted in Joe Woodward, "The End of Evolution," *Alberta Report* (December 1996), p. 33.
6. Lamoureux, "Evolutionary Creation."
7. Ibid.
8. Ibid.
9. Ibid.
10. For instance, Boyle said, quite characteristically: "God is a most free agent, and created the world not out of necessity but voluntarily, having framed it as he pleased and thought fit at the beginning of things, when there was no substance but himself and consequently no creature to which he could be obliged, or by which he could be limited. In Robert Boyle,

A Free Enquiry into the Vulgarly Received Notion of Nature, Edward B. David and Michael Hunter, eds. (Cambridge: Cambridge University Press, 1996), p. 160. See also Reijer Hooykaas's summary of Boyle's Christian empiricist epistemology in: Reijer Hooykaas, *Religion and the Rise of Modern Science*. (Grand Rapids: W. B. Eerdmans: 1972), p. 47.

11. Lamoureux, "Evolutionary Creation."
12. Stephen C. Meyer, *Signature in the Cell: DNA and the Evidence for Intelligent Design* (San Francisco: HarperOne, 2009).
13. Stephen C. Meyer, "The Origin of Biological information and the Higher Taxonomic Categories," *Proceedings of the Biological Society of Washington*, 117, no. 2 (2004), pp. 213-239; See also Stephen C. Meyer, "The Cambrian Information Explosion: Evidence for Intelligent Design," in Michael Ruse and William Dembski, eds., *Debating Design from Darwin to DNA* (Cambridge: Cambridge University Press, 2007), pp. 371–391.
14. Richard Dawkins, *River out of Eden: A Darwinian View of Life* (New York: Basic Books, 1996,) p. 17.
15. Bill Gates, *The Road Ahead* (New York: Viking, 1995), p. 188.
16. Leroy Hood, "The digital code of DNA" *Nature* 421 (2003): pp. 444–448.
17. Bernd-Olaf Küppers, *Information and the Origin of Life* (Cambridge: MIT Press, 1990), pp. 170–172.
18. Lamoureux, "Evolutionary Creation."
19. For a definition of specification; see Willia Dembski, *The Design Inference* (Cambridge: Cambridge University Press, 1998), pp. 1–66, pp. 136–174.
20. Manfred Eigen, *Steps Toward Life* (Oxford: Oxford University Press, 1992), p. 12.
21. Fred Dretske, *Knowledge and the Flow of Information* (Cambridge: MIT Press, 1981), p. 12.
22. Bruce Alberts et al., *Molecular Biology of the Cell* (New York: Garland, 1983), p. 105.
23. Bernd-Olaf Küppers, "On the Prior Probability of the Existence of Life," in *The Probabilistic Revolution*, ed. Lorenz Kruger et al. (Cambridge: MIT Press, 1987), p. 364.
24. Michael Polanyi, "Life Transcending Physics and Chemistry," *Chemical and Engineering News* 45 (1967): pp. 54–66. See also: Michael Polanyi, "Life's Irreducible Structure," *Science* 160 (1968): pp. 1308–1312, esp. p. 1309.
25. Bernd-Olaf Küppers, *Information and the Origin of Life*, pp. 170-172; also Charles Thaxton, Walter Bradley and Roger Olsen, *The Mystery of Life's Origin* (New York: Philosophical Library, 1984), pp. 24–38.
26. See, for example, Christian DeDuve, "The Beginnings of Life on Earth," *American Scientist* 83 (1995): p. 437.
27. R. A. Kok, J. A. Taylor, and W. L. Bradley, "A Statistical Examination of Self-Ordering Amino Acids in Proteins," *Origins of Life and Evolution of the Biosphere* 18 (1988), pp. 135–142.
28. Lamoureux, "Evolutionary Creation."
29. See note 13 above.
30. Gordon Mills, "Similarities and Differences in Mitochondrial Genomes: Theistic Interpretation," *Perspectives on Science and Christian Faith* 50, no. 4 (December 1998): pp. 286–292.

9. Everything Old is New Again by Denyse O'Leary

1. The Pontifical Academy of Sciences recently held a conference that addressed intelligent design, while failing to invite Michael Behe, a faithful Catholic and author of *The Edge of*

Evolution (New York: Free Press, 2007), and one of the best people to explain why Darwinism is not well supported by evidence. See information about the conference at: http://www.vatican.va/roman_curia/pontifical_academies/acdscien/.

2. See John 1:1–3.
3. Matthew 18:2 See http://bible.cc/matthew/18-2.htm For his further commentary, see: http://niv.scripturetext.com/matthew/18.htm, and compare it to the "selfish gene."
4. This is the prize announcement: http://templetonprize.org/currentwinner.html. Ayala's claim that ID is blasphemy is available at: http://www.windowview.org/sci/pgs/ayala.html. See a response view from the intelligent design community at: http://www.evolutionnews.org/2010/03/the_view_from_planet_ayala.html.
5. See, for example, Michael J. Behe, *The Edge of Evolution* and Stephen Meyer, *Signature in the Cell* (San Francisco: HarperOne, 2009).
6. See, for example: http://www.veritas-ucsb.org/library/origins/quotes/cambrian.html.
7. http://www.fossilmuseum.net/Paleobiology/CambrianExplosion.htm (Note: Many general accounts of the Cambrian period are now corrupted by an effort to defend Darwinism, and should be treated with appropriate caution.)
8. George Sim Johnston, *Did Darwin Get It Right?: Catholics and the Theory of Evolution* (Huntington: *Our Sunday Visitor,* 1998), p. 16.
9. See, for example: http://ncse.com/taking-action/ten-major-court-cases-evolution-creationism, for the Darwin lobby in action.
10. Available online at: http://www.pbs.org/wgbh/evolution/darwin/nameof/. Here, as elsewhere, it is useful to note that veneration for Darwin often prevents evaluation of his real legacy. The man who thought that black people were closer to gorillas than to white people is often carefully sheltered from a shower of abuse that anyone else would experience.
11. See, for example, the timeline at: http://civilliberty.about.com/od/gendersexuality/tp/Forced-Sterilization-History.htm; also, a forthcoming book by Jane Harris Zsovan, *Eugenics and the Firewall* (Winnipeg: J. Gordon Shillingford, 2010), which provides original document research on the Canadian history of this problem that has been systematically underreported for many decades.
12. It is difficult to find this quote, naked, due to many, many attempts to explain it away, by tax-funded Darwinists. Here it is at:

    http://books.google.ca/books?id=tvEEAAAAYAAJ&pg=PA156&dq=%22negro+or+Australian+and+the+gorilla%22+Descent+of+Man&cd=1#v=onepage&q&f=false. It has been easier to soft pedal the inevitable racism associated with Darwinism when it became unpopular, after World War II. See also *Anthropological Review* (April 1867): pp. 167, 236.
13. See, for example, http://www.fordham.edu/Campus_Resources/eNewsroom/topstories_1725.asp.
14. For a reasonable evaluation of the evidence, see Michael Behe, *The Edge of Evolution.*
15. See, for example: http://www.uncommondescent.com/darwinism/wisdom-from-your-local-zoo-introducing-the-%E2%80%9Cevolutionary-agony-aunt%E2%80%9D/.
16. See here for introduction: http://www.mercatornet.com/articles/view/wisdom_from_your_local_zoo/.
17. (Reisel: Erasmus Press, 2009). Alfred Russel Wallace (1823–1913), Darwin's co-theorist, was among the nineteenth century's most noted working English naturalists. Flannery's edition of Wallace's *World of Life* places Wallace in historical context as a non-materialist evolutionary thinker. Flannery offers a sympathetic account of what he calls Wallace's intelligent evolution, a design-based alternative to Darwin's randomness-based evolution.

Clearly, these were two very different formulations of the history of life: Wallace's design and purpose versus Darwin's randomness and chance.

18. Darwin may have witnessed considerable animal suffering in his childhood. That may have helped shape his attitudes. However, the suffering of agricultural animals, treated inhumanely, can be usefully distinguished from that which occurs in nature, as Darwin's co-theorist Alfred Russel Wallace pointed out. See, for example, Michael Flannery's re-publication of Wallace's *World of Life*. In nature in general, predators must eat their prey or see it stolen or disintegrated. Small animal predators may usually resolve the problem by simply swallowing the prey whole, which prevents theft and reduces suffering. Large animal predators generally go for the throat.
19. In the age before general public schooling, it should be remembered that many people would simply never have encountered the idea of unguided evolution, and there is no way of knowing what they "would have" thought.
20. A project aimed at getting churches on side for Darwinism. See: http://blue.butler.edu/~mzimmerm/rel_evol_sun.htm for details.
21. See, for example: http://www.aboutdarwin.com/timeline/time_07.html#0050. Note the term "radical naturalists."
22. Denyse O'Leary, "A Better Selection," *Touchstone* (November/December 2009).
23. Leonard Susskind, "Darwin's Legacy," *PhysicsWorld* (July 1, 2009), at: http://physicsworld.com/cws/article/print/39672 (registration wall).
24. Surprisingly, Nobel Prize winner Francis Crick considered this view seriously, and in the *Expelled* film Richard Dawkins did not discount it, when discussing the matter with Ben Stein. See, for example http://post-darwinist.blogspot.com/2007/10/session-two-why-is-origin-of-life-such.html.
25. See Denyse O'Leary, "A Better Selection," *Touchstone* (November/December) 2009.
26. Wallace's openness to spiritualism caused him to be ridiculed and dismissed, leaving him a comparatively obscure and misunderstood figure (although, like astrology, spiritualism was not an unreasonable belief until it came to be discredited by evidence).
27. It is difficult to find this quote, due to many, many attempts to explain it away by Darwinists, but see here: http://books.google.ca/books?id=tvEEAAAAYAAJ&pg=PA156&dq=%22negro+or+Australian+and+the+gorilla%22+Descent+of+Man&cd=1#v=onepage&q&f=false). It has been easier to soft pedal the inevitable racism associated with Darwinism when it became unpopular, after World War II. See also *Anthropological Review* (April, 1867): pp. 167, 236.
28. See, for example: http://www.touchstonemag.com/archives/article.php?id=22-08-038-b.
29. Commentator and talk show host Michael Coren has written a fine biography of Chesterton, *Gilbert, the Man Who Was G. K. Chesterton* (Vancouver, Canada: Regent College Publishing, 2001).
30. G. K. Chesterton, "Doubts about Darwin," originally published in *The Illustrated London News* (July 17, 1920), available at: http://www.gkc.org.uk/gkc/books/Doubts_About_Darwinism.html.
31. G. K. Chesterton, "On Darwinism and Mystery," originally published in *The Illustrated London News* (Aug. 21, 1920), available at: http://www.gkc.org.uk/gkc/books/On_Darwinism_and_Mystery.txt.
32. G. K. Chesterton, *The Everlasting Man* (New York: Dodd, Mead & Co., 1926), p. 14. This and further quotations are available at: www.worldinvisible.com/library/chesterton/everlasting/part1c3.htm.

33. See, for example, Judges 4:21, where a woman does that, exactly.
34. See Genesis 34: 1–34, for the general idea.
35. See, for example, Thomas Cahill's *How the Irish Saved Civilization* (New York: First Anchor, 1995). Critical ancient manuscripts were hidden in and off the coast of Ireland during ages of great and growing disorder.
36. G. K. Chesterton, "Straws in the Wind," originally published March 15, 1930, republished in *Gilbert Magazine* (2006), available online at: http://gilbertmagazine.com/page_02.html.
37. H. G. Wells, *Mr. Belloc Objects, to the Outline of History* (London: Watts & Company, 1926).
38. (London: Sheed & Ward, 1928).
39. Hilaire Belloc, *Survivals and New Arrivals: The Old and New Enemies of the Catholic Church* (London: Sheed and Ward, 1929). It is online at: http://www.ewtn.com/library/ANSWERS/SURVIV.htm
40. See, for example, Ed Pilkington, "To get a glimpse of the Ida fossil, the media make monkeys of themselves," *The Guardian* (May 19, 2009), at: http://www.guardian.co.uk/science/2009/may/19/ida-fossil-primate-media-us. See also Brandon Keim, "Bone Crunching Debunks 'First Monkey' Ida Fossil Hype," *Wired* (October 21, 2009), at: http://www.wired.com/wiredscience/2009/10/reconfiguring-ida/.
41. Ample bosoms: http://post-darwinist.blogspot.com/2007/07/evolutionary-psychology-challenge-read.html. Being gay: http://www.newscientist.com/article/dn13674-evolution-myths-natural-selection-cannot-explain-homosexuality.html. This author was once teaching a class at the University of Toronto, and pointed out that there was a very large gay community to the east of the campus, few of whose members were engaged in raising their siblings' children, as the evolutionary psychology thesis confidently proclaimed.
42. Denyse O'Leary, "Dissecting the caveman theory of psychology," *MercatorNet* (August 10, 2009), at: http://www.mercatornet.com/articles/view/dissecting_the_caveman_theory_of_psychology/. Links are provided. Note: Begley does not agree with this view and supports behavioral ecology instead—a perspective that would far better suit our traditional Catholic authors. Behavioral ecology argues that, to understand a custom, one must understand the ecology in which it arises first.
43. S. G. Mivart, "On the appendicular skeleton of primates," *Philosophical Transactions of the Royal Society of London* 157 (1867): pp. 299–429.
44. Early opponents of Darwinism, including the Catholic thinkers mentioned above, and also Wallace, were alarmed by social Darwinism. They saw human society as held together by bonds that are not governed simply by strict competition.
45. Francis Aveling, "St. George Jackson Mivart, Ph.D., M.D., F.R.S., V.P.Z.S., F.Z.S." *The Catholic Encyclopedia*, vol. 10 (New York: Robert Appleton Company, 1911), at: http://www.newadvent.org/cathen/10407b.htm.
46. Dom Paschal Scotti, "Happiness in Hell: The Case of Dr. Mivart," *Downside Review* 119 (July 2001): pp. 177–90.
47. St. George Mivart, *On the Genesis of Species* (New York: D. Appleton, 1871), available at: http://www.macrodevelopment.org/mivart/.
48. Ibid., chapter 2.
49. Stephen Jay Gould, "Not Necessarily a Wing," *Natural History* (October 1985), available at: http://www.stephenjaygould.org/library/gould_functionalshift.html.

50. Ibid. Gould, of course, thought that Darwinian theory could answer this objection. But, if anything, recent discoveries have taken the problem to an entirely new level. See, for example, Michael Behe, *Darwin's Black Box: The Biochemical Challenge to Evolution*, second edition (New York: Free Press, 2006).
51. Louis Caruana SJ, "Is Mivart still relevant?" *Thinking Faith* 24 (November 2009), available online at: http://www.thinkingfaith.org/articles/20091124_2.pdf. This article is a classic in missing the point. The attempt to defend sociobiology as doing some good overlooks the fact that its precise purpose is to reduce human relations to those of animals. That is why it is called sociobiology.
52. John D. Root, "The Final Apostasy of St. George Jackson Mivart," *The Catholic Historical Review* LXXI, no. 1 (January 1985): p. 2.
53. The doctrine is admittedly unpopular but is firmly grounded in the New Testament. See, for example, Matthew 25.
54. This Catholic view was never intended to denigrate science. Science is indispensable for determining such key policy questions for society as how embryos develop, when the process of death is irreversible, or whether a brain-damaged person is criminally responsible. However, most science deals with observed regularities, not with special events best accounted for by God's direct action.
55. Scotti, "Happiness in Hell," p. 181.
56. Ibid., p. 178–80, provides a helpful summary. He quotes Shane Leslie's apt summary (1960), "Catholicism is full of paradoxes, but this is not one of them," p. 182.
57. Scotti, p. 187.
58. Mivart's death from diabetes must be put in historical context. It was not until the 1920s that Canadian medical scientists at the University of Toronto developed a useful treatment for diabetes. The missing hormone was isolated in 1921–1922, and a way was later found to administer it to patients. Before that, most diabetes treatments were ineffectual, when not disastrous. Today, diabetes is simply a long-term chronic illness.
59. John D. Root, "The Final Apostasy of St. George Jackson Mivart."
60. See George Sim Johnston, *Did Darwin Get It Right?: Catholics and the Theory of Evolution*, p. 11.
61. Some other possibilities: Endosymbiosis (one organism moving into another, and performing new functions there) is a live possibility for bacteria, championed by Lynn Margulis. Genetic drift may account for much evolution in plants, whose pollen travels on the wind. Similarly, neoteny—the possibility that some life forms evolve into different species because their life cycle is interrupted at a certain point—is possible for animals that undergo definite life cycle stages. Convergence is another example: Only some life strategies can work in a given environment. So even if life forms are not closely related, they may show the same tendencies. For example, all tires are round because that is the only useful structure for a tire, not because they are all descended from an Ancestral Tire. Animal analogies are readily found. But there is no particular reason to believe that divine direction does not undergird these processes. That leaves Darwinism as the only theory that attempted to provide a materialist atheist account of nature, which is exactly what Darwin intended. Darwin needed to believe that natural selection created information instead of just winnowing it, which is its much more likely role. See, for example, Denyse O'Leary, "Coffee!! Which of these theories is not like the others?" *Uncommon Descent* (February 20, 2010), at: http://www.uncommondescent.com/evolution/coffee-which-of-these-theories-is-not-like-the-others/.

62. Some argue that information can never be lost from the universe, but they are referring to a highly technical definition of information. But within the human context, it can be lost for all practical purposes. If a person's great-grandfather's letters from the trenches in World War I perished in a house fire, the information is lost to that family. This problem is similar to the problem around the origin of life. Even producing life from scratch in a laboratory would not show that life actually originated that way on the early Earth.
63. Gregory W. Graffin and William B. Provine, "Evolution, Religion, and Free Will," *American Scientist* (September 2, 2008).
64. My blog, *Mindful Hack*, has featured many searchable entries on the amazing discoveries of the pseudoscience of evolutionary psychology.
65. George Sim Johnston, *Did Darwin Get It Right?: Catholics and the Theory of Evolution*, p. 16.
66. The usual time frame given is about 4.5 billion years from the formation of Earth to the present. Life started here shortly afterward, despite the extreme difficulties.
67. Mark Steyn, "Obamacare worth the price to Democrats," *Orange County Register* (March 5, 2010), at: http://www.ocregister.com/articles/health-237719-care-government.html.
68. See here, for example: Denyse O'Leary, "Polls: In Darwin's birthday year, people want to hear alternatives," *Post-Darwinist* (March 4, 2009), at: http://post-darwinist.blogspot.com/2009/03/polls-in-darwins-birthday-year-people.html#links.
69. See, for example, Susan Mazur, "Altenberg Summit," *Darwin Then and Now*, at: http://www.darwinthenandnow.com/altenberg-summit/.
70. See report on 2009 Zogby poll, "Report on 2009 Zogby Poll about Evolution and Academic Freedom," at: http://tinyurl.com/yd5qne5.

10. Can a Thomist Be a Darwinist? by Logan Paul Gage

1. *Kitzmiller v. Dover Area School District*, trial transcript Day 5 (September 30, 2005), PM Session, part 1.
2. Of course the Thomist likely denies the existence of fundamentally unguided, or purposeless processes in nature. This may be the most basic conflict between Thomism and Darwinism. While some have tried, it accomplishes nothing (except equivocation) to simply re-define Darwinism as a teleological process. It has never been seen as such by its main proponents. To take but one example among many, in 2005 thirty-eight Nobel laureates sent an open letter to the Kansas State Board of Education decrying any criticism of Darwinism in the classroom. They maintained that Darwinian "evolution is understood to be the result of an unguided, unplanned process of random variation and natural selection." See: http://media.ljworld.com/pdf/2005/09/15/nobel_letter.pdf.
3. Alzina Stone Dale (ed.), *Collected Works of G. K. Chesterton*, volume XV (San Francisco: Ignatius, 1989), p. 196.
4. Charles Darwin, *The Origin of Species: By Means of Natural Selection or the Preservation of Favored Races in the Struggle for Life* (New York: The Modern Library, 1993), pp. 78–79.
5. Thomas Aquinas, *Summa Theologica*, First Part, Question 45, Article 7.
6. Benjamin Wiker, *Moral Darwinism: How We Became Hedonists* (Downers Grove: InterVarsity Press, 2002), p. 218.
7. Stanley L. Jaki, *Chesterton, A Seer of Science* (Champaign: University of Illinois Press, 1986), p. 76.
8. Alasdair C. MacIntyre, *After Virtue: A Study in Moral Theory*, 2nd edition (South Bend: University of Notre Dame Press, 1984).
9. Personal correspondence with Lydia McGrew (June 30, 2009; July 1, 2009).

10. See Leon R. Kass, *Life, Liberty and the Defense of Dignity: The Challenge for Bioethics* (San Francisco: Encounter Books, 2002) and Peter Singer, *Animal Liberation: A New Ethics for our Treatment of Animals* (New York: Random House, 1975).
11. Johann Hari, "Peter Singer—on Killing Disabled Babies, Saving Animals, and the Dangers of Superstition," June 30, 2004, at: http://www.johannhari.com/2004/07/01/peter-singer-on-killing-disabled-babies-saving-animals-and-the-dangers-of-superstition.
12. J. Budziszewski, *The Line Through the Heart: Natural Law as Fact, Theory, and Sign of Contradiction* (Wilmington: ISI Books, 2009), p. 95. Some non-traditional Thomists have retreated to the position that there are only three kinds of form or essence in living things: vegetative, sensitive, and rational. These Thomists necessarily part ways with Darwinists regarding the formal and ontological discontinuity between these broad categories, and as such they face a similar trouble with Darwinism over essences.
13. Mark Ryland, "Applying Natural Philosophy to a Modern Controversy: The Surprisingly Difficult Case of Darwinism, Transformism, and Intelligent Design." Presentation at the 2006 American Maritain Association, November 2, 2006. It should be noted that Ryland is not an ID proponent.
14. David Berlinski, *The Devil's Delusion: Atheism and Its Scientific Pretensions* (New York: Crown Forum, 2008), pp. 155–156.
15. A. N. Wilson, "Why I Believe Again," *New Statesman* (April 2, 2009).
16. Thomas Aquinas, *Summa contra Gentiles*, Part 3, Chapter 22, quoted in Brother Benignus Gerrity, *Nature, Knowledge and God: An Introduction to Thomistic Philosophy* (Milwaukee: Bruce Publishing Company, 1947), p. 501.
17. Alfred J. Freddoso, "Medieval Aristotelianism and the Case against Secondary Causation in Nature," in Thomas V. Morris (ed.), *Divine and Human Action: Essays in the Metaphysics* of Theism (Ithaca: Cornell University Press, 1988), p. 77.
18. Gregory T. Doolan, *Aquinas on the Divine Ideas as Exemplar Causes* (Washington, DC: Catholic University of America Press, 2008), p. 43.
19. Thomas Aquinas, *Summa Theologica*, First Part, Question 44, Article 3.
20. Thomas Aquinas, *Summa contra Gentiles*, Part 2, Chapter 99.
21. Thomas Aquinas, On the Principles of Nature, chapter 3, in Ralph M. McInerny, *A First Glance at St. Thomas Aquinas: A Handbook for Peeping Thomists* (South Bend: University of Notre Dame Press, 2008), pp. 88–89.
22. Stephen C. Meyer, *Signature in the Cell: DNA and the Evidence for Intelligent Design* (San Francisco: HarperOne, 2009). Meyer does not argue, incidentally, that organisms can be reduced to their DNA, that DNA fully specifies an organism or that genetic information is the only evidence for the design of living organisms. His argument is simply that the specified complexity in the coding regions of DNA is one clear sign of intelligent design.
23. Edward Feser, *The Last Superstition: A Refutation of the New Atheism* (Huntington: St. Augustine's Press, 2008), p. 81.
24. Thomas Aquinas, *Summa contra Gentiles*, Book 3, Chapter 38, quoted in Anton C. Pegis (ed.), *Introduction to Saint Thomas Aquinas* (New York: Modern Library, 1948), pp. 454–455.
25. The fossil known as Ida is the most recent example. In a media alert, Ida was originally touted as a "revolutionary scientific find that would change everything" known about human evolution. Michael D. Lemonick, "Ida: Humankind's Earliest Ancestor! (Not Really)," *Time* (May 21, 2009). Now it appears that Ida may have left no descendants at all. See Erik R. Seiffert, Jonathan M. G. Perry, Elwyn L. Simons and Doug M. Boyer, "Convergent

evolution of anthropoid-like adaptations in Eocene adapiform primates," *Nature* 461, no. 7267 (October 22, 2009): pp. 1118–1121.

26. See J. G. Kingsolver, et al., "The Strength of Phenotypic Selection in Natural Populations," *The American Naturalist*, vol. 157, no. 3 (March 2001): pp. 245–261.
27. Avery Cardinal Dulles, "God and Evolution," *First Things* (October 2007).
28. Michael J. Behe, *The Edge of Evolution: The Search for the Limits of Darwinism* (New York: Free Press, 2007), p. 231.
29. Thomas Aquinas, *Summa contra Gentiles*, Part 2, Chapter 99.
30. Alexander R. Pruss (November 15, 2008), at: http://www.whatswrongwiththeworld.net/2008/11/dembski_frank_beckwith_finally.html#comment-40782. Note that Pruss is not an ID proponent.

11. Straining Gnats, Swallowing Camels by Jay W. Richards

1. "Contra Gentes, Part III," in William A. Dembski, Wayne J. Downs, and Fr. Justin B.A. Frederick, *The Patristic Understanding of Creation* (Riesel: Erasmus Press, 2008), p. 197. This volume contains hundreds of pages of the writings of the Churches Fathers on creation and design.
2. "Oration XXVIII—The Second Theological Oration," in Ibid., p. 277.
3. Ibid., p. 278.
4. "Intelligent Design: A Brief Introduction," in *Evidence for God*, edited by William A. Dembski and Michael R. Licona (Grand Rapids: Baker Books, 2010), p. 104.
5. In *The Privileged Planet: How Our Place in the Cosmos is Design for Discovery* (Washington, DC: Regnery, 2004).
6. See the text of the Pope's inaugural address online at: http://www.boston-catholic-journal.com/inaugural_address_of_Pope_Benedict_XVI.htm.
7. Pope Benedict XVI, "In the Beginning ...": *A Catholic Understanding of Creation and the Fall* (Grand Rapids: Eerdmans Pub., 1995), pp. ix–x. In this book he observes that the creation itself is the subject of constant attention due to the environmental debate, and yet, paradoxically, "the creation account is noticeably and nearly completely absent from catechesis, preaching, and even theology." In response to what he called "the practical abandonment of the doctrine of creation on influential modern theology," he calls for renewed attention to the doctrine.
8. Ibid., p. xi.
9. Online at: http://www.newadvent.org/library/docs_jp02tc.htm.
10. *Creation and Evolution: A Conference with Pope Benedict XVI in Castel Gandolfo*, translated by Michael Miller (San Francisco: Ignatius Press, 2008).
11. "Actus" and "potentia" in Latin. *The Catholic Encyclopedia* explains the distinction this way:

    > In general, *potentia* means an aptitude to change, to act or to be acted upon, to give or to receive some new determination. *Actus* means the fulfilment of such a capacity. So, *potentia* always refers to something future, which at present exists only as a germ to be evolved; *actus* denotes the corresponding complete reality. In a word, *potentia* is the determinable being, *actus* the determined being.

    C. Dubray, "Actus et Potentia," *The Catholic Encyclopedia* (New York: Robert Appleton Company, 1907), at: http://www.newadvent.org/cathen/01124a.htm.
12. David Sedley, *Creationism and Its Critics in Antiquity* (Berkeley: University of California Press, 2007), p. xvi.

13. Though, according to some scholars, the Demiurge does not create using pre-existing matter, but rather pre-existing space. Note that the concept of the Demiurge underwent development. For instance, the Gnostic understanding of the Demiurge in the second and third centuries AD should not be read back into Plato's thought.
14. See, for instance, Thomas Aquinas, *Summa contra Gentiles* 2:17, in which he spells out the doctrine of creation: "God's action, which is without pre-existing matter and is called creation, is neither a motion nor a change, properly speaking." Notice that he says "properly speaking." That's because he elsewhere describes God doing all sorts of things within nature, using pre-existing material. A faithful interpretation of Thomas' views cannot isolate this one statement from everything else that he says on the subject.
15. There's a complication in Thomas' view due to his understanding of how God exists in eternity. To put it basically, Thomas thinks that God's single act in eternity can manifest itself in multiple, different ways in time. While this may be hard to understand, Thomas also clearly believed, because he thought it was revealed in Scripture, that God created the world in six ordinary Earth days. So he distinguishes between God's initial act of "creation," his "distinction" of that creation (days one through three) and his "adornment" of creation (days four through seven) (I:65–74). Moreover, he thought that species have remained unchanged since their creation. Here he is following Aristotle, but he no doubt believed that this, too, was a straightforward reading of the biblical text.

    He notes the different opinion of Augustine on the days of creation, who treated the days metaphorically rather than as actual, successive days. Augustine also thought God created things potentially rather than actually on these "days." This made Augustine an outlier among the Church Fathers on these questions. Since Thomas revered Augustine, however, he seeks to give a charitable interpretation and to reconcile the great church father with the bulk of opinion on the other side—insofar as possible. But he doesn't side with Augustine.
16. See, for instance, Dembski, Downs, and Frederick, *The Patristic Understanding of Creation*.
17. From Chapter 1, "On God the Creator of All Things," and Chapter 2, "On Revelation," both available online at: http://www.papalencyclicals.net/Councils/ecum20.htm#Chapter%201%20On%20God%20the%20creator%20of%20all%20things.
18. See its treatment of creation online at: http://www.vatican.va/archive/catechism/p1s2c1p4.htm#I.
19. For a description of a significant controversy and the intrigue behind it, in the first half of the twentieth century, see Raf De Bont, "Rome and Theistic Evolutionism: The Hidden Strategies behind the 'Dorlodot Affair,' 1920–1926," *Annals of Science* 62, no. 4 (October 2005): pp. 457–478.
20. For a fairly representative discussion of the "days" of Genesis 1, see Benedict, "In the Beginning...", pp. 1–39.
21. In "Darwin's Divisions," *Touchstone* (June 2006), at: http://www.touchstonemag.com/archives/article.php?id=19-05-028-f.
22. Catechism of the Catholic Church, 390, available online at: http://www.vatican.va/archive/ENG0015/__P1C.HTM.
23. Pope Benedict XVI, "In the Beginning . . .", pp. 56–57.
24. *Creation and Evolution: A Conference with Pope Benedict XVI in Castel Gandolfo*, foreword by Christoph Cardinal Schönborn (San Francisco: Ignatius Press, 2008), p. 10.
25. Benedict, "In the Beginning . . .", pp. 50–51.
26. Quoted in Schönborn, *Chance or Purpose?*, p. 5.

27. In his introduction to ibid., Cardinal Schönborn quotes from a radio address Ratzinger gave in 1968, where he treated biological evolution as purposive. Although his language has changed over the years, Ratzinger's (now Pope Benedict's) basic views on evolution have remained quite consistent. See ibid., pp. 11–15.
28. Charles Dubray, "Teleology," *The Catholic Encyclopedia*, vol. 14 (New York: Robert Appleton Company, 1912), available online at: http://www.newadvent.org/cathen/14474a.htm.
29. It is possible, though uncommon nowadays, to be a faithful Catholic and also be a type of young earth creationist—as was Thomas Aquinas. A faithful Catholic could be an old earth creationist, a progressive creationist, or a theistic evolutionist, as long as "evolution" is understood teleologically, that is, as allowing for God to work out his purposes in the created order and in every human life.
30. George Gaylord Simpson, *The Meaning of Evolution* (Cambridge: Harvard University Press, 1967), p. 345.
31. In the first chapter of *Creation and Evolution: A Conference with Pope Benedict* XVI, (p. 27), scientist Peter Schuster explains quite assuredly: "The most important mechanism of biological evolution is the Darwinian principle of optimization through variation and selection."
32. Bryan Cones, "Why intelligent design is weird science: An interview with Francisco Ayala," *U.S. Catholic* (August 2007).
33. See, for instance, Kenneth R. Miller, *Finding Darwin's God: A Scientist's Search for the Common Ground Between God and Evolution* (New York: Cliff Street Books, 1999); and Kenneth R. Miller, *Only a Theory: Evolution and the Battle for America's Soul* (New York: Viking, 2008).
34. Michael Tkacz, "Aquinas vs. Intelligent Design," *This Rock* 19, no. 9 (November 2008), at: http://www.catholic.com/thisrock/2008/0811fea4.asp.
35. Michael J. Behe, *The Edge of Evolution: The Search for the Limits of Darwinism* (New York: Free Press, 2008).
36. I am skeptical that everything in nature can be tied back to the quite abstract, low-specificity constants of nature and cosmic initial conditions, but that is not important here.
37. See, for instance, Thomas Aquinas, *Summa contra Gentiles* 2:17, in which he spells out the doctrine of creation: "God's action, which is without pre-existing matter and is called creation, is neither a motion nor a change, properly speaking." Notice that he says "properly speaking." That's because he elsewhere describes God doing all sorts of things within nature, using pre-existing material.
38. "Aquinas Vs. Intelligent Design."
39. In his lucid summary of the history of Thomism, philosopher John Haldane observes "that the task of synthesis is promising but difficult. Thomism began as a synthesis of philosophy and theology and versions of it have ended in the tangled wreckage of unworkable combinations." John Haldane, *Faithful Reason: Essays Catholic and Philosophical* (London: Routledge, 2004), p. 4.
40. The word "intervention" is a source of much mischief in the debate not only about ID, but also about God's action in the world. It's frequently used to make the traditional view that God acts directly in the world look disreputable. For an excellent analysis of this, see Alvin Plantinga, "What is 'Intervention'?" *Theology & Science* 6, no. 4 (2008): pp. 369–401.
41. Tkacz is a so-called "River Forest" Thomist. This label refers to a suburb of Chicago, where the Albertus Magnus Lyceum for Natural Science is located. It is associated with this approach to St. Thomas. For more on this and other "brands" of Thomist, see Edward Fe-

ser, "The Thomistic tradition, Part 1," *Edward Feser* blog (October 15, 2009), at: http://edwardfeser.blogspot.com/2009/10/thomistic-tradition-part-i.html. See also his "The Thomistic tradition, Part 2," *Edward Feser* blog (October 18, 2009), at: http://edwardfeser.blogspot.com/2009/10/thomistic-tradition-part-ii.html. This school interprets Thomas in a stylized Aristotelian fashion. Other Thomists disagree with them. My own view is that Thomas' thought is "Aristotelian" compared to the patristic consensus but is still a sophisticated synthesis that resists Aristotelianism at important points. However one assesses that broader dispute, however, Tkacz seems to have subordinated Thomas to Aristotelian and perhaps other philosophies in this article, even at the expense of core truths of the faith.

42. Again, lest I be misunderstood, my point here is not that ID *per se* requires God to act directly in nature. That is a separate question. My point is that Tkacz suggests that Catholic teaching, or the "Thomistic understanding of divine causation," requires that every feature of every organism have a "natural" explanation.

12. Separating the Chaff from the Wheat by Jay W. Richards

1. Edward Feser, "'Intelligent Design Theory' and Mechanism," *What's Wrong with the World* (April 10, 2010), at: http://www.whatswrongwiththeworld.net/2010/04/intelligent_design_theory_and.html#comment-108215.
2. Etienne Gilson, *From Aristotle to Darwin and Back Again: A Journey in Final Causality, Species, and Evolution* (San Francisco: Ignatius Press, 2009), p. ix. Gilson did not write the book as a response to ID. In fact, it was first published in French in 1971 and in English in 1984. But it was recently republished by Ignatius Press with a lengthy foreword by Cardinal Schönborn. Schönborn places Gilson's book in the context of the modern debate over Darwinism and intelligent design, so that makes it especially fitting. But his foreword must be read with caution, especially in its summary of intellectual history.
3. The opening line from the *Catholic Encyclopedia* entry on "mechanism" says: "There is no constant meaning in the history of philosophy for the word Mechanism." Mark Mary de Munnynck, "Mechanism," *The Catholic Encyclopedia*, vol. 10 (New York: Robert Appleton Company, 1911), available online at: http://www.newadvent.org/cathen/10100a.htm. This is the original *Catholic Encyclopedia* published in the early twentieth century, which orthodox Catholics continue to trust. The newer *Catholic Encyclopedia* published in the 1960s is less theologically reliable from an orthodox perspective, so I won't cite it.
4. "machine," at: http://www.merriam-webster.com/dictionary/machine.
5. In the Foreword to *From Aristotle to Darwin and Back Again*, pp. ix–x.
6. Mark Mony de Munnyck "Mechanicism" in *The Catholic Enyclopedia* vol. 10 (New York: Robert Appleton Co., 1911 at: http://www.newadvent.org/cathen/10100a.htm.
7. Ed Feser, "Nothing But," *Edward Feser* (April 8, 2010), at: http://edwardfeser.blogspot.com/2010/04/nothing-but.html.
8. Edward Feser, *Aquinas* (Oxford: Oneworld, 2009), p. 39.
9. Thomas Kuhn emphasizes, even overemphasizes, the role of Neo-Platonism in the thought of Copernicus, Galileo and Kepler in *The Copernican Revolution: Planetary Astronomy in the Development of Western Thought* (Cambridge: Harvard University Press, reprint, 1992).
10. *From Aristotle to Darwin and Back Again*, p. xi.
11. Ibid., p. xii.
12. Ibid., p. 32.
13. Ibid., pp. 32–33.

14. Lenoir coined the term, though he applied it to Kant and certain German thinkers. For Newton, however, teleology was not merely a mental construct used to make sense of natural phenomena. It was a real feature of the natural world, and a legitimate basis on which to infer that the world is the product of a purposeful Creator.
15. Gilson, *From Aristotle to Darwin and Back Again*, p. 127.
16. Ibid., p. xxi.
17. In a letter to Burnet, quoted in A.J. Pyle, "Animal Generation and the Mechanical Philosophy: Some Light on the Role of Biology in the Scientific Revolution," *History and Philosophy of the Life Sciences* 9 (1987): p. 246, note 130.
18. Mark Vernon makes this charge in "Bad Science, Bad Theology, and Blasphemy," *The Guardian* (May 7, 2010), at: http://www.guardian.co.uk/commentisfree/belief/2010/may/05/intelligent-design-theology.
19. In Christoph Cardinal Schönborn, "Fides, Ratio, Scientia: The Debate About Evolution," *Creation and Evolution: A Conference with Pope Benedict XVI in Castel Gandolfo* (San Francisco: Ignatius Press, 2008), p. 85.
20. Ibid., p. 87.
21. See physicist David Snoke's provocative article, "In Defense of God-of-the-Gaps Reasoning," *Perspectives on Science & Christian Faith* 53, no. 3 (2001): pp. 152–158. He urges an important distinction:

    > We must distinguish between bad explanations for certain things *within* the theistic world view, and arguments for the theistic world view itself. People arguing that comets were signs from God or that demons caused all sickness did not argue that God *existed* because comets and demons existed. Rather, starting from a belief in God, they posited a reasonable, though ultimately falsified, theory about comets and demons.

22. For more on this distinction, see Alvin Plantinga, "What is 'Intervention'?" *Theology & Science* 6, no. 4 (2008): pp. 369–401.
23. In a justly famous article, philosopher Robert M. Adams showed that the idea of God creating the best possible world is riddled with all sorts of difficulties, and has often presupposed that there is just one such world. See his "Must God Create the Best?" *Philosophical Review* 81, no. 3 (1972): pp. 317–332.
24. This is precisely the complaint that Gottfried Leibniz had against Newton's suggestion that God tweaked planetary orbits from time to time. He charged that in Newton's view, "God Almighty wants [that is, needs] to wind up his watch from time to time; otherwise it would cease to move." To Leibniz, this suggested that God's creation is "so imperfect ... that he is obliged to clean it now and then by an extraordinary concourse, and even to mend it, as a clockmaker mends his work." Gottfried Leibniz, *The Leibniz-Clarke Correspondence*, 1715–1716, ed. H. G. Alexander (New York: Barnes & Noble, 1956) pp. 11–12. Quoted in Schönborn, "Fides, Ratio, Scientia: The Debate About Evolution," *Creation and Evolution*, p. 88.
25. But note that our modern distinction between philosophy and natural science is just that—modern. For St. Thomas and others, metaphysics and theology were also "sciences." We'll discuss this point in detail in the next chapter.
26. In paragraph 49 of *Fides et Ratio*, available online at: http://www.vatican.va/holy_father/john_paul_ii/encyclicals/documents/hf_jp-ii_enc_15101998_fides-et-ratio_en.html.
27. Indeed, Aristotle's contributions to Western thought are hard to summarize. Where would we be without Aristotelian logic? Without the concept of prudence? Without Aristotle's subtle analysis of virtue and his taxonomy of political systems? Without his con-

tributions to the concept of natural law? The mind boggles in trying to imagine Western history without him.

28. In fact, the prominence of Aristotle in the Western Church sometimes has led other philosophies, like Neo-Platonism, to get short shrift. This has widened the gulf between Western Catholics and the Eastern Rites in communion with Rome, as well as the Eastern Orthodox, who are still much more influenced by the Neo-Platonism of the early church.

    For example, if you read the entries in the (1910, 1911 version) *Catholic Encyclopedia* for "Aristotle" and "Neo-Platonism"—both written by the same author—you could easily get the impression that Aristotle was a pagan precursor to Christianity, whereas Neo-Platonism was a hostile, pagan enemy of the faith. Only at the end of the article on Neo-Platonism does the reader discover how influential Neo-Platonism was on Christianity during the first millennium. In fact, Neo-Platonism was by far the chief philosophical influence for the majority of Christian history, and shaped all the ecumenical creeds. Aristotle's influence on Christian theology during the first millennium came almost entirely through Neo-Platonism, which took up some of Aristotle's thought. William Turner, "Aristotle," *The Catholic Encyclopedia*, vol. 1 (New York: Robert Appleton Company, 1907), at: http://www.newadvent.org/cathen/01713a.htm; William Turner, "Neo-Platonism," *The Catholic Encyclopedia*, vol. 10 (New York: Robert Appleton Company, 1911), at: http://www.newadvent.org/cathen/10742b.htm.

29. For some examples, see Scott M. Sullivan, "Aquinas the Neoplatonist," available at: http://www.scottmsullivan.com/articles/AquinasNeoplatonist.pdf. See also Wayne J. Hankey, *God in Himself, Aquinas' Doctrine of God as Expounded in the Summa Theologiae* (Oxford: Oxford University Press, 1987).

30. Looking back, we can see the genius of Thomas lay in his attempt to reconcile and synthesize the long, Christian Neo-Platonic tradition with Aristotle, while striving to keep faith with the testimony of Scripture. "Thomas did not fall into either of the two opposing camps, though he was often attacked by both sides. He was intrigued by Aristotle's ideas and saw how many of them could be used in developing a Christian philosophy, but neither did he take Aristotle's side so completely that he failed to see the inadequacies of his thought. Thomas did not slavishly follow the Greek, but used him as a basis for developing his own synthesis of philosophy. In a highly original way, he used elements from Aristotle's teaching to illuminate Christian theology." Marianne Trouve, "How Aristotle Won the West," *This Rock* 9, no. 9 (September 1998), at: http://www.catholic.com/thisrock/1998/9809fea3.asp.

31. Hans Thijssen, "Condemnation of 1277," *Stanford Encyclopedia of Philosophy* (2003), at: http://plato.stanford.edu/entries/condemnation/.

32. "Bonaventure," *The Routledge Encyclopedia of Philosophy*, vol. 1, ed. Edward Craig (London: Routledge, 1998), p. 831. For discussion, see Leonard J. Bowman, "The Cosmic Exemplarism of Bonaventure," *The Journal of Religion* 55, no. 2 (April 1975): pp. 181–198.

33. See discussion in John P. Dourley, *Paul Tillich and Bonaventure* (Leiden: E. J. Brill, 1975), pp. 136–137.

34. For Thomist Ed Feser, for instance, it is what distinguishes Thomas and Aristotle's view from all "modern" views, which Feser refers to as "mechanistic." In what follows, I have Feser's arguments in mind. See his series of blog posts on the subject: "ID versus A-T Roundup," at *Edward Feser* (May 12, 2010), at: http://edwardfeser.blogspot.com/2010/05/id-versus-t-roundup.html. See also Edward Feser, *The Last Superstition: A Refutation of the New Atheism* (South Bend: St. Augustine Press, 2008).

35. "A Response to Professor Feser," *Uncommon Descent* (April 10, 2010), at: http://www.uncommondescent.com/intelligent-design/a-response-to-professor-feser/.
36. Reijer Hooykaas, *Religion and the Rise of Modern Science* (Vancouver: Regent College Publishing, 2000), p. 5. Hooykaas is a quite extreme critic of Aristotle, and seems to defend a form of occasionalism as the proper interpretation of the doctrine of creation, which, he thinks, forms the conceptual basis of modern science.
37. The details here are complicated, and involve interpretations of Aristotle that have been the perennial subject of scholarly debate. For discussion of Aristotle's concept of a Prime Mover, see David Bradshaw, *Aristotle East and West*, pp. 24–44.
38. In fact, the early Church Father Justin Martyr (c. 100–165 AD) found the similarities so striking that he thought that Plato had gotten his ideas from Moses! Justin even went so far as to say that Plato wrote "in exact correspondence with what Moses said regarding God." See his "Hortatory Appeal to the Greeks," in William Dembski, Wayne J. Downs, and Fr. Justin B.A. Frederick, *The Patristic Understanding of Creation* (Riesel, TX: Erasmus Press, 2008), pp. 10, 12–14. Unfortunately, Justin followed Plato is speaking of God as creating the universe out of chaotic matter.
39. In Donald Zehl, "Plato's Timaeus," *Stanford Encyclopedia of Philosophy* (2009), at: http://plato.stanford.edu/entries/plato-timaeus/.
40. Aristotle, *Physics* 251b14–26.
41. For a definitive study of this theme in Thomas, see Gregory T. Doolan, *Aquinas on the Divine Ideas as Exemplar Causes*. Explaining how multiple divine ideas or exemplars are to be reconciled with Thomas' doctrine of divine simplicity is a difficult task, which I do not deal with here.
42. Here's how St. Thomas puts it in *Summa Theologica*, I:103:1: "The natural necessity inherent in things that are determined to one effect is impressed on them by the Divine power which directs them to their end, just as the necessity which directs the arrow to the target is impressed on it by the archer, and does not come from the arrow itself. There is this difference, however, that what creatures receive from God is their nature, whereas the direction imparted by man to natural things beyond what is natural to them is a kind of violence. Hence, as the forced necessity of the arrow shows the direction intended by the archer, so the natural determinism of creatures is a sign of the government of Divine Providence." Available online at: http://www.newadvent.org/summa/1103.htm#article1.
43. Doolan, *Aquinas on the Divine Ideas as Exemplar Causes*, p. 245.
44. On the question of whether the ideas existed independently of individuals, Thomas sided with Plato and admitted as much, even as he saw himself as a critic of Plato (as he understood him). See *Summa contra Gentiles*, I ch. 54, 5, at: http://dhspriory.org/thomas/ContraGentiles1.htm#54. This chapter is complicated by Thomas' extraordinarily difficult doctrine of divine simplicity, but his affirmation of divine ideas is unequivocal. Moreover, Thomas mistakenly argued that Aristotle also held a doctrine of divine ideas or exemplars, but this is not relevant here. We are concerned with Thomas' views themselves, not with his beliefs about what views Aristotle held.
45. There are "forms" in God's mind—exemplars. Though God is the ultimate cause, these exemplars are also in a sense causes. Moreover, there are "forms" in organisms, which determine what an organism is. And these "forms" also get referred to as formal causes of the individual organisms. The whole business is very complex, since the term formal cause is called into a variety of services. It's no wonder that some Neo-Platonists, such as Proclus, appealed to six causes rather than to Aristotle's four. Proclus' six causes were (1) the perfective or final, (2) the paradigmatic, (3) the creative or efficient, (4) the instrumental, (5) the

formal or specifying, and (6) the material. For discussion, see Lucas Siorvanes, *Proclus: Neo-Platonic Philosophy of Science* (New Haven: Yale University Press, 1996), pp. 88–104.

46. Scott M. Sullivan, "Aquinas the Neoplatonist," available at: http://www.scottmsullivan.com/articles/AquinasNeoplatonist.pdf.
47. Andrew Haines, "Aquinas and Neo-Platonism," *Suite 101* (Feb. 5, 2009), available at: http://western-philosophy.suite101.com/article.cfm/aquinas_and_neoplatonism.
48. Ibid.
49. Wayne J. Hankey, Aquinas, Plato, and Neo-Platonism, in *The Oxford Handbook to Aquinas*, edited by Brian Davies and Eleanor Stump, available online at: http://classics.dal.ca/Files/Aquinas_Plato_and_Neo-Platonism_for_Oxford.pdf.
50. *Summa contra Gentiles* (III:100:6), at: http://dhspriory.org/thomas/ContraGentiles3b.htm#100. The title of this chapter is: "That Things Which God Does Apart From The Order Of Nature Are Not Contrary To Nature."
51. Pope Benedict XVI, *"In the Beginning . . ."*, (Grand Rapids, MI: Eerdmans, 1995), p. 93.
52. Earlier in this same chapter of the *Summa contra gentiles*, where he is speaking of miracles, Thomas points out that when "corporeal agents" act on each other, they may change what happens, but they aren't acting contrary to nature: "Therefore, it is much more impossible to say that whatever is done in any creature by God is violent or contrary to nature." That's because the "primary measure" of everything is God:

    > Now, since a judgment concerning anything is based on its measure, what is natural for anything must be deemed what is in conformity with its measure. So, what is implanted by God in a thing will be natural to it. Therefore, even if something else is impressed on the same thing by God, that is not contrary to nature.

    Ibid., III:100:4, 5. The following Question 101, On Miracles, is also pertinent.
53. R. W. Southern, *Western Society and the Church in the Middle Ages* (New York, Penguin, 1990, reprint), p. 82. Quoted in Bradshaw, *Aristotle East and West*, p. 264.
54. Corpuscularianism was actually a modification of ancient atomism, but the details are not important in the present discussion.
55. Quoted from the currently unpublished conference paper by Sveinbjorn Thordarson, "The Alliance of Christianity and Mechanistic Philosophy in 17th-Century England," *Journal of the Oxford University History Society* (2009).
56. In my view, Boyle went too far in the direction of the strong voluntarism of Ockham. Also, there is some question about just how radical Boyle's view of natural laws was. On one reading, laws don't really describe the nature of (created) matter at all, but are merely imposed on them moment-to-moment. This would mean that Boyle was effectively an occasionalist. But Boyle scholars disagree on this question. I suspect Boyle's view was somewhere between "occasionalism" and "concurrentism" since he did not attempt to develop a detailed and consistent view of the relationship between divine and natural causality. This may have been due to an overreaction to Aristotelianism. But he was emphatically not a deist. In any case, what is clear is that he wanted to remove any idea of necessity from natural laws to protect God's freedom as Creator. See Peter R. Anstey, *The Philosophy of Robert Boyle* (London: Routledge, 2000), pp. 160–180.
57. Boyle was especially keen to defend God's omnipotence and his active providential ordering of the world. When speaking of natural law and miracles, he said that by "the laws of nature [God] determin'd and bound up other Beings to act according to them, yet has not bound up his own hands by them, but can envigorate, suspend, over-rule; and reverse any of them as he thinks fit." Quoted in Ibid., p. 180.

58. See J. J. McIntosh and Peter Anstey, "Robert Boyle," in the *Stanford Encyclopedia of Philosophy* (2006), at: http://plato.stanford.edu/entries/boyle/.
59. David Sedley, *Creationism and Its Critics in Antiquity* (Berkeley: University of California Press, 2007), p. xvi.
60. Ed Feser, for instance, has written: "Abandoning Aristotelianism, as the founders of modern philosophy did, was the single greatest mistake ever made in the entire history of Western thought ... this abandonment has contributed to the civilizational crisis through which the West has been living for several centuries...." Notice he does not say the abandonment of God or the doctrine of creation or the truths in the Nicene Creed, but the abandonment of Aristotelianism, which, as a major influence on Christianity, emerged twelve hundred years after the beginning of Christian history. Obviously my appreciation of Aristotle is genuine but more reserved, as is my opposition to philosophical thought from sources other than Aristotle. Feser and others are free to maintain their opinions, of course, but those opinions should not be identified with orthodox Catholic belief. Theirs is one of many schools within Catholic philosophy.

13. Understanding Intelligent Design by Jay W. Richards

1. His seminal study is *The Design Inference: Eliminating Chance Through Small Probabilities* (Cambridge: Cambridge University Press, 1998). Dembski has refined his analysis since this initial study.
2. William A. Dembski, *The Design Revolution* (Downers Grove: InterVarsity Press, 2004), p. 152.
3. William Dembski, "Intelligent Design: Yesterday's Orthodoxy, Today's Heresy," at: http://www.designinference.com/documents/2005.04.ID_Orthodoxy_Heresy.htm. See also Dembski's discussion, and response to Ed Feser and some other Thomist critics, in "Does ID Presuppose a Mechanistic View of Nature," *Uncommon Descent* (April 18, 2010), at: http://www.uncommondescent.com/intelligent-design/does-id-presuppose-a-mechanistic-view-of-nature/. See the Gregory of Nazianzen text in "Oration XXVIII—The Second Theological Oration," William Dembski, Wayne J. Downs, and Fr. Justin B. A. Frederick, *The Patristic Understanding of Creation* (Riesel: Erasmus Press, 2008), pp. 277–278.
4. "There is a growing sense that the properties of the universe are best described not by the laws that govern matter but by the laws that govern information. This appears to be true for the quantum world, is certainly true for special relativity, and is currently being explored for general relativity." "Physicist Discovers How to Teleport Energy," *Technology Review* (February 3, 2010), at: http://www.technologyreview.com/blog/arxiv/24759/.
5. See, for instance, the forthcoming books by Richard Sternberg, *The Immaterial Genome*, and Jonathan Wells, *The Myth of Junk DNA*.
6. International Theological Commission: Communion and Stewardship: Human Persons Created in the Image of God; Vatican Statement on Creation and Evolution (July 2004), at: http://www.vatican.va/roman_curia/congregations/cfaith/cti_documents/rc_con_cfaith_doc_20040723_communion-stewardship_en.html.
7. Edward Feser, *Aquinas*, p. 45.
8. Ibid., p. 46.
9. Stephen Barr, "The End of Intelligent Design?" *First Things* (February 9, 2010), at: http://www.firstthings.com/onthesquare/2010/02/the-end-of-intelligent-design. Obscurely, after his criticism of ID, Barr admits: "None of this is to say that the conclusions the ID movement draws about how life came to be and how it evolves are intrinsically unreasonable or necessarily wrong."

10. William Dembski and Jonathan Witt, *Intelligent Design Uncensored* (Downers Grove: InterVarsity Press, 2010), p. 7.
11. I think it's problematic to speak of physical constants as "causes," but let that pass for now.
12. It's anti-teleological purpose makes Darwinism different even from other theories of biological evolution, such as endosymbiosis and convergence. For instance, no one, so far as I know, has ever said that convergence makes it possible to be an intellectually fulfilled atheist. Denyse O'Leary in "Coffee!! Which of These Theories is Not Like the Others?" *Uncommon Descent* (February 20, 2010), at: http://www.uncommondescent.com/evolution/coffee-which-of-these-theories-is-not-like-the-others/.
13. Behe discusses this issue in "God, Design, and Contingency in Nature" (Nov. 12, 2009), at: http://www.evolutionnews.org/2009/11/god_design_and_contingency_in.html.
14. Barr, "The End of Intelligent Design?"
15. For discussion of beauty as evidence of design, see Wiker and Witt, *A Meaningful World: How the Arts and Sciences Reveal the Genius of Nature* (Downers Grove: InterVarsity Press, 2006); Guillermo Gonzalez and Jay W. Richards, *The Privileged Planet: How Our Cosmos is Designed for Discovery* (Washington, DC: Regnery, 2004); Fr. Thomas Dubay, *The Evidential Power of Beauty: Science and Theology Meet* (San Francisco: Ignatius Press, 1999).
16. Del Ratzsch, *Nature, Design, and Science* (Albany: State University of New York Press, 2001).
17. Pius XII, *Humani Generis* 36.
18. The director of the Vatican Observatory, José Funes, recently made this suggestion. When asked about ID, he said:

    > The problem is when religion enters the world of science, the scientific method; that could be the problem with intelligent design. On the other side there is a danger when scientists use science outside of the scientific method, to make philosophical and religious statements—using science for a goal that science is not meant for. So, for example, you cannot use science to deny the existence of God. You can believe whatever you want but you cannot use science to prove that God does not exist.

    In Eugene Samuel Reich, "Pope's Astronomer: 'Science helps me be a priest,'" *New Scientist* (July 14, 2010), at: http://www.newscientist.com/article/mg20727684.800-popes-astronomer-science-helps-me-be-a-priest.html.
19. The late Fr. Stanley Jaki made this objection against ID. He also defined Darwinism as philosophy or metaphysics rather than science. For discussion, see Mark Cole, "Dethroning the Monkey God: Catholics, Intelligent Design & Darwin's Theory," *New Oxford Review* (Jan. 2007), at: http://www.newoxfordreview.org/article.jsp?did=0107-cole.
20. Quoted in Alvin Plantinga, "Methodological Naturalism, Part 1," *Origins & Design* 18, no. 1, at: http://www.arn.org/docs/odesign/od181/methnat181.htm.
21. See, for instance, Alvin Plantinga, "Science: Augustinian or Duhemian?" *Faith and Philosophy* 13 (July 1996): 368–94, esp. pp. 383–90; Alvin Plantinga, "Methodological Naturalism, Part 2," *Origins & Design* 18, no. 2, at: http://www.arn.org/docs/odesign/od182/methnat182.htm.
22. From the entry "Saint Bonaventure" in *The Stanford Encyclopedia of Philosophy*, at: http://plato.stanford.edu/entries/bonaventure/.
23. Examples of scientism aren't hard to find. For instance, in the *Proceedings of the National Academy of Sciences*, Anthony Cashmore argues against the notion of free will:

    > It is widely believed, at least in scientific circles, that living systems, including mankind, obey the natural physical laws. However, it is also commonly accepted that man

> has the capacity to make "free" conscious decisions that do not simply reflect the chemical makeup of the individual at the time of decision—this chemical makeup reflecting both the genetic and environmental history and a degree of stochasticism. Whereas philosophers have discussed for centuries the apparent lack of a causal component for free will, many biologists still seem to be remarkably at ease with this notion of free will; and furthermore, our judicial system is based on such a belief. It is the author's contention that a belief in free will is nothing other than a continuing belief in vitalism—something biologists proudly believe they discarded well over 100 years ago.

From: Anthony R. Cashmore, "The Lucretian swerve: The biological basis of human behavior and the criminal justice system," *Proceedings of the National Academy of Sciences USA* (February 8, 2010), at: www.pnas.org/cgi/doi/10.1073/pnas.0915161107.

24. Edward Feser, *Aquinas*, p. 2.
25. In fact, in an interesting twist on a traditional formula, he called astronomy—which he revered—the "Queen of the Sciences." Whewell was one of the most influential scientists in Victorian England. He was deeply influenced by the German philosopher Immanuel Kant.
26. Ibid., p. 18.
27. Gilson, for instance, says in passing: "The biologist perhaps does not put the question to himself as to the 'why' of the living body which he studies...." *From Darwin to Aristotle and Back Again*, p. 15. In fact, evolutionary biologists continually ask "why" questions of the things they study.
28. For example, in a Feb. 11, 2009 issue of the *Journal of the American Medical Association* dedicated to Darwin, James Evans argues that:

    > Evolutionary theory is as relevant to the teaching of medicine as to medical practice. Just as the periodic table of the elements brings structure to the study of chemistry, an evolutionary approach to medical education provides a logic to understanding the human body in health and disease. Evolution explains why humans are the way they are, ultimately answering the most fundamental questions asked by medical students, ie [sic], those that begin with why. Given the centrality of evolutionary theory to a deep understanding of the human body, it is possible to envision an entire medical curriculum built around evolution, from anatomy and molecular genetics to pathogen-host interactions.

29. Dennis Overbye, "The Joy of Physics Isn't in the Results, but in the Search Itself," *The New York Times* (Dec. 28, 2009), at: http://www.nytimes.com/2009/12/29/science/29essa.html?th&emc=th.
30. From http://evolution.berkeley.edu/evosite/nature/I3basicquestions.shtml.
31. So, in the entry for "teleology" in *The Catholic Encyclopedia*, we are told that teleology "acknowledges that the object of scientific research is to discover the laws of phenomena, and that any fact is scientifically explained when adequate causes are assigned to it, and the conditions of its occurrence are known." Charles Dubray, "Teleology," *The Catholic Encyclopedia*, vol. 14 (New York: Robert Appleton Company, 1912), available online at: http://www.newadvent.org/cathen/14474a.htm.
32. These include inductive reasoning and experiment, inference to the best explanation (also called abductive reasoning), probability calculations and statistics, all of which have been developed in recent centuries, and none of which provide absolute certainty. Even in the deductive realm, there have been insights in modal logic in recent decades—the field of logic that deals with possibility and necessity—that go beyond Aristotle.

33. These sciences sometimes involve categorizing objects, such as when a biologist classifies an animal by species, genus, and family, or when an astronomer categorizes a star or a galaxy. Such sciences also often involve the attempt to describe such objects with mathematical abstraction, such as describing the orbit of stars around the sun in terms of the law of gravity. Physicist Anthony Rizzi describes these two methods as "empirioschematic" and "empiriometric." Anthony Rizzi, *The Science Before Science: A Guide to Thinking in the 21st Century* (Baton Rouge: IAP Press, 2004).
34. See, for instance, William Dembski, *The Design Inference* (Cambridge: Cambridge University Press, 1998).
35. For a detailed explanation of this point, see Stephen C. Meyer, *Signature in the Cell* (San Francisco: HarperOne, 2009), pp. 324–358.
36. The stereotype of the chemist in his lab re-running experiments, or the physicist seeking out a simple, mathematical law, fails to capture the real diversity of natural science. The stereotype has led to a lot of skeptical mischief among philosophers of science, many of whom, upon discovering that natural science rarely renders certain conclusions, conclude that it isn't really rational at all. For a lucid and devastating critique of this tendency in the philosophy of science, see James Franklin, *What Science Knows and How It Knows it* (New York: Encounter Books, 2009), especially pp. 25–51.
37. Gilson, *From Darwin to Aristotle and Back Again*, p. 37.
38. For more on this so-called "methodological equivalence" of Darwinism and design, see Stephen Meyer, *Signature in the Cell.*
39. Christopher Baglow, *Faith, Science & Reason: Theology at the Cutting Edge* (Woodridge: Midwest Theological Forum, 2009), pp. 170–193.
40. See Meyer, *Signature in the Cell*, pp. 397–415.

14. The Maimonides Myth and the Great Heretic by David Klinghoffer

1. Evan R. Goldstein, "A Tradition's Evolution: Is Darwin Kosher?," *Wall Street Journal* (June 29, 2007).
2. Nathan Aviezer, "Intelligent Design: Why Has It Become a Battleground Between Science and Religion?" *Jewish Action* (fall 2006): pp. 12–16.
3. Speech at Stern College, August 31, 2006; reported at: http://spider.mc.yu.edu/news/articles/article.cfm?id=101231.
4. Geoffrey Cantor and Marc Swetlitz, editors, *Jewish Tradition and the Challenge of Darwinism* (Chicago: University of Chicago Press, 2006).
5. Irving Kristol, "Room for Darwin and the Bible," *New York Times* (September 30, 1986).
6. Joseph Epstein, "Ugly, Thorny Things," *Wall Street Journal* (October 22, 2006).
7. Leon Wieseltier, "Creations," *The New Republic* (August 22, 2005).
8. Moses Maimonides, *The Guide of the Perplexed*, vol. 2, translated by Shlomo Pines (Chicago: University of Chicago Press, 1963), p. 327.
9. Francis Darwin, ed., *The Life and Letters of Charles Darwin*, vol. 2 (London: John Murray, 1887), p. 227.
10. "Religion and Science: Conflict or Harmony?," Pew Forum on Religion & Public Life, May 4, 2009; transcript available at: http://pewforum.org/events/?EventID=217.
11. Joel L. Kraemer, *Maimonides: The Life and World of One of Civilization's Greatest Minds* (New York: Doubleday, 2008), p. 389.
12. *Sanhedrin* 90a, 99b.

13. Not much survives of Epicurus' own voluminous writings. Instead, his philosophy was popularized by the first century BCE Epicurean poet Lucretius, a Roman, who wrote the epic poem *On the Nature of Things*. Lucretius advised men, "Ejaculate the liquid collected in the body and do not retain it." In his book *Moral Darwinism: How We Became Hedonists* (Downers Grove: InterVarsity Press, 2002), Benjamin Wiker explains that under the Epicurean way of thinking, since there's no design in nature, there is no normative way of using and/or taking pleasure through the human body. There is nothing naturally "right" about linking sexuality with procreation. In the search for tranquility, Epicureanism sees sex as something to be avoided. Relationships expose us to the risk of pain. Rather, "it is fitting to flee images [of the beloved] and drive away fearfully from oneself the foods of love, and turn the mind elsewhere." At which point, relieve yourself in a solitary fashion. Lucretius is better known for articulating a startlingly detailed proto-Darwinian idea of life's development through random variation and natural selection.
14. Frederick Copleston, S.J., *A History of Philosophy: Book One*, vol. I (New York: Image Books, 1985), p. 405.
15. Yehudah HaLevi, *The Kuzari: In Defense of the Despised Faith*, translated by N. Daniel Korobkin (Nanuet: Feldheim, 2009).
16. The full intellectual line of descent from Epicurus to Darwin is traced with admirable clarity by Benjamin Wiker in Moral Darwinism.
17. Joseph B. Soloveitchik, Derashot Harav: *Selected Lectures of Rabbi Joseph B. Soloveitchik* (Union City: Ohr Publishing, 2003), pp. 38–39; *And from There You Shall Seek* (Jersey City, NJ: Ktav, 2008), p. 120.
18. Richard Dawkins, *The Greatest Show on Earth: The Evidence for Evolution* (New York: The Free Press, 2009), p. 362.
19. Ibid., p. 356.
20. Kenneth Miller, *Finding Darwin's God* (New York: HarperCollins, 1999), p. 127.

15. God's Image, Our Mission by David Klinghoffer

1. BioLogos Foundation, "Did Evolution Have to Result in Human Beings?" available at: http://biologos.org/questions/inevitable-humans/.
2. Kenneth Miller, comments at the "Shifting Ground" conference, Bedford, NH, March 24, 2007.
3. *The Zohar*, Pritzker Edition, vol. 3, translation and commentary by Daniel C. Matt (Palo Alto: Stanford University Press, 2006), p. 165.
4. Cf. *Pirke Avot* 5:26: "Turn it [the Torah] over and over, for everything is in it."
5. *Commentary on the Torah*, Numbers 25:3.
6. Charles Darwin, *The Descent of Man* (New York: Penguin Classics, 2004), p. 163.
7. Adolf Hitler, *Mein Kampf* (New York: Mariner Books, 1998), p. 287.
8. See Rashi on Deuteronomy 25:18.
9. Robert E. Conot, *Justice at Nuremburg* (New York: HarperCollins, 1983), p. 522.

Conclusion by Jay W. Richards

1. Richard Dawkins, *The Blind Watchmaker: Why the Evidence of Evolution Reveals a Universe Without Design* (New York: W. W. Norton and Co., 1996), p. 6.

Contributors

William A. Dembski

William A. Dembski is Research Professor in Philosophy at Southwestern Baptist Theological Seminary in Ft. Worth, where he also directs its Center for Cultural Engagement. He is a Senior Fellow at the Discovery Institute's Center for Science and Culture in Seattle and Senior Research Scientist with the Evolutionary Informatics Lab. With doctorates in mathematics and philosophy, he has published articles in mathematics, engineering, philosophy, and theology journals as well as authored and edited twenty books. In *The Design Inference: Eliminating Chance Through Small Probabilities* (Cambridge University Press, 1998), the first book on intelligent design published by a major university press, he examined the design argument in a post-Darwinian context, analyzing the connections linking chance, probability, and intelligent causation. His most comprehensive treatment of intelligent design to date, coauthored with Jonathan Wells, is titled *The Design of Life: Discovering Signs of Intelligence in Biological Systems.* In 2009, he published a book on theodicy titled *The End of Christianity: Finding a Good God in an Evil World.*

Logan Paul Gage

Logan Paul Gage received his B.A. *summa cum laude* in philosophy, history, and American studies from Whitworth College in 2004. After spending five years in Washington, DC working on a variety of public policy issues, he returned to academic life in 2009. He is currently pursuing a Ph.D. in philosophy at Baylor University. He has written for *The American Spectator, Touchstone, First Things,* and other publications.

David Klinghoffer

David Klinghoffer is a Senior Fellow at the Discovery Institute and the author of *The Lord Will Gather Me In: My Journey to Orthodox Judaism* (Free Press), *Why the Jews Rejected Jesus: The Turning Point in Western History* (Doubleday), *The Discovery of God: Abraham and the Birth of Monotheism* (Doubleday), and other books. He writes often for the *Los Angeles Times,*

National Review, The Weekly Standard, the *Forward,* and other publications. He is a former literary editor and senior editor of *National Review.* He received his A.B. *magna cum laude,* in comparative literature and religious studies, from Brown University in 1987.

Casey Luskin

Casey Luskin is an attorney with graduate degrees in both science and law. He earned his B.S. and M.S. in earth sciences from the University of California, San Diego, where he studied evolution extensively. His law degree is from the University of San Diego. He is a Program Officer in Public Policy and Legal Affairs at the Discovery Institute. He has published in both technical law and science journals, including *Journal of Church and State* and *Geochemistry, Geophysics, and Geosystems* (G3). He's a senior editor at *Salvo Magazine,* and has discussed the debate over evolution in numerous sources, including *Nature, Science,* the *New York Times,* the *Los Angeles Times,* NPR, CNN.com, *USA Today,* and Fox News. Luskin is also co-founder of the Intelligent Design and Evolution Awareness Center (IDEA) (ideacenter.org), a non-profit that helps students investigate Darwinian evolution by starting "IDEA Clubs" on college and high school campuses.

Stephen C. Meyer

Stephen C. Meyer is a Senior Fellow at the Discovery Institute and Director of Discovery Institute's Center for Science and Culture. He is the author of many articles in academic journals and books as well as editorials in magazines and newspapers such as the *Wall Street Journal,* the *Los Angeles Times,* the *Houston Chronicle,* the *Chicago Tribune, First Things,* and *National Review.* His signal contribution to ID theory is given most fully in his 2009 book *Signature in the Cell: DNA and the Evidence for Intelligent Design.* Meyer earned his Ph.D. in the history and philosophy of science from Cambridge University for a dissertation on the history of origin of life biology and the methodology of the historical sciences. Previously he worked as a geophysicist with the Atlantic Richfield Company after earning his undergraduate degrees in Physics and Geology.

Denyse O'Leary

Denyse O'Leary is a Toronto-based journalist, author, and blogger. She got involved with the intelligent design controversy when she discovered an inverted news funnel a decade ago—everyone claimed the subject was dead, yet the number of news stories was growing at a rapid rate. Finally, she wrote *By Design or by Chance?* (2004), in the process of trying to understand it. She was an Anglican at that time, but became a Catholic in 2005, when she had a difficult book to co-write (*The Spiritual Brain*), which is one long argument for non-materialist neuroscience. Meanwhile, she founded several blogs, including *The Post-Darwinist* and *The Mindful Hack*, as a sort of on-line file cabinet, intended to keep interested readers up to date. She is currently working on a book with William Dembski to expose the incompatibilities between Darwinism and Christianity.

Jay W. Richards

Jay W. Richards is a Senior Fellow at the Discovery Institute, Director of Research at the Discovery Institute's Center for Science and Culture, and a Contributing Editor of *The American* at the American Enterprise Institute. He is the author and editor of several books including *Money, Greed, and God*, and co-author of *The Privileged Planet* with astronomer Guillermo Gonzalez. Richards is also executive producer of several documentaries, including *The Call of the Entrepreneur*, *The Birth of Freedom*, and *Effective Stewardship*. He has been featured in several television-broadcast documentaries, including *The Call of the Entrepreneur*, *The Case for a Creator*, *The Wonder of Soil*, and *The Privileged Planet*, based on his book, *The Privileged Planet*. He has a B.A. with majors in political science and religion, an M.Div. from Union Theological Seminary in Virginia, a Th.M. from Calvin Theological Seminary, and a Ph.D. (with honors) in philosophy and theology from Princeton Theological Seminary.

Jonathan Wells

Jonathan Wells is a Senior Fellow of the Center for Science and Culture at the Discovery Institute. He has an A.B. in physical sciences from the University of California at Berkeley, a Ph.D. in theology from Yale University, and a Ph.D. in biology from the University of California at Berkeley. He is the author of *Charles Hodge's Critique of Darwinism, Icons of Evolution, The Politically Incorrect Guide to Darwinism and Intelligent Design,* and *The Myth of Junk DNA* (forthcoming, 2011). He is also the coauthor (with William Dembski) of *The Design of Life* and *How To Be An Intellectually Fulfilled Atheist—Or Not.* He has appeared in several documentaries, including *Unlocking the Mystery of Life, Icons of Evolution, The Case for a Creator, Expelled* and *Darwin's Dilemma.*

John G. West

John G. West is a Senior Fellow at the Discovery Institute and Associate Director of the Institute's Center for Science and Culture. He has written or edited 11 books, including *Darwin Day in America: How Our Politics and Culture Have Been Dehumanized in the Name of Science, Darwin's Conservatives: The Misguided Quest, The Politics of Revelation and Reason: Religion and Civic Life in the New Nation, The Theology of Welfare,* The *Encyclopedia of Religion in American Politics,* and *The C. S. Lewis Readers' Encyclopedia.* His current research examines the impact of social Darwinism on public policy and culture during the past century. Dr. West was previously an Associate Professor of Political Science at Seattle Pacific University, where he chaired the Department of Political Science and Geography. He has taught political science and history courses at California State University, San Bernardino and Azusa Pacific University. He holds a Ph.D. in government from Claremont Graduate University.

Jonathan Witt

Jonathan Witt is a Senior Fellow at Discovery Institute's Center for Science and Culture and a research fellow at the Acton Institute. He is the co-author of *A Meaningful World: How the Arts and Sciences Reveal the Genius of Nature* and *Intelligent Design Uncensored.* His academic writing has appeared in *Philosophia Christi, Touchstone* and *Literature and Theology*; his opinion pieces in such places as the *Seattle Times,* the *Kansas City Star, Science & Theology News,* the *American Spectator,* and *BreakPoint*; and his narrative writing in the literary journals *Windhover* and *New Texas.* Witt has written or co-written scripts for three documentaries that have appeared on PBS, including the critically acclaimed science documentary *The Privileged Planet.*

He received his Ph.D. in English from the University of Kansas where he received an honors pass for a dissertation on literary theory and aesthetics. He received his master's degree from Texas A&M University and his bachelor's degree from Abilene Christian University.

Index

E

F

G

H

20495374R00233

Made in the USA
San Bernardino, CA
12 April 2015